EXPLORING
ABSTRACT ALGEBRA
WITH
MATHEMATICA®

CD-ROM in pocket.

Allen C. Hibbard

Kenneth M. Levasseur

EXPLORING ABSTRACT ALGEBRA WITH *MATHEMATICA*®

CD-ROM Included

Springer | TELOS | THE ELECTRONIC LIBRARY OF SCIENCE

Allen C. Hibbard
Department of Mathematics and Computer Science
Central College
Pella, IA 50219
USA

Kenneth M. Levasseur
Department of Mathematics and Computer Science
University of Massachusetts Lowell
Lowell, MA 01854
USA

Library of Congress Cataloging-in-Publication Data
Hibbard, Allen C.
    Exploring abstract algebra with Mathematica / Allen C. Hibbard,
Kenneth M. Levasseur.
      p.   cm.
    Includes bibliographical references and index.
    ISBN 0-387-98619-7 (soft : alk. paper)
     1. Mathematica (Computer file)   2. Algebra, Abstract—Data
processing.   I. Levasseur, Kenneth M.   II. Title.
    QA162.H52   1998
    512′.02′028553042—dc21   98-41141

Printed on acid-free paper.

Production managed by Steven Pisano; manufacturing supervised by Jacqui Ashri.
Photocomposed pages prepared from the authors' *Mathematica* files.
Printed and bound by Hamilton Printing Co., Rensselaer, NY.
Printed in the United States of America.

9 8 7 6 5 4 3 2 1

ISBN 0-387-98619-7 Springer-Verlag New York Berlin Heidelberg SPIN 10690598

Preface

■ What is *Exploring Abstract Algebra with Mathematica*?

Exploring Abstract Algebra with Mathematica is a learning environment for introductory abstract algebra built around a suite of *Mathematica* packages entitled `AbstractAlgebra`. These packages are a foundation for this collection of twenty-seven interactive labs on group and ring theory. The lab portion of this book reflects the contents of the *Mathematica*-based electronic notebooks contained in the accompanying CD-ROM. Students can interact with both the printed and electronic versions of the material in the laboratory and look up details and reference information in the User's Guide. Exercises occur in the stream of the text of labs, providing a context in which to answer. The notebooks are designed so that the answers to the questions can either be entered into the electronic notebook or written on paper, whichever the instructor prefers. The notebooks support versions 2.2 and 3.0–4.0 and are compatible with all platforms that run *Mathematica*.

This work can be used to supplement any introductory abstract algebra text and is not dependent on any particular text. The group and ring labs have been cross-referenced against some of the more popular texts. This information can be found on our web site at `http://www.central.edu/eaam.html` (which is also mirrored at `http://www.uml.edu/Dept/Math/eaam/eaam.html`). If your favorite text isn't on our list, it can be added upon request by contacting either author. The `AbstractAlgebra` packages and the electronic documentation files are freely available for downloading from our web page, as are a variety of other resources (including useful palettes and ideas for further exploration). As we add new functionality to the packages or supplement the documentation files, updates will be made available at this web site and users are encouraged to sign up there to be notified when updates are posted.

▦ More about the `AbstractAlgebra` packages

The `AbstractAlgebra` packages provide the *Mathematica* programming code to work with structures in abstract algebra. Currently, the packages are capable of handling most of the types of objects encountered in a first-year undergraduate abstract algebra course. This includes working with (finite) groups, rings, fields, and morphisms and functions related to each of these objects. There are a large number of built-in groups (including such standard groups as $\mathbb{Z}_n$, U_n (units of $\mathbb{Z}_n$), S_n, D_n, as well as direct products and quotients of these) and rings (including $\mathbb{Z}_n$, Boolean rings, and lattice rings, as well as polynomial, matrix, and function extension rings). One can also create functions between groups or rings and investigate if these are morphisms. Documentation for these packages forms the second half of this book. After an introductory chapter, there is a chapter for working with groups, one for rings and one for morphisms. A final chapter focuses on other functions built into the packages. This portion of the book is intended to be a reference for working with the `AbstractAlgebra` packages.

▦ More about the *Mathematica* labs

Exploring Abstract Algebra with Mathematica is intended for anyone trying to learn (or teach) abstract algebra. This course is often challenging because of its formal and abstract nature. While some people are quite adept at thinking abstractly, many are helped by also thinking visually or geometrically. To this end, where possible, the *Mathematica* labs are designed to appeal to the visualization of various algebraic ideas (as pioneered by Ladnor Geissinger in his software package *Exploring Small Groups*). Additionally, the nature of the *Mathematica* notebooks encourages an exploratory environment in which one can make and test conjectures. Viewing the notebooks as interactive texts allows an environment that cannot be replicated by lecture alone. While many of the labs are designed to prepare the way for in-class discussion/lecture, they can also be used to extend examples seen in class.

There is no assumption about being able to program in *Mathematica*; users only need to know the basic concepts of using *Mathematica*, which are reviewed in Lab 0 Getting Started with *Mathematica* (found in Appendix B). Every lab starts with a set of goals as well as the prerequisites. Most labs are independent, though a few assume some experience with a previous lab. Although Part I of this book contains the 14 group labs followed by Part II containing the 13 ring labs, the ring labs can just as easily be used first by those who prefer to do rings first (as one of us does). Questions are interspersed throughout the lab at the points where it is natural to ask them. As with any text, one does not need to complete every question in every lab. While the length of the labs varies from 40 minutes to 90 minutes, they typically require about 60 minutes. (Of course, this is a function of how many of the questions are answered.) For adopters of the book, there are provisions for suggested minimal questions to be assigned, as well as listing which ones can be considered optional. Partial solutions are also available for instructors upon request (by e-mailing either author).

The enclosed CD-ROM contains all the labs (as *Mathematica* notebooks), the User's Guide, the `AbstractAlgebra` packages, and a number of palettes (for users of version 3.0 or higher—to facilitate the use of the labs and the implementation of the packages). While a copy of *Mathematica* is necessary to perform any evaluations, several versions of the read-only *MathReader* are also included on the CD-ROM. Instructions can be found both on the CD-ROM and in Appendix A for installation of the packages, documentation files, and lab notebooks. Except for the group and ring labs (and some palettes), the materials on the CD-ROM are also on our web site. (The web site also has notebooks containing just the questions that students can use as "answer sheets," as well as other resources.)

▣ Acknowledgments

We would like to thank the many people who have provided us with support for this project. This includes Stan Seltzer and Connie Elson (both at Ithaca College), whose NSF-sponsored workshop initially brought us together in 1992 to work on this project. We appreciate being involved with the Interactive Mathematics Text Project (IBM/MAA/NSF funded) which provided encouragement and support for several summers in the early years of this project. We are thankful to Wolfram Research, Inc., for providing technical tools, support staff, and the opportunity to participate in their Visiting Scholars program. In particular, Brett Barnhart, Andre Kuzniarek, and Paul Wellin were helpful on numerous occasions. We appreciate the ongoing support of our institutions, Central College and University of Massachusetts Lowell. Furthermore, we appreciate our students for their patience while testing these labs, and particularly Rochelle Rucker and Michael Thompson for their steadfast assistance in testing, editing, proofing, and converting notebooks. We are pleased with the folks at Springer-Verlag who helped bring this project to a conclusion, particularly Keisha Sherbecoe for her attention to details and Steven

Pisano for his expertise and helpfulness. We also appreciate Stan Wagon (Macalester College) for initially encouraging the publication of this project with Springer-Verlag. Our indebtedness also goes to the number of testers of the packages and labs, particularly to Eric Gossett (Bethel College) and John Kiehl (Soundtrack Recording Studios), as well as reviewers Tom Halverson (Macalester College) and Garry Helzer (University of Maryland). Finally, we would like to thank our families who have supported us throughout this project; Al particularly appreciates his daughter Christina who provided constant companionship at his desk while working, and Ken appreciates his wife Karen, and children, Joe, Kathryn, and Matt for their encouragement.

Al Hibbard
`hibbarda@central.edu`

Ken Levasseur
`Kenneth_Levasseur@uml.edu`

Contents

PART II   Ring Labs

PART III *User's Guide*

Part I

Group Labs

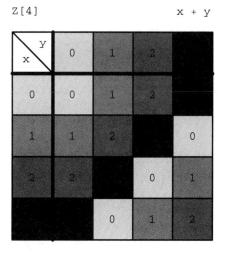

Group Lab 1

Using Symmetry to Uncover a Group

■ 1.1 Prerequisites

None.

■ 1.2 Goals

In this lab, we try to discover some of the basic properties of a group by considering the symmetries of a regular triangle.

■ 1.3 Getting started? Begin here

In most of the labs in *Exploring Abstract Algebra with Mathematica*, one needs to first read in the *Mathematica* packages necessary to provide the functionality within the lab. This is done using a `Needs` statement. (By opening the `AbstractAlgebra`Master`` package, all the functions in the `AbstractAlgebra` packages are made available.) On opening this notebook, you were probably asked, "Do you want to automatically evaluate all the initialization cells in the Notebook?" If you answered affirmatively, then you do not need to evaluate the following cell (but it doesn't hurt to do so). If you answered negatively, then you need to evaluate this cell.

```
Needs["AbstractAlgebra`Master`"];
SwitchStructureTo[Group];
```

▣ 1.4 A symmetry of an equilateral triangle

Suppose we consider an equilateral triangle with its vertices labeled with the numbers 1, 2, and 3. (Evaluate the cell, but do not worry about the parameters 3, {1, 2, 3}, and "D.")

```
triangle = ShowFigure[3, {1, 2, 3}, "D"];
```

We might think about how we can "move" this triangle so that after the "movement," it appears as if it was not moved at all, except for a new ordering of the labels on the vertices. For example, suppose we draw an imaginary line from the vertex labeled 3 perpendicularly to the side opposite and then reflect the triangle in (across) this line. (Evaluate the following cell.)

```
Show[triangle, Epilog → {Blue, Line[{{0, -1.5}, {0, 2.0}}]}];
```

After reflecting the triangle across the blue line, we have the following result (evaluate).

```
ShowPermutation[{1 → 2, 2 → 1, 3 → 3}]
```

This reflection is an example of a (geometric) transformation called a *symmetry*. Note that there are changes in the location of the vertices: vertex 1 went to the previous location of vertex 2, vertex 2 went to vertex 1, and vertex 3 went back to itself. This is called a *permutation* of the vertices. One way to represent this permutation in *Mathematica* is by

$$\{1 \to 2, \ 2 \to 1, \ 3 \to 3\}.$$

Note that even though vertex 3 went to itself, it needs to be included in the permutation list. Observe that braces are used to contain the whole list. Let's denote this first permutation by p[1]. (Yes, evaluate it.)

```
Clear[p]
p[1] = {1 → 2, 2 → 1, 3 → 3}
```

▣ 1.5 Are there other symmetries?

What other symmetries (permutations) are there? Let's label each of the others p[2], p[3], and so on and create the rules to form them. (Note that the rule symbol -> is a hyphen followed by a greater-than sign; if you hit the escape key before and after typing ->, the result is →.) As you discover each permutation, type p[k] = {1 → x, 2 → y, 3 → z}, where you supply appropriate values for *k*, *x*, *y*, and *z*. We give you one more.

> **p[2] = {1 → 2, 2 → 3, 3 → 1}**

To test if p[k] is really a valid permutation of this triangle and to see the result geometrically, type ShowPermutation[p[k]].

> **ShowPermutation[p[2]]**

Note that the permutation {1 → 2, 2 → 3, 3 → 1} can also be determined by considering only the second "coordinates" of each rule (*x* → *y*), since the first coordinates are always the same for each permutation. Here is a matrix view of this permutation.

> **PermutationMatrix[{1 → 2, 2 → 3, 3 → 1}]**

Note how each column represents a rule of the form *x* → *y*. Therefore, the list {2, 3, 1} can also be used to represent p[2] = {1 → 2, 2 → 3, 3 → 1}. So one could type

> **p[2] = {2, 3, 1}**

and then

> **ShowPermutation[p[2]]**

would give the same result. (You have been evaluating each input cell as you read along, right?)

The function ShowPermutation is set up to accept either notation.

Q1. Describe geometrically what the permutation p[2] does to the triangle.

Q2. How many other symmetries of the triangle (in permutation form) can you find? Label them p[3] and so on as indicated above, and test each one (with ShowPermutation) to see if the symmetry does what you thought. As you define each p[k], place in (* comments *) (or in a text-type cell) a geometric description of what the symmetry does to the triangle.

Q3. How many (distinct and unique) symmetries have you found altogether (including p[1] and p[2])?

■ 1.6 Multiplying the transformations

Now we would like to know the interrelationships, if any, between these various symmetries. For instance, suppose we first applied the symmetry given by permutation p[1] and then followed this with the symmetry given by p[2]. What would the figure look like? Is there a permutation in your list that would give this "product?"

> **Q4.** Take a moment and see if you can figure out what the product of p[1] followed by p[2] would be. If you have an answer, how did you come up with it?

In Group Lab 7 we will learn more of the details of how we "multiply" permutations, but for now just assume that it can be done and allow *Mathematica* to do it for you. We use the function MultiplyPermutations to do this. (If you do know how to do this, we assume the product works from right to left.) For question 4, evaluate the following to obtain the result. (First we review the definitions of the permutations p[2] and p[1].)

```
p[1]
p[2]

prod = MultiplyPermutations[p[2], p[1], Mode → Textual]
```

This yields the permutation that is the product of p[1] followed by p[2]. (Note the order of the arguments in MultiplyPermutations.) Now evaluate the following cell.

```
ShowPermutation[prod]
```

This draws the original triangle and the triangle resulting from the product of p[1] followed by p[2]. How does the second triangle compare to the first and how does the answer given for prod (which should have been {3, 2, 1}) relate to the transformation that changes the first triangle into the second?

> **Q5**. Does the permutation given by {3, 2, 1} represent a symmetry of the triangle? If so, describe the symmetry geometrically. Is it among the list of permutations you found in question 2?

■ 1.7 Are there any commuters?

Evaluate the following.

```
MultiplyPermutations[p[1], p[2], Mode → Textual]
```

> **Q6**. Did you get the same result as when you did Multiply-
> Permutations[p[2], p[1]]? Why or why not? What does this say
> about the operation of multiplying permutations? Can you think of a mathemati-
> cal term used to describe this?

■ 1.8 Is it always bad to be closed-minded?

Now we want to set up a complete "multiplication" table of all possible products, as we did with addition and multiplication when a student in elementary school. Set up your table as follows.

```
      p[1]      p[2]      p[3]    . . .
p[1]

p[2]

p[3]
  .
  .
  .
```

The top row and left column simply act as labels for the table. The body of the table is filled in by multiplying the row entries by the column entries. For example, the entry in row 3 and column 2 can be obtained by evaluating

```
MultiplyPermutations[p[3], p[2]]
(* note that this product only makes sense if
   you have already defined p[3] as a permutation *)
```

Note the order: the entry in position (i, j) (row i, column j) results from first performing the permutation corresponding to the one in column j followed by the one in row i, but when one enters this into the function MultiplyPermutations, the order is reversed. (You might wonder why this is the case. It will become clearer when you study permutations in more detail, but for now consider as a hint that a permutation is really a function and multiplying permutations is actually composing functions, usually done from right to left.)

> **Q7**. Complete the table (using p[i] notation for each product, where possible).

If you find that you obtain a product that is not in your original list, perhaps you need to look over your list and possibly make changes to it. Eventually, your list should terminate and the result of the product of any two should be one that is already in your list. In this case, the members in the body of the table should consist of only the elements listed in the column or row headings. We then say that the set of elements (symmetries of the triangle, as represented by the permutations, in this case) is said to be *closed* under this product.

▣ 1.9 We should try to find our identity

Q8. When we consider addition of real numbers, we have $r + 0 = 0 + r = r$ for all real numbers r. Similarly, with multiplication of real numbers, we have $r * 1 = 1 * r = r$. Does an analogous situation occur with the symmetries of the triangle? Is there any symmetry (say, permutation `p[j]`) which, when followed by another symmetry (`p[i]`), yields the product consisting of simply the second symmetry (`p[i]`) (i.e., `MultiplyPermutations[p[i], p[j]] = p[i]`)? If so, what is this `p[j]` and what is special about it? Do we have a name for such objects as 0, 1, and `p[j]`?

▣ 1.10 Is it perverse to not have an inverse?

In section 1.9 we found an identity (or should have found one) that takes every vertex of the triangle back to itself (i.e., leaves the triangle unchanged). Recall that with addition of real numbers, since 0 is the additive identity, for any real number r, there is another real number called $-r$ such that $r + (-r) = (-r) + r = 0$. Similarly, with multiplication of real numbers, for any $r \neq 0$, there exists $\frac{1}{r}$ such that $r * \frac{1}{r} = \frac{1}{r} * r = 1$ (and recall that 1 is the multiplicative identity).

Q9. Given a symmetry (permutation) `p[i]`, is there another symmetry `p[j]` such that `MultiplyPermutations[p[i], p[j]]` yields this identity found above? If so, then we say that `p[j]` is the inverse of `p[i]` (and `p[i]` is the inverse of `p[j]`). Does every symmetry found (when answering question 7) have an inverse? For each of the symmetries, determine the inverse (where possible). Below, list the elements and their corresponding inverses? Any observations?

▣ 1.11 Should we associate together?

Again, appealing to addition and multiplication of real numbers, recall that

$$(a + b) + c = a + (b + c) \text{ and } (a\, b)\, c = a\, (b\, c)$$

for all real numbers *a*, *b*, and *c*. These equations say that both addition and multiplication of real numbers satisfy the associative property.

Another question one might ask is whether we have associativity of permutations. In other words, is it true that `MultiplyPermutations[p[i], Multiply-Permutations[p[j], p[k]]]` = `MultiplyPermutations[MultiplyPermutations[p[i], p[j]], p[k]]` for all *i*, *j*, and *k*? This is comparable to asking if $a + (b + c) = (a + b) + c$. (Convince yourself of this.) Following is the *Mathematica* code to determine this. You have a choice of checking every possibility or you can choose to do a certain number of random checks. Do one test or the other. (If you check every possibility, there are $6^3 = 216$ tests in this case. Do you know why?)

▪ Test every possibility

To do the following test, it is assumed that you found exactly six permutations (which is how many there are) and that you have defined them in the variables `p[1], p[2], ... p[5], p[6]`.

```
assoc = True;
Do[temp = MultiplyPermutations[p[i], MultiplyPermutations[
      p[j], p[k] ] ] === MultiplyPermutations[
    MultiplyPermutations[p[i], p[j] ], p[k] ];
  assoc = And[assoc, temp], {i, 6}, {j, 6},
  {k, 6}];
assoc
```

▪ Test a random number of times

To do the following test, it is assumed that you found exactly six permutations (which is how many there are) and that you have defined them in the variables `p[1], p[2], ... p[5], p[6]`.

```
checkThisMany = 32;
randfn[_] := Random[Integer, {1, 6}]
assoc = True;
Do[index1 = randfn[1]; index2 = randfn[2];
  index3 = randfn[3]; temp = MultiplyPermutations[p[index1],
    MultiplyPermutations[p[index2], p[index3] ] ] ===
    MultiplyPermutations[ MultiplyPermutations[
      p[index1], p[index2] ], p[index3] ];
  assoc = And[assoc, temp],
{checkThisMany}];
assoc
```

■ **Now the question**

> **Q10**. Do you think that multiplication of these symmetries (permutations) is associative? Why or why not?

1.12 What else?

> **Q11**. What other observations, if any, can you make from these explorations?

1.13 Let's group it all together

The collection of the six symmetries of the triangle—you found all six, right?—with the product of one transformation followed by another (but here viewed as six permutations with multiplication of the permutations) is an example of a set with a binary operation. Below is a formal definition of what we have been investigating.

Given a set G and an operation $*$, we call G a group if it meets the following conditions.

> 1. The set is *closed* under this operation (i.e., $g * h$ is in G for all g and h in G).
>
> 2. Among the elements there exists an *identity* (i.e., there exists an e in G such that $e * g = g * e = g$ for all g in G).
>
> 3. Each element has an *inverse* among the set of elements (i.e., for all g in G, there exists h in G such that $g * h = h * g = e$).
>
> 4. The operation is *associative* (i.e., $f * (g * h) = (f * g) * h$ for all f, g, and h in G).

Furthermore, if we have $g * h = h * g$ for all g and h in G, we call G a *commutative* (or *Abelian*) group. We will be studying the properties of groups in this course.

> **Q12**. Is the set of six symmetries (permutations) with the operation of following one transformation by another (i.e., multiplying the permutations) an example of a group? Why or why not? Is it Abelian? Why or why not?

> **Q13**. Determine another example of a set with an operation that is a group?

Group Lab 2

Determining the Symmetry Group

of a Given Figure

▨ 2.1 Prerequisites

Though not absolutely necessary, it would be useful if you completed Group Lab 1 before attempting this lab.

▨ 2.2 Goals

In this lab, we continue to look at symmetries. We resume where Group Lab 1 left off. The goal is to find the complete list of symmetries (via permutations) for a variety of (more or less) random figures. When "complete," this list should be the "symmetry group" of the object.

▨ 2.3 Symmetries and how to find them

To start this lab, the following package needs to be read into *Mathematica*.

```
Needs["AbstractAlgebra`Master`"];
SwitchStructureTo[Group];
```

Consider the following array of figures.

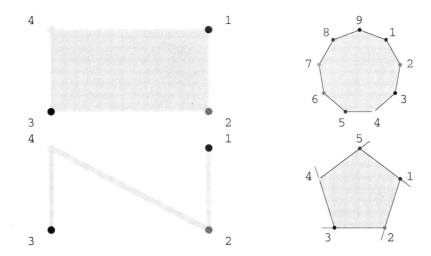

When we look for the "symmetry group" of any one of these objects, we look for a complete list of symmetries for the object. (Technically, a symmetry of an object *F* is an isometry π that maps *F* onto itself, with distances being preserved. Since an isometry is actually a function, the "natural" way of combining symmetries is to use function composition.)

Symmetries can come in several varieties. They often involve *rotations* about a certain point (the "center") through some angle. For example, if we rotate the figure at the upper left about the center of the rectangle (where the diagonals intersect) through an angle of 180°, the figure is placed back onto itself. Note that the figure does NOT include the colored and labeled dots used to mark the vertices; these are only tags to help coordinate the movements. With this rotation, observe that vertices 1 and 3 are interchanged, as are vertices 2 and 4. We can describe this by

$$\{1 \rightarrow 3, \ 2 \rightarrow 4, \ 3 \rightarrow 1, \ 4 \rightarrow 2\}$$

or, in block permutation notation,

PermutationMatrix[{1 → 3, 2 → 4, 3 → 1, 4 → 2}]

or, as you may recall from the last lab, {3, 4, 1, 2}, which constitutes the second coordinates or row of the representations above. Each of these is a representation for the permutation of the vertices that corresponds to the described rotational symmetry. Thus, we can say that the permutation {3, 4, 1, 2} represents one of the symmetries of this figure. (What are the others?)

Consider the rectangle again. If we imagine a horizontal line parallel to the base and through the center and consider it a line of reflection, then we have another common type of symmetry: the *reflection*. In this case, we could also describe the symmetry by what happens to the vertices. We would say that

$$\{1 \rightarrow 2, \ 2 \rightarrow 1, \ 3 \rightarrow 4, \ 4 \rightarrow 3\}$$

or

```
PermutationMatrix[{1 → 2, 2 → 1, 3 → 4, 4 → 3}]
```

or {2, 1, 4, 3} describes the permutation of the vertices that corresponds to this reflectional symmetry.

There are several questions you may ask yourself. Is every symmetry either a rotation or a reflection? (You might consider the case of a finite object, such as the rectangle in the figure, in contrast with an infinite object, such as an infinitely long wall-papered wall.) You might also wonder if every object has both rotational and reflectional symmetries. For this, consider the lower right figure, which has extra "wings" off the edges.

■ 2.3.1 Getting started

Evaluate the following cell to see the figure you will first consider.

```
firstfig = ShowOne[Lab2]
```

Your task is to determine the symmetry group for this figure, using permutations as the elements. In other words, as in the first lab, you are to determine the symmetries (transformations) in the form of permutations that will move the figure onto itself.

As before, you may express your permutations in either the form {$a, b, c, \ldots$} or the form {$1 \rightarrow a, 2 \rightarrow b, 3 \rightarrow c, \ldots$}. Also, as in Group Lab 1, use `p[1]` = {a,b,c, ...} or p[1] = {1 → a, 2 → b, 3 → c, ...} to define your first permutation, `p[2]` for your second, and so on.

Following is a list of functions that can be used for bookkeeping and testing while determining the permutations in the symmetry group of a figure. At the end of this section is a button that will bring up a palette to make working with these functions a bit easier.

■ 2.3.2 Functions/variables to use in this lab

From Group Lab 1, you may already be familiar with how the functions `ShowPer mutation` and `MultiplyPermutations` work; to refresh yourself, simply type ? followed by the command.

```
? ShowPermutation
```

> **? MultiplyPermutations**

What follows is a short description of the other functions that can be used in this lab.

Type `ShowOne[Lab2]` to get a figure with which to work.

> **? ShowOne**

If you are fully confident that `p[i]` is indeed a permutation of the vertices and is a symmetry of the figure, then type `AddPermToGroup[p[i]]`, which will add it to the list of all your symmetries.

> **? AddPermToGroup**

`DropPermFromGroup[p[i]]` removes the permutation `p[i]` from the accumulated list of symmetries.

> **? DropPermFromGroup**

To see the list of symmetries accumulated thus far by the previous two functions, type `MySymmetryGroup`.

> **MySymmetryGroup**

If you have a permutation `p[i]` that you believe is a symmetry, but you want to check to be sure, type `TestPermutationQ[p[i]]` and either `True` or `False` will be returned.

> **? TestPermutationQ**

To determine whether the accumulated permutations (as found in `MySymmetry-Group`) are all proper elements of the symmetry group of the figure, calling `GoodGroupElementsQ[]` answers with `True` or `False`.

> **? GoodGroupElementsQ**

When you believe that you have found all the symmetries and have added them to `MySymmetryGroup` via `AddPermToGroup`, then you may check to see if your group is correct (complete) by typing `CompleteGroupQ[]`, which will return `True` or `False` and possibly additional information.

> **? CompleteGroupQ**

If you get stuck trying to think of what other possible permutations might yield a symmetry, type `ShowPossiblePermutations[n]` (*n* being the number of vertices if a regular *n*-gon is shown, or 2 if a zee or rectangle is shown). You are given a list of all possible permutations.

? ShowPossiblePermutations

ShowPossiblePermsAsRules [n] is the same as the last function except that the permutations are listed as rules, which may be helpful in using the RestrictList function described below.

? ShowPossiblePermsAsRules

RestrictList [r] can be used to restrict the list produced by ShowPossiblePermsAsRules when you know that you want one vertex to go to another specific vertex but want to explore what can happen with the other vertices. Note that *r* should be in the form $i \to j$, $\{i \to j\}$, or as a list of either of these.

? RestrictList

It is easy to forget this list of functions. By typing AvailableFunctions (after evaluating the following), you can have the list of names always available.

```
AvailableFunctions := {ShowPermutation, MultiplyPermutations,
    ShowOne, AddPermToGroup, DropPermFromGroup,
    "MySymmetryGroup", TestPermutationQ, GoodGroupElementsQ,
    CompleteGroupQ, ShowPossiblePermutations,
    ShowPossiblePermsAsRules, RestrictList};
AvailableFunctions
```

As an alternative to typing the functions just described, there is a palette specifically designed for this lab. This also includes a tool that makes the specification of permutations easier. Click here to use this palette.

◼ 2.4 Your turn

To see the figure again for the first exercise, evaluate the following cell. (Note: You should have already viewed it once when evaluating firstfig = ShowOne [Lab2] in the section 2.3.1.)

```
firstfig[[1]] // Show;
```

> **Q1.** Using the information found in section 2.3.2, determine the complete list of all the symmetries (in the form of permutations) for the given figure. Use p[i] notation for each one (or ppp[i] if used by one of the functions described above). Show all work in establishing this list. (Or show your work with the Permutation Specifier, found in a palette for this lab.)

```
AvailableFunctions
(* evaluate to be reminded what functions are available *)
```

Q2. Use the function `MultiplyPermutations` as in the Group Lab 1 to determine the Cayley table for the symmetry group of this figure. (`Multiply-Permutations[p[i], p[j]]` determines the permutation of `p[j]` followed by `p[i]`.)

Q3. Repeat the steps in questions 1 and 2 with `secondfig`.

```
Clear[p, ppp]
secondfig = ShowOne[Lab2]

AvailableFunctions
(* evaluate to be reminded what functions are available *)
```

Q4. Repeat the steps in questions 1 and 2 again.

```
Clear[p, ppp]
thirdfig = ShowOne[Lab2]
```

Group Lab 3

Is This a Group?

▣ 3.1 Prerequisites

To complete this lab, you should have already seen the definition of a group and become familiar with the basic group properties: being closed, having an identity, inverses, and associativity (and commutativity).

▣ 3.2 Goals

Does a set of elements and some operation on the elements form a group? This lab explores a number of pairs consisting of a set and an operation. A Cayley table for the set and operation is presented and the user is asked which of the defining properties of a group hold for this pair.

▣ 3.3 When do we have a group?

We know that before we can call a set with an operation a *group*, it must satisfy certain properties. Namely, the set must be *closed* under the operation, there must be an *identity* element, each element must have an *inverse*, and the operation needs to be *associative*. Our goal is to determine which, if any, of the above properties hold when presented a set with an operation. Before we begin, we need to read in the *Mathematica* code used in this lab.

```
Needs["AbstractAlgebra`Master`"];
SwitchStructureTo[Group];
```

When you evaluate ShowOne[Lab3] to obtain the first problem, you will be shown something like this.

ShowOne[Lab3]

```
The alleged group consists of the set of elements
{4, 0, 1, 3, 2} and the operator is myPower. This is case
15. Below is the Cayley table for this
alleged group.
```

^	2	4	1	3	0
2	4	16	2	8	1
4	16	256	4	64	1
1	1	1	1	1	1
3	9	81	3	27	1
0	0	0	0	0	Ind

Note that the Cayley table may have the elements (in the headings) in a different order than that in the written description. (Before making the table, they are randomized—for this lab only.) The properties of closure, identity, and inverses should be readily observable from this Cayley table. Associativity is a little more difficult. We will use a function that randomly tests for associativity, testing 25 triples by default. Since what is shown is identified as "case 15," the argument to our function is LG[15] (for Lab Groupoid 15.). Evaluate the following cell.

RandomAssociativeQ[LG[15]]

If you wish to change the number of tests, add this number as the second argument:

RandomAssociativeQ[LG[15], 30]

▣ 3.4 Your turn

Q1. Use `ShowOne[Lab3]` to bring up an alleged group. With careful examination of the Cayley table, you should be able to answer the following. (Also indicate the "case number" of the alleged group.)

(a) Determine if the set is closed under the given operation. If so, why is it, and if not, why not?

(b) Determine if there is an identity element. If so, state it, and if not, why not?

(c) Determine the inverse of each element, where one exists.

(d) Use the associativity test to make a conjecture whether the operation is associative or not. (Remember to specify the particular case number in the argument of `LG[]`.)

(e) Although not a required property to be a group, determine if this operation is commutative.

(f) Finally, determine whether the set with the operation is indeed a group. (Guess, if you are not fully convinced about the associativity.)

`ShowOne[Lab3]`

Q2. Repeat the instructions in question 1. (Again, indicate the "case number" of the alleged group for this question.)

`ShowOne[Lab3]`

Q3. Repeat the instructions in question 1. (Again, indicate the "case number" of the alleged group for this question.)

`ShowOne[Lab3]`

Q4. Repeat the instructions in question 1. (Again, indicate the "case number" of the alleged group for this question.)

`ShowOne[Lab3]`

Group Lab 4

Let's Get These Orders Straight

▣ 4.1 Prerequisites

To complete this lab you should be familiar with the basic definition of a *group*. You should also be familiar with the definition of the order of an element in a group. (Recall that the *order* of an element g of a finite group G is the least positive integer k such that g^k is equal to the identity of G.)

▣ 4.2 Goals

In this lab, we look at issues regarding the order of groups and their elements. First we consider the relationship between the order of g and the order of the inverse of g. We then look at the distribution of the orders of elements in $\mathbb{Z}_n$, followed by an inspection of which elements share a common order. We then begin an exploration regarding the probability that an arbitrary element of $\mathbb{Z}_n$ will generate the whole group. Finally, we consider the order of the group U_n (the multiplicative units of $\mathbb{Z}_n$) and try to find an expression for this order in terms of n.

▣ 4.3 Order of g and its inverse

Suppose we consider any group G and take a random element g from G. The issue we would like to consider is how the order of g, denoted $|g|$, compares with the order of its inverse, $|g^{-1}|$.

For this section of the lab, the groups we consider come from the following list: $\mathbb{Z}_n$ ($1 < n < 31$), U_n ($2 < n < 41$), D_n ($1 < n < 8$), GaussianUnits ($\{\pm 1, \pm i\}$, under multiplication), and IntegerUnits ($\{\pm 1\}$, under multiplication).

The function ShowOne[Lab4] presents a random group G from the preceding list and then chooses a random element g from the chosen group. First, we read in the *Mathematica* code needed for this lab; evaluate the following two cells.

```
Needs["AbstractAlgebra`Master`"];
SwitchStructureTo[Group];
SetOptions[ListPlot, PlotStyle → RGBColor[0, 0, 1]];

{G, g} = ShowOne[Lab4]
op = Operation[G];
```

This gives a random group G and an element g in G; we have also defined op as a variable for the group operation.

We now wish to determine the order of g in G. We can do this by successively applying the operation to g.

```
op[g, g]
```

This calculates g^2. To calculate g^3, apply op to g and the last result (indicated by %).

```
op[g, %]
```

To calculate g^4, apply op to g and the last result (%), and so on.

```
op[g, %]
```

Q1. For the group G given above, determine the order of the element g. Also record your group G and element g.

Let's pick another group and element.

```
{G, g} = ShowOne[Lab4]
```

As an alternative, one can use the ElementToPower function that calculates g^n for any integer n.

```
ElementToPower[G, g, 2]
```

The following calculates the first 6 powers; adjust the range of the table accordingly to determine $|g|$.

```
Table[ElementToPower[G, g, k], {k, 1, 6}]
```

> **Q2**. What is the order of *g* in this case? (Record your group *G* and element *g*.)

We can also use the following function to calculate the order. Since it is important to know how to calculate orders, be sure you answer questions 1 and 2 "by hand" before confirming your answer with this function.

```
Order[G, g]
```

> **Q3**. For the given group *G*, determine the inverse of the element *g* in the group. (You are expected to do this without using *Mathematica*.)

Just as it is important to be able to determine orders through calculations, it is also important to be able to determine ("by hand") the inverse of an element. The following command can be used to confirm your answer to the last question, but evaluate it only after you have found an answer.

```
invg = GroupInverse[G, g]
```

Now let's ask *Mathematica* to help us calculate the order of the inverse.

```
Order[G, invg]
```

> **Q4**. In this case, what is the relationship between the order of *g* and the order of its inverse? Record *g* and *g*'s inverse.

Let's try this again. First we pick out a group and an element.

```
{G, g} = ShowOne[Lab4]
```

Next determine the order of *g* in *G* (without *Mathematica*) and then use the following to confirm your answer.

```
Order[G, g]
```

Now determine the inverse of *g* and confirm with the following.

```
invg = GroupInverse[G, g]
```

Finally, calculate the order of the inverse and compare it to the order of *g* itself.

```
Order[G, invg]
```

> **Q5**. In this case, what is the relationship between the order of *g* and the order of its inverse? Since the pair (G, g) you investigated was randomly generated, include in your answer the group *G*, elements *g* and g^{-1}, and the orders of *g* and g^{-1}.

Let's generate some more data. The output from the following cell might take a little time to compute. It will consist of a list of several groups, a random element from each group, the inverse of the element, and the orders of the element and its inverse.

```
TableForm[Table[{G, g} = ShowOne[Lab4, Verbal → False];
  {GroupoidName[G], g, invg = GroupInverse[G, g],
   Order[G, g], Order[G, invg]}, {20}], TableHeadings →
  {None, {"group", "g", "g⁻¹", "|g|", "|g⁻¹|\n"}},
  TableSpacing → {0.5, 3}]
```

> **Q6**. Make a conjecture about the relationship between the order of an element in a group and the order of its inverse.

> **Q7**. Try to prove your conjecture.

▣ 4.4 Distribution of the orders of elements in $\mathbb{Z}_n$

We now pick a random index n between 20 and 60 and consider the group $\mathbb{Z}_n$.

```
n = Random[Integer, {20, 60}]
G = Z[n]
```

Let's take a random element from $\mathbb{Z}_n$ and find its order. Making a table, we will do this experiment with 20 trials.

```
numTrials = 20;
TableForm[Table[g = RandomElement[G];
   {g, Order[G, g]}, {numTrials}],
  TableHeadings → {None, {"g", "|g|\n"}},
  TableSpacing → {0.5, 3}]
```

Record, either here or on your paper, both the index n used and all the *different* orders (appearing in the second column); at this point, we only want to record *which* orders occur for a given index n.

Now repeat this experiment with a new group by evaluating the following cell. *In the unlikely event that you get the same index, evaluate the cell again to get a new value for n.*

```
n = Random[Integer, {20, 60}];
G = Z[n]
numTrials = 20;
```

```
TableForm[Table[g = RandomElement[G];
  {g, Order[G, g]}, {numTrials}],
 TableHeadings → {None, {"g", "|g|\n"}},
 TableSpacing → {0.5, 3}]
```

Record both the index *n* and the orders that appear. Keep evaluating this input cell until you can answer question 8.

Q8. Given a positive integer *n*, what can you say about the orders of the elements of $\mathbb{Z}_n$? State your conclusion formally in the form of a conjecture.

4.5 Another look at orders

In section 4.4, we considered the relationship between $|g|$ and $|G|$ for elements *g* from some group $G = \mathbb{Z}_n$. Now that the relationship is determined, we wish to consider the frequencies of various orders. In other words, how frequently does a particular order occur, once we know that it does occur? For the moment, we will still consider $G = \mathbb{Z}_n$. Let's pick an arbitrary index for this group.

```
n = Random[Integer, {8, 50}]
G = Z[n]
```

Evaluate the following command.

```
ShowGroupOrders[Z[n]];
```

If these graphics are hard to read, you may wish to enlarge them. (To do so, click once on the graphic to select it and then drag from the lower right corner until the graphic is sufficiently enlarged.)

A word of explanation is in order. The first graph, a `ListPlot`, consists of a graphical representation of what was accomplished in section 4.4. Along the horizontal axis are the elements in $\mathbb{Z}_n$ and along the vertical axis are the orders the elements can take. For each element in *G*, there is a dot at the height corresponding to the order of that element.

The second graph is a bar chart showing the frequencies with which the orders occur. Along the bottom are the different orders of elements in the group, and the height of the bar corresponds with how many elements in the group have that particular order.

Consider another example.

```
n = Random[Integer, {8, 50}]
G = Z[n]
ShowGroupOrders[G];
```

Let's now change the form of what we observe. Instead of simply making a list of the orders, we divide the order of the group by the index n and then plot these values.

```
data = MapIndexed[ #1/First[#2] &, data]
ListPlot[data, AxesOrigin → {0, 0}, PlotRange → {0, 1}];
```

Q17. Does this look at all familiar? Have you seen this before?

Let's continue our investigation into the order of U_n by considering various types of integers n. For example, when n is prime we get

```
data1 = Table[n = Prime[p];
    {n, Size[U[n]]}, {p, 1, 40}]
data2 = Table[n = Prime[p];
    {n, Size[U[n]] / n}, {p, 1, 40}];
Show[GraphicsArray[{ListPlot[data1, DisplayFunction →
    Identity, Ticks → {Table[50 i, {i, 1, 4}], Automatic}],
    ListPlot[data2, DisplayFunction →
    Identity, PlotRange → {0, 1},
    Ticks → {Table[50 i, {i, 1, 4}], Automatic}]}],
    DisplayFunction → $DisplayFunction];
```

The first plot reflects the orders of U_n for prime n (horizontal axis is n and vertical is $|U_n|$). The second plot has the vertical axis being $|\frac{U_n}{n}|$ (again, for prime n). We use just the latter type of plot in what follows.

When n is a power of two we get

```
max = 8;
data1 = Table[n = 2^p;
    {n, Size[U[n]]}, {p, 1, max}]
data2 = Table[n = 2^p;
    {n, Size[U[n]] / n}, {p, 1, max}]
ListPlot[data2, PlotRange → {0, 1}, AxesOrigin → {0, 0}];
```

When n is a power of three we get

```
max = 8;
data1 = Table[n = 3^p;
    {n, Size[U[n]]}, {p, 1, max}]
data2 = Table[n = 3^p;
    {n, Size[U[n]] / n}, {p, 1, max}]
ListPlot[data2, PlotRange → {0, 1}, AxesOrigin → {0, 0}];
```

> **Q18.** The following code shows similar results for powers of 4. Evaluate it. Then change it to powers of 5, 6, 7, 8, 9, and so on until you can make some kind of conjecture. Can you prove it? (max is the number of terms to compute; you may wish to reduce max to 5 for larger values.)

```
powersOf = 4;
max = 6;
data1 = Table[n = powersOf^p;
    {n, Size[U[n]]}, {p, 1, max}]
data2 = Table[n = powersOf^p;
    {n, Size[U[n]] / n}, {p, 1, max}]
ListPlot[data2, PlotRange → {0, 1}, AxesOrigin → {0, 0}];
```

Now consider multiples of three.

```
data1 = Table[n = 3 p;
    {n, Size[U[n]]}, {p, 1, 40}]
data2 = Table[n = 3 p;
    {n, Size[U[n]] / n}, {p, 1, 40}];
Show[GraphicsArray[{ListPlot[data1, DisplayFunction →
    Identity,
     PlotStyle → {RGBColor[0, 0, 1], PointSize[0.025]}],
    ListPlot[data2, DisplayFunction → Identity,
     AxesOrigin →
    {0, 0}, PlotRange → {0, 1},
     PlotStyle → {RGBColor[0, 0, 1],
    PointSize[0.025]}]}], DisplayFunction →
    $DisplayFunction];
```

If the graphs appear a little small, click on the graph, then move the mouse to the lower right corner. When the cursor is a double arrow (facing NW and SE), then press down and drag to an appropriate size.

> **Q19.** Can you see any dichotomy in the immediately preceding graphs? Describe it and try to explain why it exists. Look at the data used to generate the plots. (In particular, consider factoring some of the indices.)

Let's try multiples of two.

```
multiplesOf = 2;
max = 60;
data1 = Table[n = multiplesOf * p;
    {n, Size[U[n]]}, {p, 1, max}]
data2 = Table[n = multiplesOf * p;
    {n, Size[U[n]] / n}, {p, 1, max}];
Show[GraphicsArray[{ListPlot[data1, DisplayFunction →
    Identity,
     PlotStyle → {RGBColor[0, 0, 1], PointSize[0.025]}],
```

```
ListPlot[data2, DisplayFunction → Identity,
 AxesOrigin →
{0, 0}, PlotRange → {0, 1},
 PlotStyle → {RGBColor[0, 0, 1],
PointSize[0.025]}]}], DisplayFunction →
$DisplayFunction];
```

Q20. Change `multiplesOf` to 5, 6, 7, 8, 9, and so on until you can make some kind of conjecture. (Again, `max` indicates the number of terms that are calculated; you may wish to modify it.) Can you provide any proof?

Q21. What is the order of U_n? Give as complete an answer as possible, even if you don't have all the cases covered.

Group Lab 5

Subversively Grouping Our

Elements

▦ 5.1 Prerequisites

To complete this lab, you should be familiar with the definition of a subgroup of a group.

▦ 5.2 Goals

What constitutes a subgroup? What elements are necessary before a set can be considered a subgroup? What do the subgroups of $\mathbb{Z}_n$ look like? What about the subgroups of U_n? What is the probability that a randomly chosen subset of elements from $\mathbb{Z}_n$ will actually be a subgroup? What elements of $\mathbb{Z}_n$ guarantee *closure* to the full group? These are some of the questions that are explored in this lab.

▦ 5.3 When do we have a subgroup?

In Group Lab 3, we considered when a set with an operation on the set forms a group. In this lab, we consider when a subset of a group is a group in its own right (when using the operation from the parent set).

First let's consider a random group $\mathbb{Z}_n$ for n in [6, 20]. To define this group, we need to first read in the *Mathematica* package that defines $\mathbb{Z}_n$ and the other functions that we will be using.

```
Needs["AbstractAlgebra`Master`"];
SwitchStructureTo[Group];

n = Random[Integer, {6, 20}]
G = Z[n]
```

Next we pick a random integer m, less than n, and then choose this many elements from G and put them in a set that we call H.

```
m = Random[Integer, {1, Floor[N[√n]]}]
H = RandomElements[G, m, Replacement → False]
```

The question we would like to pursue first is whether this set H forms a subgroup of G, and if not, how we can make one with it.

Q1. In this case, is H a subgroup of G? Justify your answer. (Indicate the group G and subset H that were chosen.)

Now use the following command to confirm your answer to question 1.

```
SubgroupQ[H, G]
```

Simply knowing whether it is true or false if H is a subgroup of G is of limited value. We would also like to know how to make H become a subgroup (by adding certain elements, if necessary). Let's look at a Cayley table where we focus on the elements of H.

```
SubgroupQ[H, G, Mode → Visual]
```

Note that the elements colored red are in G but not in H.

Q2. The presence of red elements (if any) indicates that the set H does not satisfy which property relative to the operation in G?

What would happen if we modified H to include some (or all) of the elements that were sums of elements in H, but not already in H (namely, the red elements)? In the variable labeled `ElementsToAdd`, add the elements (between {}) you would like to join to H to see if you can make H a subgroup of G.

```
ElementsToAdd = {};
H = Union[Join[H, ElementsToAdd]]
SubgroupQ[H, G, Mode → Visual];
```

> **Q3**. By deleting the previous elements in the list `ElementsToAdd` and replacing them with new ones, keep modifying the code until you have enlarged *H* to become a subgroup. You will know that you are done if there are no longer any red elements. (You may have enlarged *H* to become *G* itself.) What is your subgroup *H*? (Also, what was your group *G*?)

Let's try this again.

```
n = Random[Integer, {6, 20}];
G = Z[n]

m = Random[Integer, {1, Floor[N[√n]]}]
H = RandomElements[G, m, Replacement → False]
SubgroupQ[H, G, Mode → Visual];
```

> **Q4**. As before, keep modifying the following code until you have enlarged *H* to become a subgroup. What is your subgroup *H*? (Also, what was your group *G* and the original *H*?)

```
ElementsToAdd = {};
H = Union[Join[H, ElementsToAdd]]
SubgroupQ[H, G, Mode → Visual];
```

Once you feel comfortable knowing how to enlarge *H* to make it a subgroup, you can ask *Mathematica* to do that part and you can focus on related issues. We call the new subgroup of *G* formed from the set *H* the *closure of H in G* and we can use the `Closure` command.

Let's try this with a new group.

```
n = Random[Integer, {6, 20}];
G = Z[n]

m = Random[Integer, {1, Floor[N[√n]]}]
H = RandomElements[G, m, Replacement → False]
```

If all we want is to determine the closure of *H*, we use the following.

```
Closure[G, H]
```

(Apply `SortGroupoid` to this result if you want to see the elements ordered.) If we ever want to know what else we can do with a function, it is often useful to ask for information about the function:

```
?Closure
```

Let's try a few of these variations.

```
Closure[G, H, ReportIterations → True]
```

Note that this returns the closure of *H* first, followed by the number of iterations and the results of each iteration. Let's consider another option.

```
Closure[G, H, Mode → Visual]
```

This simply shows the same information visually. These graphics can now be animated, if desired. (To do so, double-click on one of the graphics and adjust the motion with the arrow keys.) If we do not want to see all the graphics at once, we can try the following.

```
Closure[G, H, Mode → Visual, Staged → True];
```

To see the next stage, evaluate the following.

```
NextStage[Closure];
```

Or to see a previous stage, try

```
PreviousStage[Closure];
```

Either of these last two commands can be repeatedly cycled. Test yourself one more time; evaluate the following.

```
n = Random[Integer, {6, 20}]
G = Z[n]
m = Random[Integer, {1, Floor[N[√n]]}]

H = RandomElements[G, m, Replacement → False]
Closure[G, H, Mode → Visual, Staged → True];
```

You should know which elements need to be added. Now predict which elements will be colored red, if any, in the next iteration. When you think you know, evaluate the following. Keep doing this until you have found the closure of *H*.

```
NextStage[Closure];
```

Q5. You may have noticed that sometimes the closure becomes the whole group. There are many questions related to this to think about, some of which we consider in this lab. Here is one with which to start. If we let $H(n)$ be the size of the closure of *H* at the *n*th iteration (so $H(1) = |H|$), how big does $H(n)$ have to become before we can be certain that the closure of *H* will be all of *G*? You may wish to evaluate the following cell a number of times to gain some insights.

```
n = Random[Integer, {6, 30}];
G = Z[n]
m = Random[Integer, {1, Floor[N[√n]]}];

H = RandomElements[G, m, Replacement → False]
Closure[G, H, ReportIterations → True] // Last // Last //
  ColumnForm
```

▣ 5.4 Subgroups of $\mathbb{Z}_n$

Let's pick a random group $\mathbb{Z}_n$, where n is in [6, 30].

```
n = Random[Integer, {6, 30}]
G = Z[n]
```

What are the subgroups of this group $G = \mathbb{Z}_n$? One naive way of exploring this would be to pick a random set of elements and look at the closure, which we have seen always results in a subgroup. If we repeat this enough times, we might find all the subgroups of the group. Let's try it. Evaluate the following three to five times.

```
m = Random[Integer, {1, Floor[N[√n]]}];
H = RandomElements[G, m, Replacement → False]
Closure[G, H, Sort → True]
```

You may notice that the full group is often returned. Suppose we try restricting the number of elements in H to one or two. Try evaluating the following a number of times.

```
m = Random[Integer, {1, 2}];
H = RandomElements[G, m, Replacement → False]
Closure[G, H, Sort → True]
```

Q6. What do you think are the subgroups of $\mathbb{Z}_n$ for the n with which you have been working? (Also indicate what n you were given.)

Let's get a new group and try this again.

```
n = Random[Integer, {6, 30}]
G = Z[n]

m = Random[Integer, {1, 2}];
H = RandomElements[G, m, Replacement → False]
Closure[G, H, Sort → True]
```

> **Q7**. What group did you get this time? What do you think are the subgroups for this group?

▣ 5.5 $P(H < G)$ for a random subset H of $G = \mathbb{Z}_n$

Suppose we consider the group $\mathbb{Z}_{12}$. Recall that if H is a subgroup of G, we sometimes denote this by $H < G$. If we choose a random set of elements, H, from the elements of G, what is the probability that H is indeed a subgroup of G (denoted $P(H < G)$)? In this section, we pursue this question (and modifications of it, using other indices n in $\mathbb{Z}_n$).

```
G = Z[12]
```

First we start with $|H| = 1$. Evaluate the following to determine the results of randomly choosing one element 30 different times to see if it forms a subgroup of G.

```
sizeOfH = 1;
TableForm[
  Table[{H = RandomElements[G, sizeOfH, Replacement → False],
    SubgroupQ[H, G]}, {30}], TableDepth → 2,
  TableHeadings → {None, {"H", "(H < G)?\n"}}]
```

> **Q8**. Out of the 30 attempts, how many yielded a subgroup? What did you expect to happen? What would you expect to happen if 100 people did this experiment and each ran the loop for 1000 times instead of 30? Justify your answer. You should have an answer for $P(H < \mathbb{Z}_{12})$ for $H = \{g\}$ for some g in $\mathbb{Z}_{12}$.

Next we consider the case when $|H| = 2$. Evaluate the following to determine the results of choosing two elements (40 times) to see if the subset forms a subgroup of G.

```
sizeOfH = 2;
TableForm[
  Table[{H = RandomElements[G, sizeOfH, Replacement → False],
    SubgroupQ[H, G]}, {40}], TableDepth → 2,
  TableHeadings → {None, {"H", "(H < G)?\n"}}]
```

If you didn't get a `True`, try evaluating this cell again (which will not guarantee a `True` but may be worth trying, in some cases).

> **Q9.** How many successes did you have? (That is, how many times did you get `True`?) Which pair of elements yielded a subgroup, if any? Is there any (other) subset of size two that will (also) be a subgroup? Why or why not? Given a random set H of two elements from $\mathbb{Z}_{12}$, what do you think is the probability that H will be a subgroup (i.e., $P(H < \mathbb{Z}_{12})$)?

Next we consider the case when $|H| = 3$. Evaluate the following to determine the results of choosing three elements (40 times) to see if the subset forms a subgroup of G.

```
sizeOfH = 3;
TableForm[
  Table[{H = RandomElements[G, sizeOfH, Replacement → False],
    SubgroupQ[H, G]}, {40}], TableDepth → 2,
  TableHeadings → {None, {"H", "(H < G)?\n"}}]
```

If you didn't get a `True`, try evaluating this cell again. (Again, no guarantee.)

> **Q10.** How many successes did you have? Which triple of elements yielded a subgroup, if any? Is there any (other) subset of size three that will (also) be a subgroup? Why or why not? Given a random set H consisting of three elements of $\mathbb{Z}_{12}$, what is $P(H < \mathbb{Z}_{12})$?

Next we consider the case when $|H| = 4$. Evaluate the following to determine the results of choosing four elements (40 times) to see if the subset forms a subgroup of G.

```
sizeOfH = 4;
TableForm[
  Table[{H = RandomElements[G, sizeOfH, Replacement → False],
    SubgroupQ[H, G]}, {40}], TableDepth → 2,
  TableHeadings → {None, {"H", "(H < G)?\n"}}]
```

If you didn't get a `True`, evaluate again until you do, keeping track of how many attempts were made. If you get tired of doing this and think you know what you should expect, you can quit.

Next we consider the case when $|H| = 5$. Evaluate the following to determine the results of choosing five elements (40 times) to see if the subset forms a subgroup of G.

```
sizeOfH = 5;
TableForm[
  Table[{H = RandomElements[G, sizeOfH, Replacement → False],
    SubgroupQ[H, G]}, {40}], TableDepth → 2,
  TableHeadings → {None, {"H", "(H < G)?\n"}}]
```

Keep trying to get True or stop when you think you know what is likely to happen. Keep increasing the order of *H* and evaluating the cell until you can answer the following question.

Q11. Given $G = \mathbb{Z}_{12}$, what are the different possible orders of the subgroups of *G*? Also, how many subgroups are there of each order?

Q12. Now suppose we have $G = \mathbb{Z}_{10}$. What are the orders of the subgroups of *G* and how many subgroups are there of each order? Use the following cell if you want to do some experimenting.

```
G = Z[10];
sizeOfH = 1;
TableForm[
  Table[{H = RandomElements[G, sizeOfH, Replacement → False],
    SubgroupQ[H, G]}, {40}], TableDepth → 2,
  TableHeadings → {None, {"H", "(H < G)?\n"}}]
```

Q13. Now suppose we have $G = \mathbb{Z}_{11}$. What are the orders of the subgroups of *G* and how many are there of each order? Use the following cell if you want to do some experimenting.

```
G = Z[11];
sizeOfH = 1;
TableForm[
  Table[{H = RandomElements[G, sizeOfH, Replacement → False],
    SubgroupQ[H, G]}, {40}], TableDepth → 2,
  TableHeadings → {None, {"H", "(H < G)?\n"}}]
```

Q14. Summarize your findings by writing a conjecture about the subgroup structure of $\mathbb{Z}_n$. How might you prove your answer?

Q15. Given $G = \mathbb{Z}_n$ and a subset *H* of *G* with $|H| = m$, what is $P(H < G)$?

Q16. Look back at the subgroups you found for $\mathbb{Z}_{12}$. Starting with the subgroup(s) of order 2 and working up, what can you say about the relationship(s), if any, between the *order* of the subgroup, the elements of $\mathbb{Z}_{12}$, and the actual elements in the subgroup?

Q17. Do you think the results summarized in question 14 pertain only to $\mathbb{Z}_n$ or are they valid for other groups (either some or all) as well? Try the following to help you think about this question.

```
G = U[20]
sizeOfH = 2;
TableForm[
  Table[{H = RandomElements[G, sizeOfH, Replacement → False],
    SubgroupQ[H, G]}, {40}], TableDepth → 2,
  TableHeadings → {None, {"H", "(H < G)?\n"}}]
```

■ 5.6 Necessary elements for full closure

Suppose we focus on the group $\mathbb{Z}_{10}$ as an example in thinking about the question of what elements must be in a set H to guarantee that we have the closure of H be the entire group.

First we define G.

```
G = Z[10]
```

Then we look at a table of random sets H with one or two elements, with their closure.

```
TableForm[Table[m = Random[Integer, {1, 2}];
  {H = RandomElements[G, m, Replacement → False],
    Elements[Closure[G, H, Sort → True]]}, {25}],
  TableHeadings → {None, {"H", "closure of H\n"}},
  TableSpacing → {0.5, 3}, TableDepth → 2]
```

> **Q18.** Partition the elements of $\mathbb{Z}_{10}$ into three classes: (1) those whose presence in H cause the closure of H to be the full group, (2) those whose presence in H do NOT cause the closure of H to be the full group, and (3) the elements you are not sure about.

Consider another example, $\mathbb{Z}_8$.

```
G = Z[8]
TableForm[Table[m = Random[Integer, {1, 2}];
  {H = RandomElements[G, m, Replacement → False],
    Elements[Closure[G, H, Sort → True]]}, {25}],
  TableHeadings → {None, {"H", "closure of H\n"}},
  TableSpacing → {0.5, 3}, TableDepth → 2]
```

> **Q19.** Repeat question 18 with the results of $\mathbb{Z}_8$.

Finally, consider another example, $\mathbb{Z}_{12}$.

```
G = Z[12]
ColumnForm[Table[m = Random[Integer, {1, 2}];
    {H = RandomElements[G, m, Replacement → False],
        Elements[Closure[G, H, Sort → True]]}, {25}]]
```

Q20. Consider the results of the last three examples. If $H = \{g, h\}$ is a subset of $\mathbb{Z}_n$ and the closure of H is all of $\mathbb{Z}_n$, what can you conclude about the relationship between at least one of g or h and the number n?

5.7 Subgroups of U_n

For a quick review of the group U_n, let's view the first 20 groups where n runs from 1 to 20.

```
ColumnForm[Table[{n, Elements[U[n]]}, {n, 1, 20}]]
```

From this it should be clear how many elements are in each group listed here (and if you answered all the questions from Group Lab 4, you perhaps know the order of U_n as a function of n for any n). What about the subgroups of U_n? Since we know that the trivial subgroup consisting of the identity is always a subgroup, as is the full group, we can ignore them. Therefore, the first group to consider for nontrivial subgroups is U_5. (Why?) Furthermore, we know that any subgroup must have the identity, so we can be sure that 1 will be in any subgroup.

Let's look at the possible subgroups of U_5 by first considering all the nonidentity elements. (Recall what the complement of a set is and that 1 is the identity.)

```
els = Complement[Elements[U[5]], {1}]
```

These are the elements from which we need to consider all possible subsets. The function KSubsets (from DiscreteMath`Combinatorica) does this job.

```
? KSubsets
```

We will make a table of all possible subsets of length 1, 2, ... up to one less than the length of els, which is 2 (= 3 − 1) in this case. (We have deleted the identity 1 from els, and we know that the full group is a subgroup, so we search for sets of length up to two less than the size of the group.)

```
Table[KSubsets[els, i], {i, Length[els] - 1}]
```

There are too many levels of braces, so we remove one level.

```
Flatten[%, 1]
```

Now we want to join the identity back into each one. These are now *candidates* for being (proper) subgroups.

```
Hsets = Map[Join[{1}, #] &, %]
```

The next step is to test them by mapping the `SubgroupQ` function on each one.

```
subgroups = Map[SubgroupQ[#, U[5]] &, Hsets]
```

It is easier to see which are subgroups if we match up the sets with these results.

```
Transpose[{Hsets, subgroups}] // MatrixForm
```

> **Q21**. What can you say about the order of the element 2? What about 3? What about 4? Do you think U_5 is cyclic or not? Justify your answer.

Now let's put all the steps above into one compact function that gives the final output.

```
FindNontrivialSubgroupsOfUn[n_Integer?Positive] :=
        Module[{els, Hsets, subgroups},
    els = Complement[Elements[U[n]], {1}];
    Hsets = Flatten[
    Table[KSubsets[els, i], {i, Length[els] - 1}], 1];
    Hsets = Map[Join[#, {1}] &, Hsets];
    subgroups = Map[SubgroupQ[#, U[n]] &, Hsets];
    Transpose[{Hsets, subgroups}] // MatrixForm
]
```

It is time to test it on the next index, $n = 6$.

```
FindNontrivialSubgroupsOfUn[6]
```

> **Q22**. What happened? Was there a mistake made in the coding? Think about this. Why was there no output?

What about $n = 7$?

```
FindNontrivialSubgroupsOfUn[7]
```

We see many `False` conclusions. Why? Can we be more efficient in our search for subgroups? How many elements are there in U_7?

```
Elements[U[7]] // Length
```

Perhaps through some previous experiences, in this lab, a previous lab, or classwork, you have become aware that if we have a subgroup, its order must be a divisor of the order of the group. This is an important result, called Lagrange's Theorem, which will be proven later.

What are the divisors of 6? It is easy in this case, but here is how *Mathematica* can be asked.

```
divs = Divisors[6]
```

Now let's gather all the actual "HSets" our function generates.

```
temp =
 FindNontrivialSubgroupsOfUn[7][[1]] // Transpose // First
```

Since we really have questions about only those of orders 2 and 3 (why?), let's select just them.

```
Select[temp, MemberQ[{2, 3}, Length[#1]] &]
```

Now we can test them as subgroups. The last several steps are implemented in a new version of our function.

```
Clear[FindNontrivialSubgroupsOfUn];

FindNontrivialSubgroupsOfUn[n_Integer?Positive] :=
        Module[{els, Hsets, subgroups, divs},
    els = Complement[Elements[U[n]], {1}];
    Hsets = Flatten[
    Table[KSubsets[els, i], {i, Length[els] - 1}], 1];
    Hsets = Map[Join[#, {1}] &, Hsets];
    Hsets =
    Select[Hsets, MemberQ[Complement[Divisors[Order[U[n]]],
        {1, n}], Length[#]] &];
    subgroups = Map[SubgroupQ[#, U[n]] &, Hsets];
    Transpose[{Hsets, subgroups}] // MatrixForm
  ]
```

Now we try it again.

```
FindNontrivialSubgroupsOfUn[7]
```

Q23. Do you think U_7 is cyclic? Why or why not?

```
FindNontrivialSubgroupsOfUn[8]
```

Q24. Do you think U_8 is cyclic? Why or why not?

```
FindNontrivialSubgroupsOfUn[9]
```

Q25. Do you think U_9 is cyclic? Why or why not?

```
FindNontrivialSubgroupsOfUn[10]
```

Q26. Do you think U_{10} is cyclic? Why or why not?

If you are willing to wait a little while, try the following.

```
FindNontrivialSubgroupsOfUn[11]
```

Q27. Why does this take so long?

We can also use a more general function that determines the actual subgroups for any (finite) group (given enough time, memory, and disk space). Here we try it on U_{20}.

```
Subgroups[U[20]]
```

Q28. Do you think U_{20} is cyclic? Why or why not?

Group Lab 6

Cycling Through the Groups

6.1 Prerequisites

Other than familiarity with the basic definitions related to a group, there are no prerequisites.

6.2 Goals

We first look at what it means for a group to be cyclic and try to determine the generators when it is. We try to classify the cyclicity of the groups $\mathbb{Z}_n$, D_n, and U_n and determine the set of generators when they are cyclic. Next we consider the case when the direct product of $\mathbb{Z}_m$ and $\mathbb{Z}_n$ yields a cyclic group. Finally, we look at some of the cyclic subgroups of the infinite additive group of integers, $\mathbb{Z}$.

6.3 What, when, how, and why about cyclic groups

We need to read in the `Master` package inside the `AbstractAlgebra` directory to conduct this lab.

```
Needs["AbstractAlgebra`Master`"];
SwitchStructureTo[Group];
```

Recall that we say a group G is *cyclic* if there exists an element g in G such that $G = \{g^k \mid k \in \mathbb{Z}\}$. In this case, we call g a *generator* of G and denote this relationship by $\langle g \rangle = G$. Which groups, if any, are cyclic? First let's consider a random group $\mathbb{Z}_n$, for n in [6, 14].

```
n = Random[Integer, {6, 14}]
G = Z[n]
```

Is this group cyclic? How can we find out? Recalling the definition of being cyclic (which was written for multiplicative groups), we need to see if there is an element g in G such that the set of *multiples* of g constitutes the whole group. Let's generate multiples of each element and find out if G is cyclic. Note that since our group G is finite with order n, we need to consider only a finite number of multiples.

Q1. Why is it true that we need to consider only a finite number of multiples? Explain your answer.

So let's take a look at some multiples of each of the elements in G.

```
els = Elements[G]
TableForm[
  multiples = Table[j els[[i]], {i, n}, {j, n}], TableHeadings →
    {Table["multiples of " <> ToString[i] <> ":", {i, 0, n - 1}],
      None}, TableSpacing → {0.5, 1}]
```

Whoops! These don't all look like elements in G. What did we forget?

```
TableForm[Map[Mod[#1, n] &, multiples], TableHeadings →
    {Table["multiples of " <> ToString[i] <> ":", {i, 0, n - 1}],
      None}, TableSpacing → {0.5, 1}]
```

The first list of numbers represents multiples of 0, the second list represents multiples of 1, the third list has the multiples of 2, and so on; the last row represents multiples of $n - 1$.

Q2. Do any of these lists represent G? Is G cyclic? What had you expected? If G is cyclic, what are the generators? Does this list of generators surprise you? (Record the group G that was given to you.)

We can get similar results by using a function called `SubgroupGenerated`.

```
Table[Elements[SubgroupGenerated[G, g]],
    {g, 0, n - 1}] // ColumnForm
```

```
? SubgroupGenerated
```

Let's focus on $G = \mathbb{Z}_{12}$ for the moment. (If this is the group that was randomly generated for you, pardon the redundancy.)

```
G = Z[12]
```

Does the element 5 generate the whole group? Let's find out.

```
SubgroupGenerated[G, 5]
```

The list, as given, is the order in which the elements are generated. To see this with a little more explanation, try the following:

```
SubgroupGenerated[G, 5, Mode → Textual]
```

A visual perspective might also be useful. Evaluate the following cell and then double-click on any graphic cell (perhaps closing the enclosing cell bracket first) and then adjust the speed by typing a number from 1 (slow) to 9 (fast). Or you can step through the animation with the (up/down) arrow keys.

```
SubgroupGenerated[G, 5, Mode → Visual];
```

What follows is another visual way of seeing the subgroup generated by 5. In this case, the colors indicate the order in which the multiples (powers) occur, following the rainbow; the key at the bottom helps by listing the element in the colored box and the multiple (power) above it.

```
SubgroupGenerated[G, 5, Mode → Visual2]
```

These illustrations show that indeed $\mathbb{Z}_{12}$ is cyclic. If there is a Thomas in the crowd, he/she can also try

```
CyclicQ[Z[12]]
```

It is interesting to know not only *whether* a group is cyclic but what elements generate the whole group. Try the following.

```
OrderOfAllElements[G, Mode → Textual]
```

Q3. What are the generators of $\mathbb{Z}_{12}$? How do you know? Justify your answer.

We can also approach the question visually:

```
OrderOfAllElements[G, Mode → Visual]
```

Q4. Look at this table. What observations can you make? Can you explain any of these observations?

Now let's consider the group $\mathbb{Z}_{15}$.

```
G = Z[15]
```

> **Q5.** Determine if $\mathbb{Z}_{15}$ is cyclic. If so, specify the generators; if not, explain why not.

> **Q6.** What about $\mathbb{Z}_{17}$?

Let's check out the cyclicity of some other $\mathbb{Z}_n$.

```
TableForm[Table[{n, CyclicQ[Z[n]]}, {n, 2, 25}],
  TableSpacing → {0.5, 2},
  TableHeadings → {None, {"n", "cyclic?\n"}}]
```

> **Q7.** What conclusion can you infer about the cyclicity of $\mathbb{Z}_n$? Prove your statement, if you can.

> **Q8.** For an arbitrary n, what elements are the generators of $\mathbb{Z}_n$? You should be able to be very specific with a description here. Can you also specify (as a function of n), the *number* of generators for $\mathbb{Z}_n$?

Let's think about some other groups. What about the dihedral family?

```
CyclicQ[Dihedral[1]]
```

```
CyclicQ[Dihedral[2]]
```

Or if you wish to see this visually, try

```
OrderOfAllElements[Dihedral[2], Mode → Visual]
```

```
OrderOfAllElements[Dihedral[3], Mode → Visual]
```

```
OrderOfAllElements[Dihedral[4], Mode → Visual]
```

Table allows us to quickly look at a few more examples

```
TableForm[Table[{n, CyclicQ[Dihedral[n]]}, {n, 5, 8}]]
```

> **Q9.** What do you suppose is true about D_n being cyclic? Justify your answer. Can you prove it?

What about the group U_n? Let's consider a few examples.

```
OrderOfAllElements[U[15], Mode → Visual]
```

```
OrderOfAllElements[U[14], Mode → Visual]
```

```
OrderOfAllElements[U[13], Mode → Visual]
```

Q10. Is U_n cyclic for all n? Why or why not?

Considering only whether U_n is cyclic or not, we can use `Table` and `CyclicQ`. The following is already generated—do not evaluate the cell again.

```
TableForm[
  Partition[Table[{n, CyclicQ[U[n]]}, {n, 3, 52}], 10] //
    Transpose,
      TableSpacing → {0.5, 1.5}, TableDepth → 2]
  (* already evaluated - simply open up *)
```

{3, True}	{13, True}	{23, True}	{33, False}	{43, True}
{4, True}	{14, True}	{24, False}	{34, True}	{44, False}
{5, True}	{15, False}	{25, True}	{35, False}	{45, False}
{6, True}	{16, False}	{26, True}	{36, False}	{46, True}
{7, True}	{17, True}	{27, True}	{37, True}	{47, True}
{8, False}	{18, True}	{28, False}	{38, True}	{48, False}
{9, True}	{19, True}	{29, True}	{39, False}	{49, True}
{10, True}	{20, False}	{30, False}	{40, False}	{50, True}
{11, True}	{21, False}	{31, True}	{41, True}	{51, False}
{12, False}	{22, True}	{32, False}	{42, False}	{52, False}

Here is another list that is also already generated—do not evaluate the cell again.

```
TableForm[
  Partition[Table[{n, CyclicQ[U[n]]}, {n, 53, 104}], 10] //
    Transpose,
      TableSpacing → {0.5, 1.5}, TableDepth → 2]
  (* already evaluated - simply open up *)
```

{53, True}	{63, False}	{73, True}	{83, True}	{93, False}
{54, True}	{64, False}	{74, True}	{84, False}	{94, True}
{55, False}	{65, False}	{75, False}	{85, False}	{95, False}
{56, False}	{66, False}	{76, False}	{86, True}	{96, False}
{57, False}	{67, True}	{77, False}	{87, False}	{97, True}
{58, True}	{68, False}	{78, False}	{88, False}	{98, True}
{59, True}	{69, False}	{79, True}	{89, True}	{99, False}
{60, False}	{70, False}	{80, False}	{90, False}	{100, False}
{61, True}	{71, True}	{81, True}	{91, False}	{101, True}
{62, True}	{72, False}	{82, True}	{92, False}	{102, False}

Q11. For what values of n is U_n cyclic? (Think about the answer—you do not need to list the values.) Can you see any patterns? What conclusions can you draw?

6.4 Cyclicity of $\mathbb{Z}_m \oplus \mathbb{Z}_n$

We are familiar with the $\mathbb{Z}_n$ groups. Suppose we consider pairing up two (possibly the same) $\mathbb{Z}_n$ groups and making a new group from the pairs. For example, consider $\mathbb{Z}_2$ and $\mathbb{Z}_3$ and form all pairs (x, y) where the x comes from $\mathbb{Z}_2$ and the y comes from $\mathbb{Z}_3$: $G = \{(x, y) \mid x \in \mathbb{Z}_2 \wedge y \in \mathbb{Z}_3\}$. What are the elements in G? This is called the direct sum (also called direct product) of $\mathbb{Z}_2$ and $\mathbb{Z}_3$, denoted here simply by $\mathbb{Z}_2 \oplus \mathbb{Z}_3$.

```
Elements[DirectSum[Z[2], Z[3]]]
```

Note that all the first elements in the pairs are either 0 or 1 and the second elements are 0, 1, or 2, exactly as specified. So what operation do we use? There is a "natural" one to consider. Suppose (x, y) and (a, b) are generic elements. Then we say $(x, y) + (a, b) = ((x + a) \bmod 2, (y + b) \bmod 3)$. In other words, we treat each dimension (component or coordinate) as we did before joining them, using the operation of the contributing group for that part.

Q12. Do some calculations in this setting.
   a. $(0, 1) + (1, 0) = ?$
   b. $(1, 1) + (1, 1) = ?$
   c. $(1, 0) + (1, 2) = ?$
   d. $(0, 1) + (0, 2) = ?$

Consider the Cayley table of the group.

```
CayleyTable[DirectSum[Z[2], Z[3]], Mode → Visual];
```

Since the elements do not fit very well in the limited space in the table, we use the Key to guide us.

Q13. Check your answers to the previous question. Were they right? Is this group Abelian? Why? What is the identity? Is the group cyclic? How do you know?

So when is $\mathbb{Z}_m \oplus \mathbb{Z}_n$ cyclic? The following produces a table for various m and n (and is already evaluated—simply open it up).

```
Flatten[Table[G = DirectSum[Z[m], Z[n]];
    {m, n, CyclicQ[G]}, {m, 2, 7}, {n, 2, 7}], 1] //
  Partition[#, 4] & // TableForm[#,
    TableHeadings → {None, {"{m,n, cyclic?}\n"}},
    TableSpacing → {0.5, 2}, TableDepth → 2] &
(* already evaluated - simply open it up *)
```

```
{m,n, cyclic?}

{2, 2, False}    {2, 3, True}     {2, 4, False}    {2, 5, True}
{2, 6, False}    {2, 7, True}     {3, 2, True}     {3, 3, False}
{3, 4, True}     {3, 5, True}     {3, 6, False}    {3, 7, True}
{4, 2, False}    {4, 3, True}     {4, 4, False}    {4, 5, True}
{4, 6, False}    {4, 7, True}     {5, 2, True}     {5, 3, True}
```

```
{5, 4, True}      {5, 5, False}     {5, 6, True}      {5, 7, True}
{6, 2, False}     {6, 3, False}     {6, 4, False}     {6, 5, True}
{6, 6, False}     {6, 7, True}      {7, 2, True}      {7, 3, True}
{7, 4, True}      {7, 5, True}      {7, 6, True}      {7, 7, False}
```

Q14. Give a conjecture about the *necessary* conditions for $\mathbb{Z}_m \oplus \mathbb{Z}_n$ to be cyclic. How might a proof go?

▓ 6.5 Structure of intersections of subgroups of $\mathbb{Z}$

For any integer m, the subgroup $\langle m \rangle$ is a cyclic subgroup of the integers $\mathbb{Z}$. Given two such subgroups, $\langle m \rangle$ and $\langle n \rangle$, we know that the intersection of these two subgroups is another subgroup (as is the intersection of any two subgroups from the same group). Furthermore, we know that it is another subgroup of the form $\langle p \rangle$.

Q15. Why must this intersection look like $\langle p \rangle$ for some integer p?

Let's look at some examples. Since these subgroups are all infinite, all we can do is look at a finite swath, but this should be sufficient to build our intuition. Consider $H = \langle 4 \rangle$ and $K = \langle 10 \rangle$:

```
leftAndRight = 15;
H = Table[4 i, {i, -leftAndRight, leftAndRight}]

K = Table[10 i, {i, -leftAndRight, leftAndRight}]

H ∩ K
```

Q16. What does the intersection appear to be? What connection, if any, is there between your answer and the fact that $H = \langle 4 \rangle$ and $K = \langle 10 \rangle$?

Consider $H = \langle 4 \rangle$ and $K = \langle 11 \rangle$:

```
leftAndRight = 22;
H = Table[4 i, {i, -leftAndRight, leftAndRight}]

K = Table[11 i, {i, -leftAndRight, leftAndRight}]

H ∩ K
```

Q17. What does the intersection appear to be? What connection, if any, is there between your answer and the fact that $H = \langle 4 \rangle$ and $K = \langle 11 \rangle$?

Consider $H = \langle 4 \rangle$ and $K = \langle 12 \rangle$.

```
leftAndRight = 15;
H = Table[4 i, {i, -leftAndRight, leftAndRight}]

K = Table[12 i, {i, -leftAndRight, leftAndRight}]

H ∩ K
```

Q18. What does the intersection appear to be? What connection, if any, is there between your answer and the fact that $H = \langle 4 \rangle$ and $K = \langle 12 \rangle$?

Q19. Can you give a conjecture? Test it with a few more examples. Try to prove it.

Group Lab 7

Permutations

🗐 7.1 Prerequisites

To complete this lab, you should have a good understanding of functions, including "right to left" composition. You do not need to complete any previous labs to attempt this one.

🗐 7.2 Goals

We look at the notion of a permutation and how a group can be formed with permutations. Additionally, we look at properties of permutations and consider different ways of rewriting a permutation to gain insights regarding products and orders.

🗐 7.3 What is a permutation?

We need to read in the following *Mathematica* code to work through this lab.

```
Needs["AbstractAlgebra`Master`"];
SwitchStructureTo[Group];
```

What is a permutation? Suppose we had five colored squares labeled 1 through 5.

```
RandomColoredSquares[5];
```

Suppose further that we decide we wish to change the order in which these squares appear. Let's say we want square 1 to go to the second location, square 2 to the third location, square 3 to the first, square 4 to the fifth, and square 5 to the fourth.

```
PermuteColoredSquares[{1 → 2, 2 → 3, 3 → 1, 4 → 5, 5 → 4}];
```

Note how the first square went to the second location ($1 \to 2$), the second square went to the third position ($2 \to 3$), the third square went to the first position ($3 \to 1$), and so on. Here we say that the second row of squares is a *permutation* of the squares in the first row. Technically, a permutation of a set X is a function g from X to X that is both one-to-one and onto. In the case above, we could define $g : X \to X$ by $g(1) = 2$, $g(2) = 3$, $g(3) = 1$, $g(4) = 5$, and $g(5) = 4$, where $X = \{1, 2, 3, 4, 5\}$. This is often represented by a matrix.

```
PermutationMatrix[{1 → 2, 2 → 3, 3 → 1, 4 → 5, 5 → 4}]
```

Note that the domain occurs in the first row and the corresponding range elements in the second row. (Note also that the range, or second row in the matrix, is NOT the same as the numbers in the second row of labeled squares. There is a connection, however; can you see it?)

Since permutations are functions, we can combine two permutations by using function composition; this is shown in section 7.4. Using this binary operation, the set of all permutations of a set X is called the *permutation group of the set X*. If X is the set of integers $\{1, 2, \dots n\}$, then we call this group the *symmetric group of degree n* (denoted S_n). If the set X is some other collection of n elements, we can simply (re)label the elements 1 through n and still consider the group of permutations as the symmetry group S_n.

7.4 Computations with permutations

Suppose we have two permutations.

```
p = {1 → 2, 2 → 4, 3 → 3, 4 → 5, 5 → 1}
q = {1 → 5, 2 → 4, 3 → 1, 4 → 2, 5 → 3}
```

We can view each permutation as a matrix.

```
PermutationMatrix[p]
PermutationMatrix[q]
```

To perform the product of p followed by q (note the order here and below), it may help to think of p and q as side-by-side matrices.

```
SideBySideMatrices[q, p]
```

Observe that *q* is on the left and *p* is on the right. We start by first considering the right-hand side. Observe that *p* maps 1 to 2. Since *q* follows *p*, we note that *q* takes 2 to 4. Therefore, the composition of *p* followed by *q* takes 1 to 2 to 4, or simply 1 to 4. Let's try another. Note that *p* takes 5 to the element 1 and that *q* takes the element 1 to the element 5. Therefore the composition of *p* followed by *q* takes 5 to 1 to 5 (and thus 5 is said to be *fixed* under this composition). The complete product is shown by

```
PermutationMatrix[MultiplyPermutations[q, p]]
```

> **Q1**. Suppose that we are given two permutations *s* and *t* defined by $s = \{1 \to 2, 2 \to 3, 3 \to 1, 4 \to 4\}$ and $t = \{1 \to 2, 2 \to 3, 3 \to 4, 4 \to 1\}$. Determine the two compositions, *s* followed by *t* and *t* followed by *s* (indicating which is which). First do your work on paper and then verify it using the *Mathematica* functions. What property does S_4 *not* have?

When we viewed the permutation *p* in matrix form, it looked like

```
PermutationMatrix[p]
```

You may have noticed that this form always has the top row as the first consecutive *n* integers, and the bottom row captures the images of the top row under the function defined by the permutation. Therefore, it is really the bottom row that holds the important information. Consequently, we frequently refer to a permutation by simply using the bottom row.

```
MultiplyPermutations[{2, 3, 1, 4}, {4, 2, 3, 1}, Mode → Textual]
```

> **Q2**. Determine the product of {2, 3, 4, 1} followed by {1, 3, 2, 4}.

Consider the following permutation and its illustration using colored squares.

```
p = {3, 1, 2, 5, 4, 6}
ShowColoredPermutation[p];
```

Study the permutation (and/or its colored representation). Note that elements in the set {1, 2, 3} permute among themselves, as do the elements in the sets {4, 5} and {6}. In particular, note that we have $p(1) = 3, p(p(1)) = p(3) = 2$, and $p(p(p(1))) = p(p(3)) = p(2) = 1$. Therefore, these elements cycle through as $1 \to 3 \to 2 \to 1$. We say that the (ordered) subset {1, 3, 2} is a *cycle* of the permutation *p*. Furthermore, {4, 5} and {6} are also cycles. The standard way to denote these cycles is by the notation (1, 3, 2), (4, 5), and (6) respectively. (Often the commas are dropped, if only single digits are used.) Since *Mathematica* allows parentheses to be used only as grouping symbols (and not delimiters), we use `Cycle[1, 3, 2]` to denote the cycle (1, 3, 2); the leading word `Cycle` should make it clear that a cycle is under discussion, not a permutation. Frequently it is convenient to rewrite permutations into disjoint cycles. (The three cycles (1, 3, 2), (4, 5) and (6) are said to be *disjoint* since the intersection of any pair is empty.) The following command does this.

```
ToCycles[p]
```

The following takes us back.

```
FromCycles[%]
```

The cycle (3, 2, 5, 4, 1) is different from the permutation {3, 2, 5, 4, 1}. Study the following until you see how these differ.

```
PermutationMatrix[{3, 2, 5, 4, 1}]
(* viewing the permutation *)

PermutationMatrix[FromCycles[{Cycle[3, 2, 5, 4, 1]}]]
(* viewing the cycle;
 the output is the cycle converted to a permutation *)
```

Read your text for further details about working with cycles; in particular, determine how to multiply permutations represented in cycle notation.

Q3. Suppose the permutation p is given as a product of disjoint cycles as $p = (1, 2, 5)(3, 6)(4)$ and q is given by $q = (2, 5, 6)(3, 1, 4)$. Determine the product of p followed by q.

7.5 Applications of permutations

You may recall working with permutations in Group Labs 1 and 2. There, we had figures such as the following.

```
ShowFigure[4, {1, 2, 3, 4}, "D"];
```

In those labs, you were supposed to find all the symmetries of a figure and determine the group of symmetries. You may realize now that this figure's group of symmetries consists of four rotations and four reflections, the dihedral group D_4. Using Rot to represent the lowest-order rotation (90° in this case) and Ref for any reflection, the group is given by

```
Dihedral[4]
```

This representation of D_4 is useful in some contexts. However, viewing it in terms of the permutations of the four vertices to accomplish these symmetries might be more useful at the moment, as well as a reminder of what we did in earlier labs.

```
G = Dihedral[4, Form → Permutations]
```

Let's look at just the elements.

```
els = Elements[G]
```

> **Q4**. As we already know, D_4 consists of four rotations and four reflections. Match up the elements when using the Rot and Ref form with the elements corresponding to permutations.

Now let's randomly choose one of these permutations and show the result of applying it to our figure.

```
ShowPermutation[els[[Random[Integer, {1, 8}]]]]
```

Consider the permutation

```
p = {1, 2, 4, 3}
```

and its effect on our square:

```
ShowPermutation[p]
```

What symmetry is this? Note that vertices 1 and 2 both stayed fixed. A rotation fixes only one point (which one?), so this cannot be a rotation. A reflection moves all points except those on the line of reflection. But using a line of reflection through vertices 1 and 2 would not land this square back onto itself again. Furthermore, it would not transpose vertices 3 and 4. Therefore, this is not a symmetry of the square. So what is it? It is just a permutation of the vertices that can not be obtained by a symmetry.

Recall that we use S_4 to denote the group of all permutations of four objects. This group is given below.

```
Symmetric[4]
```

How many elements are there in S_4? The *order* of a group is precisely that number.

```
Order[Symmetric[4]]
```

> **Q5**. For any given n, what is the order of S_n? Why?

7.6 Questions about permutations

Consider the following permutation.

```
p1 = {3, 1, 2, 5, 4, 6}
```

We might want to know how to write this as a product of disjoint cycles. This is something you want to learn how to do *without* the computer, but for now, let's use *Mathematica*.

```
ToCycles[p1]
```

Since there are three cycles, let's call them *a*, *b*, and *c*.

```
{a, b, c} = ToCycles[p1]
```

Recall that cycle notation is not to be read in the same way as a permutation. We can determine the permutation for any one of these.

```
ToPermutation[a, 6]
ToPermutation[b, 6]
ToPermutation[c, 6]
```

The second parameter (6) is used to indicate that we want to think of these cycles as permutations living inside S_6 (i.e., permutations of length 6). Without this we get slightly different results.

```
ToPermutation[a]
ToPermutation[b]
```

As usual, when a new function is encountered, it is good to learn more about it.

```
? ToPermutation
```

Since the cycles *a*, *b*, and *c* are really permutations in S_6, it makes sense to multiply them.

```
MultiplyCycles[a, b]
MultiplyCycles[b, a]
```

Here's a shortcut for multiplying cycles.

```
a @ b
b @ a
```

What happens if we multiply two cycles from another permutation? Evaluate the following lines of code until you get a cycle representation consisting of at least two cycles, each of which has length at least 2.

```
p2 = RandomPermutation[9]
ToCycles[p2]
```

Q6. Multiply the two cycles by hand. Confirm your work with *Mathematica*. Record the cycles you multiplied.

Consider the following permutation in S_8.

```
q = {1, 2, 3, 4, 5, 7, 6, 8}
```

Let's view this in matrix form.

```
PermutationMatrix[q]
```

It is clear that the only "action" in the permutation is that 6 goes to 7 and 7 goes to 6; everything else goes to itself. What happens when we view this using cycles?

```
ToCycles[q]
```

Q7. Give a reasonable explanation why the cycles (1), (2), (3), (4), and (5) are omitted in this list of cycles. Additionally, explain why (8) is included.

When do two cycles commute? Try the following.

```
Cycle[2, 3, 4] @ Cycle[5, 6]
Cycle[5, 6] @ Cycle[2, 3, 4]

Cycle[2, 3, 4] @ Cycle[3, 6]
Cycle[3, 6] @ Cycle[2, 3, 4]
```

The following picks a random permutation from S_6, converts it to cycles, and then grabs the first cycle and calls it *a*. This is repeated to obtain *b*. These are then multiplied, in both orders ($a * b$ and $b * a$).

```
a = First[ToCycles[RandomPermutation[6]]]
b = First[ToCycles[RandomPermutation[6]]]
MultiplyCycles[a, b]
MultiplyCycles[b, a]
```

Q8. Keep evaluating this code until you can make a conjecture regarding when two cycles will commute. (Evaluate it five times at a minimum.)

Let's pick a cycle from some permutation in S_6.

```
a = First[ToCycles[RandomPermutation[6]]]
```

What is the order of this element? (Recall that the order of an element g in a finite group G is the least positive integer n such that g^n is the identity of the group. In S_6, the identity is $\{1, 2, 3, 4, 5, 6\}$.)

```
OrderOfElement[S[6], ToPermutation[a, 6]]
```

Let's try it again.

```
a = First[ToCycles[RandomPermutation[6]]]
OrderOfElement[S[6], ToPermutation[a, 6]]
```

> **Q9**. Keep evaluating this code until you can make a conjecture regarding the order of a cycle. How might a proof go?

Now that we know how to find the order of any cycle, and we know that all permutations can be written as a product of disjoint cycles, let's see if we can determine the order of an arbitrary permutation.

```
p = RandomPermutation[7]
ToCycles[p]
OrderOfElement[S[7], p]
(* read the following question *)
```

> **Q10**. Why do you think this first warning message is given?

We try again, following the instructions.

```
p = RandomPermutation[7]
ToCycles[p]
OrderOfElement[S[7, IndexLimit → 7], p]
```

> **Q11**. Keep evaluating this code until you can make a conjecture regarding the order of a permutation. Make sure you get some permutations with two or more cycles before forming a conjecture.

Now we want to think about permutations in terms of 2-cycles (not to be confused with bicycles—here we mean cycles of length two), which are also called *transpositions*. Before continuing on that thread, an observation needs to be made, if not already observed. Note that when a permutation is given in cycle notation, sometimes singleton cycles (cycles of length 1) show up.

```
ToCycles[{2, 1, 3}]
```

In this case, the permutation can really be thought of just as the transposition (1, 2) since the 1-cycle (3) really acts like the identity permutation—the three goes to itself and everything is fixed! Keep this in mind as we continue.

First we pick a random permutation, convert it to cycles, and pick one of the cycles at random. The code in the next cell might look a little confusing, but it is set up to give a cycle with length at least three.

```
long = False;
While[! long,
 a = First[Randomize[ToCycles[p = RandomPermutation[7]]]];
 long = Length[a] ≥ 3]
a
```

Now we convert this cycle to a product of transpositions. It is important to note that these transpositions are *not* disjoint.

```
transp = ToTranspositions[a]
```

If we multiply these cycles together,

```
MultiplyCycles[transp]
```

we get a permutation as the output, so we can convert it back to cycle notation.

```
ToCycles[%]
```

This should be where we started out. Therefore, converting to transpositions is a safe operation and one that can be undone by multiplying them out. What can be obtained by writing a cycle as a product of transpositions? Let's pick two random cycles from two random permutations, convert them to a product of transpositions, and look at the number of transpositions obtained from each cycle.

```
a = First[Randomize[ToCycles[RandomPermutation[7]]]]
b = First[Randomize[ToCycles[RandomPermutation[7]]]]

transpa = ToTranspositions[a]
transpb = ToTranspositions[b]

Length[transpa]
Length[transpb]
```

Next, we form the product of the two cycles we found and convert it back to cycle notation. We want to convert this list of cycles to transpositions, so we apply the function ToTranspositions to each cycle in the list (this is what the second line does). Since this result has extra levels of {}'s, we "flatten" out the unnecessary levels and then count how many are in the list.

```
ToCycles[MultiplyCycles[a, b]]
Map[ToTranspositions, %]

Flatten[%, 1]
Length[%]
```

Compare this length with the two lengths above it. We look at this process again, this time suppressing all output except the counts.

```
a = First[Randomize[ToCycles[RandomPermutation[7]]]];
b = First[Randomize[ToCycles[RandomPermutation[7]]]];
transpa = ToTranspositions[a];
transpb = ToTranspositions[b];
Length[transpa]
Length[transpb]

prodCyc = ToCycles[MultiplyCycles[a, b]];
trans = Map[ToTranspositions, prodCyc];
Flatten[trans, 1] // Length
```

The first value is the number of transpositions in the first cycle, the second value is the number of transpositions in the second cycle, and the third value is the number of transpositions in the product of the two cycles.

Q12. Keep evaluating this code until you can make a conjecture regarding a relationship between the number of transpositions in the product (the last number) and the number of transpositions in the two factors. Hint: Do not try to be too specific.

There is another function we can apply to permutations, the `Parity` function. Let's see how this works.

```
p = RandomPermutation[7]
ToTranspositions[p]

Length[%]
Parity[p]
```

Q13. Keep evaluating this code until you can make a conjecture regarding a relationship between the number of transpositions in a permutation and its parity.

Now let's resume our pursuit that preceded question 12. Here we pick random cycles *a* and *b*, form the product, and call it *p*. We convert each of these to transpositions and count them; the first triple is the number of transpositions of each in the list {*a*, *b*, *p*}. Then we also calculate the parity of each, reflected in the second triple.

```
a = First[Randomize[ToCycles[RandomPermutation[7]]]];
b = First[Randomize[ToCycles[RandomPermutation[7]]]];
transpa = ToTranspositions[a];
transpb = ToTranspositions[b];
p = MultiplyCycles[a, b];
ToTranspositions[p];
```

```
{Length[transpa], Length[transpb], Length[%]}
{Parity[ToPermutation[a, 7]],
  Parity[ToPermutation[b, 7]], Parity[p]}
```

The output is {# transpositions in *a*, # transpositions in *b*, # transpositions in $a * b$} followed by {parity of *a*, parity of *b*, parity of $a * b$}.

Q14. Keep evaluating this code until you can make a conjecture regarding a relationship between the parity of a product and the parity of its factors. Now go back to questions 12 and 13 and see if you have any changes to make; if so, state them here.

A permutation with parity 1 is called an *even permutation* and if the parity is -1, we call it an *odd permutation*. Two functions test evenness and oddness of permutations.

```
p = RandomPermutation[6]

OddPermutationQ[p]
EvenPermutationQ[p]
```

Q15. Using your knowledge about sums of odd and even integers and how products result from various combinations of -1 and 1, do you see any connection between parity and the number of transpositions? If you do, state it.

Group Lab 8

Isomorphisms

🔳 8.1 Prerequisites

To complete this lab, you should be familiar enough with the basic properties of groups to be able to compare the various pairs of groups that you will be asked to examine. No previous labs are necessary.

🔳 8.2 Goals

This lab explores the notion of isomorphisms. First we define an *isomorphism* and then we see how one can be constructed. Next we explore when two groups are isomorphic.

🔳 8.3 What is an isomorphism?

Before beginning, we read in the necessary code for this lab.

```
Needs["AbstractAlgebra`Master`"];
SwitchStructureTo[Group];
```

Let's consider two groups, G_1 and G_2.

```
Clear[G]
G₁ = U[10]
G₂ = Z[4]
```

An important question in abstract algebra (and in mathematics in general) is "When are two objects the 'same' by some appropriate measurement?" In this case, are groups G_1 and G_2 the same? Clearly they have different elements and different operations, so they are not identical. They do, however, both have four elements and in this way they are the same. Therefore, we can construct a one-to-one, onto function (bijection) from G_1 to G_2. In fact, there are 24 different ways of setting up such functions.

Q1. Explain how one comes up with 24 different bijections from G_1 to G_2. How many functions (bijections or not) are there from G_1 to G_2?

If we consider only the *number* of elements (four in each, in our case), we ignore a large portion of the richness of groups and we ignore the operation altogether. Since each group has a special element, the identity, it may be reasonable to want to match these up.

```
GroupIdentity[G₁]
GroupIdentity[G₂]
```

Thus, if we want to define a function $f : G_1 \to G_2$ that somehow illustrates "sameness," we might want to define $f(1) = 0$.

```
Clear[f]
f[1] = 0
```

Q2. With this assumption, how many different bijections are there from G_1 to G_2?

We need three more matches to make a bijection. Consider the element 3 in G_1. To what should it be mapped? Is there any special property related to 3 that should also exist with the element in G_2 to which it will be mapped? What about the order of 3?

```
OrderOfElement[G₁, 3]
```

It would seem reasonable to want to map 3 to an element in G_2 of the same order. What are the orders of the elements in G_2?

```
OrderOfAllElements[G₂]
```

If the output is not clear, ask about the function:

```
? OrderOfAllElements
```

So we see that the two elements 1 and 3 in G_2 both have order 4. (Of course, you may have already known this; we are illustrating a general process we might take.) Suppose we map 3 in G_1 to 1 in G_2.

```
f[3] = 1
```

Q3. With this assumption, *now* how many different bijections are there from G_1 to G_2?

Since we are mapping 3 to 1, it might be reasonable to map the inverse of 3 in G_1 to the inverse of 1 in G_2. (Again, you should already know the results of the following; we are illustrating a general procedure.)

```
GroupInverse[G₁, 3]
GroupInverse[G₂, 1]
```

So we map 7 in G_1 to 3 in G_2.

```
f[7] = 3
```

Since $|g| = |g^{-1}|$, we know that the orders in this assignment match. The two remaining elements are put together by default.

```
f[9] = 2
```

Do the orders match in this case?

```
OrderOfElement[G₁, 9]
OrderOfElement[G₂, 2]
```

Indeed. If all we care about is whether orders match up, we are done and we might call these two groups the same. There might be other issues to consider, however. In G_1, when we multiply 7 by 9 we get 3.

```
Clear[op]
op₁ = Operation[G₁];
op₁[7, 9]
```

Let's review our definition of f:

```
? f
```

We see that 7 is mapped to 3 and 9 is mapped to 2. What happens if we add 3 (the image of 7) and 2 (the image of 9) in G_2? Surely we get 1.

```
op₂ = Operation[G₂];
op₂[3, 2]
```

More important, observe that $1 = 3 + 2 = f(7) + f(9)$ but also $1 = f(3) = f(7 * 9)$. In other words,

$$f(7 * 9) = f(7) + f(9).$$

Note that on the left-hand side, the operation inside the parentheses is taking place inside G_1 (the domain of f), while the operation on the right-hand side is taking place inside G_2. In this case, we say f is "operation preserving" for the elements 7 and 9. What about other pairs? We can make a table and check all possibilities:

```
els = Elements[G₁]
Table[f[op₁[els〚i〛, els〚j〛]] == op₂[f[els〚i〛], f[els〚j〛]],
  {i, 4}, {j, 4}]
```

Or if you want to see more details:

```
ListOperationPreservingElements[f, G₁, G₂]
```

This shows that the function f is operation preserving for all the elements in G_1. This is when algebraists are satisfied with calling two groups G_1 and G_2 the same: there exists a bijection $f : G_1 \rightarrow G_2$ that is operation-preserving in the sense that

$$f(x * y) = f(x) * f(y)$$

for all elements x and y in G_1 (with the understanding that on the left-hand side the operation taking place between x and y is the operation in G_1 and the operation taking place on the right-hand side between $f(x)$ and $f(y)$ is taking place in G_2). In such a case, the two groups are said to be *isomorphic* and the defining function f is said to be an *isomorphism*.

Q4. Change G_1 to U_8 and determine whether U_8 and $\mathbb{Z}_4$ are isomorphic. If so, provide the isomorphism; if not, indicate why not. (Clear[f] before beginning.)

▥ 8.4 Creating Morphoids

In previous labs, we have used a structure called a Groupoid when we worked with groups. Now we use a structure called a Morphoid to work with morphisms, including homomorphisms and isomorphisms. Recall from the previous section that an isomorphism f between two groups is a bijective (one-to-one and onto) mapping such that $f(x * y) = f(x) * f(y)$ for all elements x and y in the domain. If we relax the condition that f is bijective and require only the "operation-preserving" part, we have a *homomorphism*. A Morphoid is a structure that has the potential of being a homomorphism (and therefore potentially also an isomorphism).

Recall that our function f was defined as follows.

```
Clear[f]
f[1] = 0;
f[3] = 1;
f[7] = 3;
f[9] = 2;
```

We can use `FormMorphoid` to create a `Morphoid` based on f.

```
func1 = FormMorphoid[f, U[10], Z[4]]
```

Alternatively we can set up a list of rules that are equivalent to f.

```
rules = Map[Rule[#, f[#]] &, Elements[U[10]]]
```

The `FormMorphoid` function can also handle a list of rules to create a Morphoid.

```
func1alt = FormMorphoid[rules, U[10], Z[4]]
```

The two `Morphoids` `func1` and `func1alt` appear to be different, but they are really the same. `Morphoids` can be defined either by a list of rules or a function. If the underlying correspondences between the elements of the `Groupoids` are the same, the `Morphoids` are considered to be equal.

```
EqualMorphoidQ[func1, func1alt]
```

We already know that `func1` is an isomorphism. Let's set this knowledge aside for the moment (for the sake of learning how to use some of the available tools in situations where we don't know if we have an isomorphism). What we want to know is whether, for any given pair of elements in G_1, the operations are preserved?

First we will take a specific pair, 9 and 3.

```
PreservesQ[func1, {9, 3}, Mode → Visual]
```

In this case, we see that $f(9*3)$ does equal $f(9) + f(3)$, since $9*3$ is 7 in G_1 and the image of 7 is 3, which happens to be the sum of 2 (the image of 9) and 1 (the image of 3). Note that the operation in G_1 is multiplication (hence, $*$), whereas it is addition in G_2 (and so + is used).

Let's pick two random elements and check out whether the operations of G_1 and G_2 are preserved under the `Morphoid` `func1` (based on f).

```
{g, h} = RandomElements[G₁, 2]
PreservesQ[func1, {g, h}, Mode → Visual]
```

Q5. Evaluate the two lines of code again several times. Do you find any case where the operations were *not* preserved? Why or why not?

Note that `PreservesQ` is used when one wishes to explore whether a function is operation preserving for a *particular* pair of elements. More generally, we are interested in doing this for *all* possible pairs. The function `MorphismQ` does this for us. Here we use it in the `Visual` mode; pairs that are preserved are colored at their point of intersection in the table.

> `MorphismQ[func1, Mode → Visual]`

We can also check to see if `func1` is an isomorphism as well as a homomorphism.

> `IsomorphismQ[func1]`

In case one begins to think that having a homomorphism happens naturally, let's take a look at an example where not all the elements preserve the operation. We define a `Morphoid` `func2` that has $\mathbb{Z}_4$ as its domain and $\mathbb{Z}_5$ for its codomain, and the rule to get from the domain to the codomain is to take an element from $\mathbb{Z}_4$, add 1, and then reduce this result mod 5.

> `func2 = FormMorphoid[Mod[#1 + 1, 5] &, Z[4], Z[5]]`

Now let's check out which elements preserve the operations.

> `MorphismQ[func2, Mode → Visual]`

We see that some do and some do not. In particular, since the entry at the intersection of the row headed by 1 and the column headed by 2 is not colored, `func2` is not operation-preserving for the pair (1, 2). Let's look at this in detail.

> `PreservesQ[func2, {1, 2}, Mode → Visual]`

Q6. Explain in your own words how and why this failed. Why were the operations not preserved?

As you work with homomorphisms in greater detail, you will learn about the kernel and the image of a homomorphism. For the two `Morphoids` created here, we illustrate how to obtain the kernel and image.

> `K₁ = Kernel[func1]`
> `I₁ = Image[func1]`
>
> `K₂ = Kernel[func2]`
> `I₂ = Image[func2]`

The operations of these `Groupoids` are as follows.

> `Map[Operation, {K₁, I₁, K₂, I₂}]`

> **Q7**. Which of the four (the two kernels and the two images) are groups? Justify your answer.

▣ 8.5 Seeing isomorphisms

Often, by looking at appropriate data, one can almost literally "see" an isomorphism at hand. Here we look at some examples.

▪ 8.5.1 Example 1

Consider the example with which we first started.

```
G₁ = U[10]
G₂ = Z[4]
```

Let's look at the Cayley tables of these two groups.

```
CayleyTable[{G₁, G₂}, Mode → Visual];
```

We have shown that these two groups were isomorphic. Shouldn't their Cayley tables be identical? Think about why they are not. Perhaps if we ordered the elements in the second table in the order corresponding to the first (via our isomorphism), we might obtain different results.

```
CayleyTable[{{G₁}, {G₂, TheSet → {0, 1, 3, 2}}}, Mode → Visual];
```

This looks better. Now the isomorphism is clear.

▪ 8.5.2 Example 2

We now consider U_{10} and another group of order four.

```
G₁ = U[10]
G₂ = U[12]
```

Are these two isomorphic?

```
CayleyTable[{G₁, G₂}, Mode → Visual];
```

Certainly the two tables do not look alike, but we know that this does not necessarily mean that the two groups are not isomorphic. Below we rearrange the elements of U_{10}.

```
CayleyTable[{{G₁, TheSet → {1, 9, 3, 7}}, {G₂}}, Mode → Visual];
```

Q8. Try other arrangements for the elements in U_{10} until you show that the two groups are isomorphic or conclude that they are not isomorphic. If they are isomorphic, describe how the isomorphism map works; if they are not isomorphic, explain why not.

■ 8.5.3 Example 3

We now consider the group D_2 and its Cayley table.

```
G₁ = Dihedral[2]
CayleyTable[G₁, Mode → Visual];
```

We have seen several other groups of order four. Find one that is isomorphic to D_2. Define it and then show the Cayley tables side by side.

```
G₂ =

CayleyTable[{G₁, G₂}, Mode → Visual]
```

If the elements in G_2 are not arranged to properly reflect the isomorphism, do so below.

```
CayleyTable[{{G₁}, {G₂, TheSet → {}}}, Mode → Visual]
```

Q9. What is the other group to which D_2 is isomorphic and what are the details of the isomorphism? (Specify which elements are mapped to which.)

■ 8.5.4 Example 4

Following are several groups of order six.

```
SL22 = FormGroupoid[
    {{{0, 1}, {1, 0}}, {{0, 1}, {1, 1}}, {{1, 0}, {0, 1}},
     {{1, 0}, {1, 1}}, {{1, 1}, {0, 1}}, {{1, 1}, {1, 0}}},
    Mod[#1.#2, 2] &, WideElements → True,
    GroupoidName → "SL[2, 2]"];
(* two-by-two matrices with determinant
   1 and entries from ℤ₂ *)
G₁ = U[9]
G₂ = Z[6]
G₃ = Symmetric[3]
G₄ = Dihedral[3]
G₅ = SL22
G₆ = DirectSum[Z[2], Z[3]]
```

Are any of these isomorphic to one another? Here are some tools you may wish to use.

```
CayleyTable[{G₁, G₂, G₃, G₄, G₅, G₆}, Mode → Visual];
(* After evaluating this,
 you may wish to enlarge the graphic by
   selecting it and then dragging from a corner. *)

AbelianQ[{G₁, G₂, G₃, G₄, G₅, G₆}]

CyclicQ[{G₁, G₂, G₃, G₄, G₅, G₆}]

TableForm[
  OrderOfAllElements[{G₁, G₂, G₃, G₄, G₅, G₆}], TableDepth → 1]

OrderOfAllElements[{G₁, G₂, G₃, G₄, G₅, G₆}, Mode → Textual];

? GroupCenter

Map[GroupCenter, {G₁, G₂, G₃, G₄, G₅, G₆}] // ColumnForm

? CommutatorSubgroup

Map[Elements[CommutatorSubgroup[#]] &,
    {G₁, G₂, G₃, G₄, G₅, G₆}] // ColumnForm
```

Q10. If groups H_1 and H_2 are isomorphic, we denote this by $H_1 \cong H_2$. For every pair (G_i, G_j) of groups in $\{G_1, G_2, G_3, G_4, G_5, G_6\}$, determine whether $G_i \cong G_j$. Note: There are "six-choose-two"—$\binom{6}{2}$— such pairs to consider.

Q11. You probably found that D_3 was isomorphic to at least one of the other groups. Below, we use the variable `otherGroup` to denote the group you had in mind; change G_4 to the group you have in mind and evaluate the following cell. You now see the list of elements in your chosen group. Between the {} on the right-hand side of `orderedElementsOfOtherGroup`, place the elements of your chosen group matched up according to the elements of D_3. After double-checking your ordering, evaluate the cell and the cell that defines the rules for this isomorphism, as well as the `Morphoid` f itself. Now use the tools illustrated earlier and verify that f is indeed an isomorphism (assuming that you correctly found an isomorphic group).

```
otherGroup = G₄
Elements[otherGroup]
Elements[Dihedral[3]]
```

```
orderedElementsOfOtherGroup = {}

rules = Transpose[{Elements[Dihedral[3]],
    orderedElementsOfOtherGroup}] /. {x_, y_} :> x → y
f = FormMorphoid[rules, Dihedral[3], otherGroup]
```

Q12. Consider the set $H = \{1, \text{Rot}, \text{Rot}^2\}$, which is a subgroup of D_3. What is the image of H under f? (You may wish to use the following line.) Is this also a subgroup of the image of D_3?

```
Image[f, {1, Rot, Rot²}]
```

Group Lab 9

Automorphisms

9.1 Prerequisites

To complete this lab, you should have completed Group Lab 8.

🔳 **9.2 Goals**

This lab continues the exploration of isomorphisms begun in Group Lab 8. In this lab, we look at a collection of isomorphisms from a group to itself and ask what kind of structure, if any, might be present.

🔳 **9.3 Automorphisms on $\mathbb{Z}_n$**

Before beginning, we read in the necessary code for this lab.

```
Needs["AbstractAlgebra`Master`"];
SwitchStructureTo[Group];
```

In Group Lab 8 we considered questions such as "Is U_{12} isomorphic to U_{10} (or other groups besides U_{10})?" Consider the group $\mathbb{Z}_{12}$. We never asked "Is $\mathbb{Z}_{12}$ isomorphic to $\mathbb{Z}_{12}$ (i.e., itself)?" Why not ask this? Is the answer obvious? On the one hand, since isomorphisms indicate that two groups are the "same" in some sense, if two groups are identical (as in this case), we should certainly expect them to be isomorphic.

Indeed, $\mathbb{Z}_{12}$ is isomorphic to itself. Which isomorphism (map) will show this? Try the identity function that takes every element to itself. All the properties hold. Since we are mapping from $\mathbb{Z}_{12}$ to itself, we call this kind of isomorphism an ***automorphism***. We illustrate the use of the identity function.

```
Clear[f]
f₁ = FormMorphoid[Identity, Z[12], Z[12]]
IsomorphismQ[f₁]
```

We can see the details of this `Morphoid` by looking at the rules used in its definition.

```
ToRules[f₁]
```

A more important question to ask at this time is whether there are any other automorphisms besides the identity map. We know that we must map 0 to 0. (Why?) What about the element 5? Since it is a generator and has order 12, to what must it be mapped? Clearly it needs to go to another generator, if we wish to preserve orders. So it could be mapped to 1, 5, 7, or 11.

Let x be any other element of $\mathbb{Z}_{12}$. Since 5 is a generator, $x = n\,5 = 5 + 5 + \cdots + 5$ (n summands) for some integer n. To what should we map x? Suppose our function is f. Then,

$$f(x) = f(5 + 5 + \cdots + 5) = f(5) + f(5) + \cdots + f(5)$$

with n summands in both instances. This should indicate that once we know where 5 is mapped, we know where every element is mapped.

Q1. Explain this last statement (in such a way that would convince a classmate who may not have heard this yet).

Since we know we can send the generator 5 to any one of the four generators, let's make a function for each possibility. Below we use a function called `Automorphism` that allows us to specify a rule as a generator and build the complete morphism from that single piece of information. Asking for more information about a function is always a good idea.

```
? Automorphism
```

We can also get further information by specifying the `Textual` mode of `Automorphism`.

```
f₂ = Automorphism[Z[12], 5 → 1, Mode → Textual]
```

Instead of checking to see if f_1 and f_2 are morphisms, we can check whether they are isomorphisms directly with `IsomorphismQ`.

```
IsomorphismQ[f₁]
IsomorphismQ[f₂]
```

Let's create our last two automorphisms and check them.

```
f₃ = Automorphism[Z[12], 5 → 7]
IsomorphismQ[f₃, Cautious → True]

f₄ = Automorphism[Z[12], 5 → 11]
IsomorphismQ[f₄, Cautious → True]
```

We now have four automorphisms. What can we do with these functions? Is this set of four functions anything special? What would happen if we follow one by another (i.e., compose the functions)? In particular, what do we get if we follow f_2 by f_3?

First, define Els to be the elements of the group $\mathbb{Z}_{12}$ under consideration.

```
Els = Elements[Z[12]]
```

We want the output of f_2 to be the input for f_3, so let's evaluate f_2 at the set Els.

```
f₂[Els]
```

This maps f_2 over each element in Els. We could also have f_2 mapped onto each element by using the Map command.

```
f2Output = Map[f₂, Els]
```

Now we use this as input for f_3.

```
Map[f₃, f2Output]
```

Now the question is which function, if any, has the same output (in this order) when mapped over Els? Let's try each one.

```
Map[f₁, Els]
Map[f₂, Els]
Map[f₃, Els]
Map[f₄, Els]
```

Great! The function f_4 has the same images as the images of f_2 followed by f_3. But what does this mean?

Q2. What is the relationship, if any, between f_2, f_3, and f_4?

Q3. Modify the steps above and determine the result of taking f_2 followed by f_4. Which function, if any, does this yield?

> **Q4**. Where are we going with all this? What seems to be lurking around the corner (i.e., *Mathematica* computation)?

Suppose we try to automate all this work. First we define our list of automorphisms but redefine f_1 using `Automorphism`, converting it into a rules-based `Morphoid`. (This is necessary to form a `Groupoid` of `Morphoids`.)

```
f₁ = Automorphism[Z[12], 5 → 5]
automorphisms = {f₁, f₂, f₃, f₄}
```

We can operate on these automorphisms with the function `MorphoidComposition`.

```
? MorphoidComposition
```

If we want to know f_4 followed by f_3, we use the following. (Notice that the ordering is from right to left.)

```
MorphoidComposition[f₃, f₄]
```

Or f_1 followed by f_4:

```
MorphoidComposition[f₄, f₁]
```

The next question we might ask concerns a Cayley table. First we need to turn this set and operation into a `Groupoid`. We will call it `Automorphisms`.

```
Automorphisms =
  FormGroupoid[automorphisms, MorphoidComposition[#1, #2] &,
    WideElements → True, KeyForm → OutputForm]
```

Now for the moment for which we have all been waiting:

```
CayleyTable[Automorphisms, Mode → Visual];
```

> **Q5**. Does this appear to be a group? Why? Is there an identity? If so, what is it; if not, why not? Since this is of order four, to what familiar group is this isomorphic?

All the work above is automated in the function `AutomorphismGroup`.

```
? AutomorphismGroup
```

Let's change our modulus from 12 to 10 and see what kind of group we get.

```
G = AutomorphismGroup[Z[10]]
```

Now take a look at its Cayley table.

```
CayleyTable[G, Mode → Visual];
```

> **Q6.** To what familiar group is this isomorphic?

Let's try one more.

```
G = AutomorphismGroup[Z[14]];
CayleyTable[G, Mode → Visual];
```

> **Q7.** To what familiar group is this isomorphic?

9.4 Inner automorphisms

We have seen how we obtain all the automorphisms on a cyclic group. If the group is not cyclic, the question is a bit harder. There are certain automorphisms, however, that are easy to generate for any group.

Let G be any group and g by any element in G. Consider the function $f_g : G \to G$ defined by $f_g(h) = g\,h\,g^{-1}$. We call $g\,h\,g^{-1}$ the *conjugate of h by g* and the process of applying f_g to an element is called *conjugation.*

> **Q8.** If the group G is Abelian, what happens when we conjugate h by g? What is the map f_g in this case?

Consider the group $G = D_4$ and let's try conjugating some elements using the function `ElementConjugate`.

```
Dihedral[4]
(* recall Rot is the element
   corresponding to the lowest order rotation,
 90° in this case, and Ref is any reflection *)

?ElementConjugate

ElementConjugate[Dihedral[4], Rot³, Rot]

ElementConjugate[Dihedral[4], Ref, Rot]
```

We can speed this up by looking at all the conjugates at once.

```
ConjugatingElement = Rot;
TableForm[Transpose[{els = Elements[Dihedral[4]], Map[
    ElementConjugate[Dihedral[4], #1, ConjugatingElement] &,
    els]}], TableHeadings → {None,
{"h", "h conjugated by " <> ToString[ConjugatingElement] <>
    "\n"}}, TableSpacing → {0.5, 5}]
```

Q9. Which elements were not changed when they were conjugated by `Rot`? Is there anything special about this set?

What if we conjugate by `Ref`?

```
ConjugatingElement = Ref;
TableForm[Transpose[
    {Elements[Dihedral[4]], (ElementConjugate[Dihedral[4], #1,
        ConjugatingElement] &) /@Elements[Dihedral[4]]}],
    TableHeadings → {None, {"h", "h conjugated by " <>
        ToString[ConjugatingElement] <> "\n"}},
    TableSpacing → {0.5, 5}]
```

For each element g in a group G, there is a conjugation function f_g (that takes an element to its conjugation by g). Note that f_g is a function from G to G and therefore a candidate for being an automorphism. In fact, each of these functions is an automorphism and is called an *inner automorphism of G induced by g*. We can use *Mathematica* to help us construct them.

```
? InnerAutomorphism

Clear[f]
f₁ = InnerAutomorphism[Dihedral[4], Rot²]
```

Let's look at some details of this function. One way of seeing the action is to map f_1 onto the elements of the group.

```
Map[f₁, Elements[Dihedral[4]]]
```

Is f_1 really an automorphism?

```
IsomorphismQ[f₁]
```

Suppose we call f_2 the inner automorphism induced by Rot^3.

```
f₂ = InnerAutomorphism[Dihedral[4], Rot³]
Map[f₂, Elements[Dihedral[4]]]
```

Since f_1 and f_2 are functions, it is legitimate to ask about the composition. Will this be another inner automorphism? If so, which one? Perhaps we should try to find all the inner automorphisms by mapping the `InnerAutomorphism` function onto the elements of the group.

```
funcs = Map[InnerAutomorphism[Dihedral[4], #] &,
    Elements[Dihedral[4]]]
```

Now let's map each of these onto the elements of the group to see if any of these functions are the same.

```
images = Map[Map[#, Elements[Dihedral[4]]] &, funcs]
```

Q10. By looking at this list, would you say that any of these functions are the same? If so, which ones?

Sorting this list, we remove any duplicates.

```
Union[images]
```

At this stage, you should be able to match up each of these lists with those in `images` and likewise determine the inner automorphisms with which they are associated. We put all of this (and more) together in one function.

```
G = InnerAutomorphismGroup[Dihedral[4]]
```

The elements of this `Groupoid` are suppressed until we ask for them.

```
Elements[G]
```

Let's look at the details of the second element.

```
MorphoidRules[Elements[G][[2]]]
```

Here is the Cayley table for this group.

```
CayleyTable[G, Mode → Visual];
```

Q11. You now know about the different groups of order four. To what (common) group is *G* isomorphic?

Consider the following groups.

```
K = InnerAutomorphismGroup[Dihedral[5]];
CayleyTable[K, Mode → Visual];

L = InnerAutomorphismGroup[Dihedral[6]];
CayleyTable[L, Mode → Visual, ShowKey → False];
```

Q12. Make a conjecture about the inner automorphism group of D_n?

Group Lab 10

Direct Products

10.1 Prerequisites

To complete the last section of this lab, you should have completed the lab on isomorphisms (Group Lab 8).

10.2 Goals

This lab explores the direct product of two groups. First we define the concept of a direct product and how to determine its order. Next we determine the order of an element in a direct product. We also consider when a direct product might be cyclic, given that its factors are cyclic. Finally, we consider when U groups are isomorphic to direct products of other U groups.

10.3 What is a direct product?

In abstract algebra, when an object (such as a group) is being studied, there are a couple of natural questions that are often asked: (1) When does a subset still have the algebraic properties of the parent set? (2) Can we take two (or more) objects and combine them to build a new one with similar properties? (We pursue a third natural question in Group Lab 12.)

Since we have already spent some time looking at subgroups, we now consider the second question. Let's consider two groups, G_1 and G_2, as defined below. First we read in code from the AbstractAlgebra packages.

```
Needs["AbstractAlgebra`Master`"];
SwitchStructureTo[Group];

G1 = U[10]
G2 = Z[4]
```

How can we combine them to form a new group where both G_1 and G_2 can be viewed as "subgroups" in some sense? From linear algebra, we know what it means to view vectors as ordered pairs and then use component-wise addition, so perhaps we can implement a similar strategy.

What we want to do is form all ordered pairs of the form (x, y) where x is obtained from G_1 and y is obtained from G_2. The set of all these ordered pairs can be given a group structure by considering the operation defined by $(x, y)(a, b) = (x\,a, y\,b)$, where the operation in the first component on the right-hand side takes place in G_1 and in G_2 in the second component. We call this group the *direct product of G_1 and G_2*. We denote this direct product by $G_1 \times G_2$.

```
G = DirectProduct[G1, G2]
```

> **Q1.** Here we see there are 16 elements. If $|G_1| = n$ and $|G_2| = m$, what is the order of the direct product $G_1 \times G_2$? Why?

Let's look at another direct product.

```
H = DirectProduct[U[10], Z[2]]
```

As we have seen in the past, studying the Cayley table of a group can reveal significant information. Study the following.

```
CayleyTable[H, Mode → Visual];
```

> **Q2.** What observations can you make about the group H? Can you "see" any subgroups? Can you identify the order of any elements? Is this Abelian? What other observations can you make?

▣ 10.4 Order of an element in a direct product

We continue to use the same G as in section 10.3. (Evaluate the following only if you have changed your definition of G.)

```
G = DirectProduct[U[10], Z[4]]
```

Let's pick a random element from G.

```
? RandomElement
(* if you are not sure how RandomElement works, try this *)
```

```
g = RandomElement[G, SelectFrom → NonIdentity]
```

What is the order of this element?

```
OrderOfElement[G, g]
```

Now let's determine the order of each component in its respective group. That is, we would like to know the order (in G_1) of the first component of g and the order (in G_2) of the second component of g. In *Mathematica*, one uses $g[[k]]$ to obtain the kth component of a list, so here we use $g[[1]]$ and $g[[2]]$ to find the coordinates of g in G_1 and G_2 respectively.

```
OrderOfElement[{{G1, g[[1]]}, {G2, g[[2]]}}]
```

Recall that the first number is the order of the first coordinate of g in the group G_1 and the second number is the order of the second coordinate of g in the group G_2. Let's look at a table and see if we can come to any conclusions.

```
TableForm[
  Table[g = RandomElement[G, SelectFrom → NonIdentity];
   {g, OrderOfElement[G, g],
    OrderOfElement[{{G1, First[g]}, {G2, Last[g]}}]}, {8}],
  TableDepth → 2, TableHeadings →
   {None, {"g = {x, y}", "|g|", "{|x|, |y|}\n"}},
  TableSpacing → {0.5, 3}]
```

> **Q3**. Evaluate the cell at least one more time. Make a conjecture about a relationship between the order of the element $g = \{x, y\}$ in $G_1 \times G_2$ and the orders $|x|$ and $|y|$.

Change the groups G_1 and G_2 to some other groups (use Z, U, Symmetric, Dihedral or any other groups you might know).

```
Clear[G1, G2]
G1 = (* <--- fill in your group here *)

G2 = (* <--- fill in your group here *)

G = DirectProduct[G1, G2];
Elements[G]
```

```
TableForm[
 Table[g = RandomElement[G, SelectFrom → NonIdentity];
  {g, OrderOfElement[G, g],
   OrderOfElement[{{G1, First[g]}, {G2, Last[g]}}]}, {5}],
  TableDepth → 2, TableHeadings →
  {None, {"g = {x, y}", "|g|", "{|x|, |y|}\n"}}]
```

> **Q4**. Test your conjecture from question 3. Does it still hold? What are your groups G_1 and G_2 in your direct product? Try this again by changing groups (again listing your component groups). State your final conjecture. How might you prove it?

■ 10.5 When is a direct product of cyclic groups cyclic?

Consider the following two groups and their direct product.

```
G1 = Z[6]
G2 = Z[4]
G = DirectProduct[G1, G2]
```

We know that both G_1 and G_2 are cyclic; what about G?

```
CyclicQ[{G1, G2, G}]
```

In this case, we see that G is not cyclic. Let's make a table of some random indices for $\mathbb{Z}_m$ and $\mathbb{Z}_n$ and then consider the cyclicity of the direct product $\mathbb{Z}_m \times \mathbb{Z}_n$. (The following is already evaluated; just open up the cell.)

```
numberToSample = 6;
TableForm[Table[n = Random[Integer, {2, 10}];
  m = Random[Integer, {2, 11}]; {m, n,
   CyclicQ[DirectProduct[Z[m], Z[n]]]}, {numberToSample}],
  TableHeadings → {None, {"m", "n", "Zm × Zn cyclic?\n"}}]
```

m	n	$\mathbb{Z}_m \times \mathbb{Z}_n$ cyclic?
7	10	True
2	2	False
6	3	False
7	3	True
10	3	True
8	3	True

Q5. By either repeating the preceding code or substituting values for m and n below, try enough examples until you can make a conjecture about a relationship between m and n that will guarantee that the direct product $\mathbb{Z}_m \times \mathbb{Z}_n$ is cyclic.

```
m = 7
n = 8
CyclicQ[DirectProduct[Z[m], Z[n]]]
```

Q6. Give a reasonable explanation why your conjecture makes sense.

- **What are the generators of a cyclic direct product?**

One of the examples you may have tested, $\mathbb{Z}_4 \times \mathbb{Z}_5$, turns out to be cyclic. A natural question to ask concerns the generators of this cyclic group. *Before* you evaluate the following cell, think about what the list of generators might be.

```
? CyclicGenerators

CyclicGenerators[DirectProduct[Z[4], Z[5]]]
```

Q7. What is another example of indices m and n for which $\mathbb{Z}_m \times \mathbb{Z}_n$ is cyclic? Without using *Mathematica*, determine the generators for this group. Test yourself using *Mathematica*.

Fill in values for n and m for which the direct product is cyclic.

```
m = 4
n = 5
CyclicGenerators[DirectProduct[Z[m], Z[n]]]
```

Q8. When is $\mathbb{Z}_m \times \mathbb{Z}_n$ isomorphic to $\mathbb{Z}_{mn}$?

10.6 Isomorphisms among U_n groups

Consider the following, where we form the groups U_m, U_n, U_{mn} and the direct product $U_m \times U_n$.

```
m = 5;
n = 4;
G1 = U[m]
G2 = U[n]
```

```
G = DirectProduct[G1, G2]
```

```
U[m * n]
```

Note that U_{mn} and $U_m \times U_n$ have the same number of elements.

```
Length[Elements[U[m * n]]] ==
 Length[Elements[DirectProduct[U[m], U[n]]]]
```

Suppose we define a function from U_{mn} to $U_m \times U_n$. To do so, we need to make sure that the image of any element lands in $U_m \times U_n$, which means that the first component must be in U_m and the second in U_n. The following function is one way to accomplish this.

```
Clear[f]
f[x_] := {Mod[x, m], Mod[x, n]}
```

What is the image of this function on U_{mn}?

```
Map[f, Elements[U[m * n]]]
```

How does this compare to the elements of $U_m \times U_n$?

```
SameSetQ[%, Elements[DirectProduct[U[m], U[n]]]]
```

Thus, we see that we have the same set of elements. Could this possibly be an isomorphism? First, let's set up a `Morphoid`.

```
func = FormMorphoid[f, U[m * n], DirectProduct[U[m], U[n]]]
```

Since we see that the map is onto (and one-to-one), what we really need to do is check to see if it is a morphism.

```
MorphismQ[func, Mode → Visual]
```

Indeed. We have an isomorphism. Is there anything special about the *m* and *n* we chose? Let's try some random values and see what happens.

```
numberToSample = 6;
Off[Morphoid::dff];
TableForm[Table[m = Random[Integer, {2, 10}];
  n = Random[Integer, {2, 10}]; G₁ = U[m]; G₂ = U[n];
  G = DirectProduct[G₁, G₂]; H = U[m n]; Clear[f];
  f[x_] := {Mod[x, m], Mod[x, n]}; {m, n, IsomorphismQ[
    FormMorphoid[f, U[m n], DirectProduct[U[m], U[n]]],
    Cautious → True]}, {numberToSample}],
  TableHeadings → {None, {"m", "n", "isomorphic?\n"}}]
On[Morphoid::dff];
```

Q9. By either repeating the preceding code or substituting values for n and m below, try enough examples until you can make a conjecture about a relationship between n and m that will guarantee that the direct product of U_n and U_m is isomorphic to U_{mn}.

```
n =
m =
Clear[f];
f[x_] := {Mod[x, m], Mod[x, n]};
IsomorphismQ[
    FormMorphoid[f, U[m n], DirectProduct[U[m], U[n]]]]
```

Q10. Is U_{30} isomorphic to $U_2 \times U_3 \times U_5$? Why or why not? If so, set up an isomorphism; if not, explain why not.

Q11. Is $\mathbb{Z}_{30}$ isomorphic to $\mathbb{Z}_2 \times \mathbb{Z}_3 \times \mathbb{Z}_5$? Why or why not? If so, set up an isomorphism; if not, explain why not.

Q12. Is $\mathbb{Z}_{mn}$ isomorphic to $\mathbb{Z}_m \times \mathbb{Z}_n$? Set up an isomorphism when it is and explain why not when it is not.

Group Lab 11

Cosets

11.1 Prerequisites

This lab is self-contained. No prior labs need to be completed to attempt this one, but familiarity with the basic groups such as $\mathbb{Z}_n$, U_n and D_n is helpful.

11.2 Goals

This lab explores the notion of cosets. We will look at how cosets are determined, the different types of cosets, and some of the properties of cosets.

11.3 Cosets, left and right

Suppose we start with some group G, say the dihedral group of order 8, D_4. First we read in the necessary *Mathematica* code.

```
Needs["AbstractAlgebra`Master`"];
SwitchStructureTo[Group];

G = Dihedral[4]
```

Now consider a subgroup of G, say $H = \{1, \text{Rot}^2\}$, consisting of the identity and the 180° rotation (when we view D_4 as the symmetries of the square).

```
H = {1, Rot²}
```

We would like to take another element *g* from *G* and multiply it by all the elements in *H*. Suppose we choose one of the reflections, say `Ref`. Before multiplying, we need to first choose on which side to multiply `Ref`, since this group is not Abelian. We also need to find a way of calculating this product in *Mathematica*.

If we have a function *f* of two variables (such as $f(x, y) = x^2 + y^2$) and we want to hold one variable fixed and let the other roam over some fixed set, we can use the `Map` function as follows.

```
Clear[f, x, y]
f[x_, y_] := x^2 + y^2
Map[f[2, #] &, {-2, -1, 0, 1, 2}]
```

The expression `f[2, #]&` indicates that all the elements in the list {−2, −1, 0, 1, 2} should, one at a time, be placed in for the `#` and then *f* can be properly evaluated with the two arguments (the first always being 2, in this example). Think about what the following will return; don't evaluate it until you have made some (educated) guess.

```
Map[f[#, 1] &, {-2, -1, 0, 1, 2}]
```

We will use this mapping principle to calculate our product of `Ref` times the elements in *H*. First we need the group's operation, which is a function of two variables. The following allows us to use `op` as a short name for the operation. (The semicolon suppresses the actual definition, since its details are not important.)

```
op = Operation[G];
(* the ; suppresses output, so do not expect any *)
```

Now we will use `op` to do the calculations, placing `Ref` on the left (as the first operand).

```
Map[op[Ref, #] &, H]
```

This result is called the *left coset of H in G containing g* (where $H = \{1, \text{Rot}^2\}$ and *g* = Ref). In general, we denote the left coset of *H* in *G* containing *g* by the notation *gH*. This coset is the set {*g* ∗ *h* | *h* ∈ *H*}. We define the right coset of *H* in *G* containing *g* in a similar fashion: *Hg* = {*h* ∗ *g* | *h* ∈ *H*}. (Note that when a group is written additively, we denote these cosets by *g* + *H* and *H* + *g*.) We can calculate the right coset in a similar fashion.

```
Map[op[#, Ref] &, H]
(* note that Ref is now the second, or right, operand *)
```

Observe that, in this case, the right coset and the left coset are the same. Suppose we use a new element, $g = \text{Rot}^2 \text{ Ref}$.

A technical *Mathematica* note might be useful here, since in *Mathematica* we enter $Rot^2 Ref$ as `Rot^2 ** Ref`. In *Mathematica*, multiplication is denoted either by juxtaposition, 6 7, or by using the asterisk, $6*7$.

```
6 7
6 * 7

Clear[g, h]
g * h
h * g
```

Note that both $g*h$ and $h*g$ return $g\,h$. In other words, by default *Mathematica* assumes that the multiplication $*$ is commutative and the elements are returned in some canonical order (alphabetically in this case). To force a *noncommutative* multiplication we use $**$.

```
g ** h
h ** g
```

Now let's try calculating the left coset of H in G containing $Rot^2 Ref$.

```
Map[op[Rot^2 ** Ref, #1] &, H]
```

Note that this is the same result (as sets they don't need to be in the same order) as the left coset of H in G containing Ref.

> **Q1**. Calculate both the left and right cosets of H in G containing Rot. Are they the same?

Since this is a somewhat awkward way of calculating cosets (though it conveys exactly what is happening), let's use another method. We have functions `Left-Coset` and `RightCoset` that give us what we want.

```
LeftCoset[G, H, Rot^2 ** Ref]
RightCoset[G, H, Rot^2 ** Ref]

? LeftCoset
```

Suppose we consider another group, say $\mathbb{Z}_8$, and let $H = \{0, 2, 4, 6\}$.

```
G = Z[8]
H = {0, 2, 4, 6}
SubgroupQ[H, G]
```

(`SubgroupQ` is a function that one can use to test to see if a set is truly a subgroup or not. We want to make sure that H is a subgroup before we try making cosets.)

Let's try some cosets.

```
LeftCoset[G, H, 0]
LeftCoset[G, H, 1]
```

For a little more information, try the `Textual` mode.

```
LeftCoset[G, H, 1, Mode → Textual]
```

Here is the last coset from another perspective:

```
LeftCoset[G, H, 1, Mode → Visual]
```

To get all the cosets, we can use our `Map` function and map the `LeftCoset` function across the whole set of elements of *G*.

```
Map[LeftCoset[G, H, #] &, Elements[G]]
```

This can be understood better, perhaps, by viewing both the coset and the element that generates the coset; try the following.

```
Map[{#, LeftCoset[G, H, #]} &, Elements[G]] // ColumnForm
```

> **Q2.** What observations can you make about these cosets? If you can't think of anything, try the following (but try to think of something first).

```
Map[{#, Sort[LeftCoset[G, H, #]]} &, Elements[G]] //
  ColumnForm
```

Now compare the left cosets and the right cosets of $H = \{0, 2, 4, 6\}$ in $G = \mathbb{Z}_8$.

```
TableForm[
  Map[{#, LeftCoset[G, H, #], RightCoset[G, H, #]} &,
   Elements[G]], TableHeadings →
   {None, {"g", "g+H", "H+g\n"}}, TableDepth → 2]
```

In each row, the first element is the element that we are multiplying by (adding to, in this case) the set *H*. (This element is usually called a *coset representative*.) The second item is the left coset and the third is the corresponding right coset.

> **Q3.** What observations can you make about the relationship between these left and right cosets?

Lest you jump to a false conclusion, let's consider another example.

```
G = Symmetric[3]
H = {{1, 2, 3}, {1, 3, 2}}
SubgroupQ[H, G]
```

Now that we know that *H* is a subgroup of *G*, we can consider the various left and right cosets.

```
TableForm[
  Map[{#, LeftCoset[G, H, #], RightCoset[G, H, #]} &,
    Elements[G]], TableHeadings →
    {None, {"g", "gH", "Hg\n"}}, TableDepth → 2]
```

Q4. What observations can you make about the relationship between these left and right cosets?

▦ 11.4 Properties of cosets

Let's look for some common properties among the three collections of cosets is section 11.3.

```
G = Dihedral[4];
H = {1, Rot^2};
TableForm[
  Map[{#, LeftCoset[G, H, #], RightCoset[G, H, #]} &,
    Elements[G]], TableHeadings →
    {None, {"g", "gH", "Hg\n"}}, TableDepth → 2]

G = Z[8];
H = {0, 2, 4, 6};
TableForm[
  Map[{#, LeftCoset[G, H, #], RightCoset[G, H, #]} &,
    Elements[G]], TableHeadings →
    {None, {"g", "gH", "Hg\n"}}, TableDepth → 2]

G = Symmetric[3];
H = {{1, 2, 3}, {1, 3, 2}};
TableForm[
  Map[{#, LeftCoset[G, H, #], RightCoset[G, H, #]} &,
    Elements[G]], TableHeadings →
    {None, {"g", "gH", "Hg\n"}}, TableDepth → 2]
```

For each of the following questions, consider the preceding examples and see if you can provide an answer. (Note that in the following questions we use gH and Hg generically, independent of whether the group G writes the cosets additively or multiplicatively.)

Q5. Recall that when we talk about the left (or right) coset of H in G containing g, we call the element g the coset representative of gH (or Hg). Does it appear that g is one of the elements in gH and Hg? Why or why not?

Q6. As you look at the cosets above, when is it the case that $g H$ or $H g$ actually is the set H? Why is this the case?

Q7. Pick any two elements x and y in G. What can you say about the two cosets $x H$ and $y H$ (or, for that matter, the corresponding right cosets)? Justify your answer.

Q8. How many (different) cosets are there for a given group G and subgroup H? Why is this the case?

Q9. Although there is only limited evidence (you could produce more, however), what can you say about when the left coset $g H$ might be equal to the right coset $H g$?

Q10. When is the left coset $g H$ (or right coset $H g$) a subgroup of G? Are there any conditions that guarantee it?

Let's consider two more examples of cosets. Instead of using the Map function to get cosets, we can use the functions LeftCosets and RightCosets

```
? LeftCosets

G = U[35];
H = {4, 16, 29, 11, 9, 1};
SubgroupQ[H, G]

LeftCosets[G, H]
RightCosets[G, H]

G = Z[35];
H = {5, 10, 15, 20, 25, 30, 0};
SubgroupQ[H, G]

LeftCosets[G, H]
RightCosets[G, H]
```

Q11. For both of the preceding groups, count the number of cosets (either left or right), and calculate the orders of H and G. Now offer a conjecture.

Q12. Suppose $G = \mathbb{Z}$ and $H = 2\,\mathbb{Z}$, the even integers. Verify that H is a subgroup of G. Describe the coset $3 + H$.

Q13. Suppose $G = \mathbb{Z}$ and $H = 5\,\mathbb{Z}$, the multiples of 5. Suppose x and y are the scores of two opponents playing table tennis. There is a change of service whenever the sum of the scores x and y is a multiple of 5 (and so in H). Each score belongs to one of the following cosets: H, $1 + H$, $2 + H$, $3 + H$ or $4 + H$. (Why?) Is there any relationship between the coset to which x belongs and the one to which $-y$ belongs at the time of the change of service? Can you give a reasonable explanation for it?

Q14. Suppose $G = U_5 \times \mathbb{Z}_4$ and $H = \{\{2, 2\}, \{4, 0\}, \{3, 2\}, \{1, 0\}\}$ is a subgroup. Determine the other left cosets by hand. See if you can use *Mathematica* to confirm your answer.

Group Lab 12

Normality and Factor Groups

▣ 12.1 Prerequisites

Before attempting this lab you should complete Group Lab 11 (on cosets).

▣ 12.2 Goals

After defining a *normal* subgroup, we try to make some sense of finding an operation to act on cosets. This leads to the development of the *quotient* or *factor* group. We conclude by illustrating why normality is important in constructing the factor group.

▣ 12.3 Normal subgroups

Suppose we start out, as we did in the last lab, with the dihedral group of order 8, D_4. First we read in the code needed for this lab.

```
Needs["AbstractAlgebra`Master`"];
SwitchStructureTo[Group];

G = Dihedral[4]
```

Now consider a subgroup of G, say $H = \{1, \text{Rot}^2\}$, consisting of the identity and the 180° rotation.

```
H = {1, Rot²}
```

In Group Lab 11 we saw that on some occasions a left coset was the same as a right coset. Will this happen in this case? Following is a table consisting of a column of the elements of G followed by the elements in the left and right cosets of H in G containing g.

```
TableForm[
  Map[{#, LeftCoset[G, H, #], RightCoset[G, H, #]}&,
    Elements[G]], TableDepth → 2,
  TableHeadings → {None, {"g", "gH", "Hg\n"}}]
```

> **Q1**. Would you say that every left coset $g\,H$ is equal to the right coset $H\,g$? Why or why not?

Consider another example. Suppose we have $G = U_{40}$.

```
G = U[40]
```

Let H be the subgroup $\langle 13 \rangle$.

```
H = SubgroupGenerated[G, 13]
```

Let's check out the left cosets $g\,H$ and the right cosets $H\,g$ (same g). This time we will let *Mathematica* do the checking of equality. (Note that we sort the two cosets before comparing for equality. Why is this done?)

```
TableForm[
  Map[{#, LeftCoset[G, H, #], RightCoset[G, H, #], Sort[
      LeftCoset[G, H, #]] === Sort[RightCoset[G, H, #]]}&,
    Elements[G]], TableDepth → 2, TableHeadings →
    {None, {"g", "gH",
  "Hg", "= ?\n"}}]
```

We repeat this with $G = S_3$ and $H = \langle\{1, 3, 2\}\rangle$.

```
G = Symmetric[3]
H = SubgroupGenerated[G, {1, 3, 2}]

TableForm[
  Map[{#, LeftCoset[G, H, #], RightCoset[G, H, #], Sort[
      LeftCoset[G, H, #]] === Sort[RightCoset[G, H, #]]}&,
    Elements[G]], TableDepth → 2, TableHeadings →
    {None, {"g", "gH",
  "Hg", "= ?\n"}}]
```

Here we note that four of the left cosets are not equal to their corresponding right cosets. So we have evidence that equality of left and right cosets is *not* a general property for all groups or subgroups. This leads to a definition: Given a subgroup H of a group G, if $g\,H = H\,g$ for all g in G, we say H is a *normal* subgroup of G. We can use the function `NormalQ` to test for normality of a subgroup.

```
NormalQ[H, G]
```

Note that if we change *H* to a different subgroup of *G* we may obtain normality.

```
NormalQ[SubgroupGenerated[G, {2, 3, 1}], G]
```

Furthermore, note the *index* of this subgroup in *G* (i.e., the number of cosets).

```
Index[G, SubgroupGenerated[G, {2, 3, 1}]]
```

An exercise in group theory is to show that any subgroup *H* whose index in *G* is 2 will always be normal in *G*. Let's try some other groups.

```
NormalQ[{0, 4}, Z[8]]
NormalQ[SubgroupGenerated[U[52], 15], U[52]]
NormalQ[SubgroupGenerated[Dihedral[4], Ref], Dihedral[4]]
```

> **Q2**. Suppose *G* is an Abelian group. What can you say about the normality of any subgroup *H* of *G*? Why?

> **Q3**. Suppose $G = \mathbb{Z}_5 \times U_6$. Let $H = \langle (2, 5) \rangle$. Is *H* normal in *G*? Give an answer and justification and then check using *Mathematica* (recalling that `Direct-Product` is used to obtain the desired direct product).

▣ 12.4 Making a new group

Let's consider the group $G = \mathbb{Z}_8$ and $H = \{0, 4\}$ and investigate the cosets that arise.

```
G = Z[8]
H = {0, 4}
```

Consider the following four cosets. (Recall that the coset *g H* is written *g + H* when the group is written additively.)

```
0 + H
1 + H
2 + H
3 + H
(* if you think Mathematica
   knows what it is doing here,
   understanding cosets, you are mistaken - try 5 + H *)
```

Q4. Why are these the only cosets considered? In other words, why aren't both left cosets and right cosets considered? Additionally, what about the cosets $4 + H, 5 + H, 6 + H$ and $7 + H$?

Suppose we consider these four cosets and think about a way in which we might combine (operate on) any two of them. What might it mean to add two cosets, say $2 + H$ and $3 + H$? As alluded to, *Mathematica* doesn't really (yet) know how to add these cosets.

```
(2 + H)
(3 + H)
(2 + H) + (3 + H)
Mod[(2 + H) + (3 + H), 8]
```

Even reducing these mod 8 is not right. What might it mean to "add" {2, 6} and {3, 7}? Suppose we try all possible sums and then reduce mod 8.

```
{2 + 3, 2 + 7, 6 + 3, 6 + 7}
```

Reduce mod 8:

```
Mod[%, 8]
```

Finally, remove duplicates.

```
Union[%]
```

Note that this is really $5 + H$ (as well as $1 + H$). Note also that $2 + 3 = 5$. Since $2 + H$ could have been written as $6 + H$, one might wonder if interchanging them would affect how we "add" these cosets. As will likely be proved in your class, it does not. In fact, given any group G and any normal subgroup N of G, the set of all (left) cosets $\{g N \mid g \in G\}$ forms a group under the operation $(g N)(h N) = (g h) N$ (shown multiplicatively here—additively it would be written $(g + N) + (h + N) = (g + h) + N$). This new group is called the *factor group* of G by H, or the *quotient group* of G by H. This is frequently denoted G / H.

```
FactorGroup[G, H]
QuotientGroup[G, H]

?QuotientGroup
```

Q5. Try adding a few of these cosets (as elements of the factor group). On scratch paper, you may wish to make the Cayley table for the group G / H. To what group is this isomorphic?

▥ 12.5 Factor groups

Maybe you thought about making the Cayley table with *Mathematica*. (If not, why not?) Let's do so.

```
G = Z[8];
H = {0, 4};
CayleyTable[QuotientGroup[G, H]]
```

Of course, there is also the pretty version.

```
gr1 = CayleyTable[QuotientGroup[G, H],
    Mode → Visual, Output → Graphics];
```

Observe that the elements in the group table are the left cosets of $H = \{0, 4\}$ in $\mathbb{Z}_8$. Just as a reminder, here they are.

```
LeftCosets[G, H]
```

Q6. To what group is $\mathbb{Z}_8 / \{0, 4\}$ isomorphic?

Here is a visual way of seeing how these cosets interact with each other.

```
gr2 = LeftCosets[G, H, Mode → Visual, Output → Graphics];
```

Putting these two graphics together, we obtain the following.

```
Show[GraphicsArray[{gr1, gr2}]];
```

Q7. What observations can you make from considering these two tables? Include an accounting of the coloring of both tables and the "blocking" (coloring by "chunks") in the second table.

Recall that the subgroup $H = \langle\{1, 3, 2\}\rangle$ is *not* normal in S_3.

```
NormalQ[SubgroupGenerated[S[3], {1, 3, 2}], S[3]]
```

How do the left cosets interact in this case?

```
LeftCosets[S[3],
    SubgroupGenerated[S[3], {1, 3, 2}], Mode → Visual]
```

> **Q8**. Before we can form the factor group, the subgroup must be normal. What goes wrong when the subgroup is not normal? Consider the preceding visualization and answer this question. Focus, in particular, on the last four columns and try to determine why the coloration does not occur in square blocks there. Give specific explanations for the coloring in the last four columns.

Next we consider the same group and subgroup but focus on the right cosets.

```
RightCosets[S[3],
    SubgroupGenerated[S[3], {1, 3, 2}], Mode → Visual]
```

> **Q9**. Explain why the last four rows occur as they do. Provide explicit details.

> **Q10**. Determine another subgroup H of some group G that is not normal in G. Define them below. Now consider the left and right cosets visually, using the cell below your definitions.

```
G =
H =
NormalQ[G, H]

LeftCosets[G, H, Mode → Visual]
RightCosets[G, H, Mode → Visual]
```

Now let's consider two more examples of normal subgroups and the corresponding products of cosets.

```
G = U[16]
H = SubgroupGenerated[G, 3]
NormalQ[H, G]

LeftCosets[G, H, Mode → Visual]

G = Z[6]
H = {0, 3}
NormalQ[H, G]

LeftCosets[G, H, Mode → Visual]
```

> **Q11**. For these normal subgroups, why did we only look at the visual representation of the left cosets (and not the right cosets)?

> **Q12**. Make a summary statement regarding why normality is a requirement for a quotient group to be a coherent, well-defined structure.

Group Lab 13

Group Homomorphisms

🖼 13.1 Prerequisites

The reader should be familiar with isomorphisms (Group Lab 8) and cosets (Group Lab 11).

🖼 13.2 Goals

This lab explores the concept of group homomorphisms. The ultimate goal is an understanding of the relationship between the domain, kernel, and image of a homomorphism through the Fundamental Theorem of Group Homomorphisms.

🖼 13.3 What is a group homomorphism?

Note on terminology: The terms *homomorphism* and *morphism* are used synonymously in algebra. Since the former is somewhat more prevalent in textbooks, we use it here, but in the interest of using shorter names where possible, we use the latter in our *Mathematica* code.

A *homomorphism* from a group $(G, *)$ into a group $(H, \#)$ is a function from G into H that preserves the operations between the two groups ($*$ and $\#$ respectively). That is, if f is a homomorphism, then for all values of x and y in G,

$$f(x * y) = f(x) \# f(y).$$

You should be familiar with the one-to-one, onto variety of these functions, called *isomorphisms*. Recall that the existence of an isomorphism between two groups establishes that the two groups are "equal" in an algebraic sense. The automorphisms on a group (isomorphisms from a group to itself) help us to categorize groups. Homomorphisms provide us with yet another tool in exploring groups.

Before continuing, we first read in the code used in this lab.

```
Needs["AbstractAlgebra`Master`"];
SwitchStructureTo[Group];
```

To review the notion of preserving an operation, we start by looking at two functions from the group of integers modulo 12 into the group of integers modulo 6. Again we use objects called `Morphoids` and the function `FormMorphoid` to create them. The first function adds 3 to each element of $\mathbb{Z}_{12}$ and then takes the residue modulo 6 (reduces it mod 6). The second adds 6 before taking the residue modulo 6.

```
f = FormMorphoid[Mod[#1 + 3, 6] &, Z[12], Z[6]]
```

```
g = FormMorphoid[Mod[#1 + 6, 6] &, Z[12], Z[6]]
```

Next we see if either function preserves the operations between the two groups for a specific pair of elements, say 5 and 9. We use the `Visual` mode of `PreservesQ` to help visualize the process. First f.

```
PreservesQ[f, {5, 9}, Mode → Visual]
```

The operations are not preserved: $f(5 + 9) \neq f(5) + f(9)$. So, with this single example, we see that f is not a homomorphism. Now we look at g. (Note: The visual mode displays the name of any function generically as f, but here g is the function we are examining.)

```
PreservesQ[g, {5, 9}, Mode → Visual]
```

For the pair $(5, 9)$, g acts like a homomorphism. Does g satisfy this property for all pairs in $\mathbb{Z}_{12}$? We can use `MorphismQ` to test all pairs.

```
MorphismQ[g, Mode → Visual]
```

If g is replaced with f in this cell, an opposite color pattern results.

> **Q1.** There is a simpler way to define g than the function we used. Can you describe it?

13.4 The kernel and image

The kernel and image of a homomorphism are special subsets worthy of consideration. The kernel is a subset of the domain and the image is a subset of the codomain.

13.4.1 The kernel

The kernel of a group homomorphism is the set of elements in the domain that map onto the identity of the codomain. Recall that g is a homomorphism.

```
Kernel[g]
```

The `Kernel` function returns a `Groupoid`, which we know by now does not in itself imply that it is a group (or a subgroup of the domain)

> **Q2.** In this case, is the kernel a group (i.e., a subgroup of $\mathbb{Z}_{12}$)?

Let's look at some more kernels. Consider the following `Morphoids`.

```
f2 = FormMorphoid[Mod[#1, 5] &, Z[10], Z[5]]
f3 = FormMorphoid[Mod[#1, 2] &, Z[10], Z[2]]
f4 = FormMorphoid[2 → 1, U[25], Z[20]]
f5 = FormMorphoid[2 → 2, U[25], Z[20]]
f6 = FormMorphoid[Mod[#1, 3] &, Z[10], Z[5]]
```

Let's determine which of these are homomorphisms and compute the kernels.

```
Map[MorphismQ, {f2, f3, f4, f5, f6}]
```

```
Map[Kernel, {f2, f3, f4, f5, f6}]
```

> **Q3.** For which of these five functions is the kernel a group (i.e., a subgroup of the domain)?

> **Q4.** If h is a homomorphism and a and b are in $K = \text{Kernel}[h]$, explain why $a * b$ must also be in K. (In other words, show that K is closed.) Note: Closure of a (nonempty) subset of a finite group implies that it is a subgroup.

The *Mathematica* function `Kernel` acts on any `Morphoid`. If the codomain has an identity, it returns a `Groupoid` (it returns `$Failed` otherwise). Recall that f was not a homomorphism.

```
Kernel[f]
ClosedQ[%]
```

Q5. True or false: If ϕ is a `Morphoid` from some group G into some group H and Kernel[ϕ] is a subgroup of the G, then ϕ is a homomorphism. (See the hint at end of the section, if needed.)

In some sense, the size of the kernel indicates how much the domain is "reduced" when a homomorphism is applied to it.

▪ Hint

Consider $G = H = \mathbb{Z}_3$ and the function that squares its input (mod 3).

▪ 13.4.2 The image

The image of a homomorphism is the range of the function and thus is a subset of the codomain. We want to know when this subset is a subgroup. For our first homomorphism, g, the image is the whole codomain, and so it is clearly a subgroup. This is a common situation.

```
Image[g]
```

Consider the five `Morphoids` defined in section 13.4.1, in addition to f defined at the outset of the lab. Are the images subgroups of the codomain?

```
{f, f2, f3, f4, f5, f6} // ColumnForm

Map[Image, {f, f2, f3, f4, f5, f6}]
```

Q6. Which of these images are subgroups of the given codomain?

Q7. Suppose π is a function from the group G into the group H. Consider the following two statements: (1) If π is a homomorphism, then the image of π is a subgroup of H, (2) If the image of π is a subgroup of H, then π is a homomorphism. Determine whether these statements are true or false. If true, give a proof; if false, provide an example of a `Morphoid` that demonstrates it is false.

▦ 13.5 Properties that are preserved by homomorphisms

The following results are proven in many abstract algebra texts, so we simply report them and give a few illustrations.

Let f be a homomorphism from G into H. Then

    1. f(the identity of G) = the identity of H.
    2. $f(x^{-1}) = f(x)^{-1}$ for all x in G.
    3. If S is a subgroup of G, then $f(S) = \{f(s) \mid s \in S\}$ is a subgroup of H.

We use one homomorphism to illustrate these three properties, mapping the integers modulo 16 into the fourth roots of unity.

```
G = Z[16]
H = RootsOfUnity[4, Mode → Visual]
```

The visual form of `RootsOfUnity[4]` is given to illustrate that this group is the set of complex numbers (on the unit circle) that satisfy the equation $x^4 = 1$. The `Morphoid` maps each integer to that power of i, so $x \mapsto i^x$.

```
Clear[f];
f = FormMorphoid[(I^#) &, G, H]
```

The identities here are 0 and 1, respectively. Therefore, by property 1, the element 0 in G should be mapped to the element 1 in H.

```
f[0] == 1
```

Notice below how the equation is much more explicit in *Mathematica* than in property 2. To test the property we need to specify the groups in which each inversion is taking place. We pick a random element x out of the domain and then test property 2.

```
x = RandomElement[G]
f[GroupInverse[G, x]] == GroupInverse[H, f[x]]
```

Let's generate a few more examples and place them in a table. The third and fifth columns should match if our `Morphoid` is a morphism.

```
TableForm[
  Table[x = RandomElement[G]; xinv = GroupInverse[G, x];
   fxinv = f[xinv]; fx = f[x]; invfx = GroupInverse[H, f[x]];
   {x, xinv, fxinv, fx, invfx}, {10}], TableHeadings →
   {None, {"x", "x⁻¹", "f(x⁻¹)", "f(x)", "f(x)⁻¹\n"}}]
```

For property 3, note that the even elements of $\mathbb{Z}_{16}$ are a subgroup of G.

```
Y = Range[0, 15, 2]
SubgroupQ[Y, G]
```

Now consider its image.

```
W = Union[Map[f, Y]]
SubgroupQ[W, H]
```

This too is a subgroup.

> **Q8**. Recall the `Morphoid` *f5* defined earlier. Verify that these three properties hold for this morphism.

```
f5 = FormMorphoid[2 → 2, U[25], Z[20]]
```

13.6 The kernel is normal

The kernel of a homomorphism is not just an ordinary subgroup; it is always a normal subgroup. One proof of this statement is based on the definition of a normal subgroup.

H is a normal subgroup of *G* if and only if $g^{-1}h\,g$ is in *H* for all *h* in *H* and *g* in *G*.

In other words, every conjugate of an element of *H* is also in *H*. (Recall that $g^{-1}h$ *g*, for some *g* in *G*, is a conjugate of *h*.)

> **Q9**. Verify that if *f* is a homomorphism and *k* is in *K* = Kernel[*f*], then any conjugate of *k* is also in *K*.

Since every subgroup of an Abelian group is normal (Do you know why?) we need to consider a nonabelian group to illustrate the normality of the kernel. Consider the cyclic subgroup generated by `Ref` in D_3.

```
H = SubgroupGenerated[Dihedral[3], Ref]
```

We can choose *g* to be any element of the group—here we pick `Rot` and look at its conjugates.

```
Map[Conjugate[Dihedral[3], #, Rot] &, Elements[H]]
```

This result shows that *H* is not a normal subgroup. (Why?) It also implies something we can not easily illustrate: *H* is not the kernel of any homomorphism with domain D_3. Let's see what kernels we *can* get from homomorphisms on D_3. Suppose we consider the exponent of the symbol `Rot` in any element of D_3. Recall the elements of D_3.

```
Elements[Dihedral[3]]
```

Clearly the exponent is an integer in {0, 1, 2}, so let's consider this as a `Mor`-phoid into $\mathbb{Z}_3$. The *Mathematica* expression here is a bit complicated because the elements in D_3 involve the function `NonCommutativeMultiply`.

```
xifunction =
  Exponent[# /. {NonCommutativeMultiply → Times}, Rot] &
```

We put the `Morphoid` into rules-form to make it easier to read.

```
ξ = FormMorphoid[xifunction, Dihedral[3], Z[3]] // ToRules
```

Now consider the kernel.

```
Kernel[ξ]
```

> **Q10.** Since the kernel of ξ is {1, Ref}, what immediate conclusion should we draw regarding ξ? You might want to use `PreservesQ` or `MorphismQ` to verify your answer.

Let's do the same thing, but with the exponent of `Ref`.

```
taufunction =
  Exponent[#1 /. {NonCommutativeMultiply → Times}, Ref] &
```

```
τ = FormMorphoid[taufunction, Dihedral[3], Z[2]] // ToRules
```

We get a different kernel here.

```
K = Kernel[τ]
```

Is the kernel normal? We can use the function `NormalQ`.

```
NormalQ[K, Dihedral[3]]
```

Alternatively, we can ask whether τ is a homomorphism.

```
MorphismQ[τ]
```

> **Q11.** Explain why checking the value of `MorphismQ[τ]` gives us more information than does `NormalQ[K, Dihedral[3]]`.

▣ 13.7 The First Homomorphism Theorem

Recall that normality is exactly the condition on a subgroup required in order for its left (or right) cosets to form a group. If f is a homomorphism from G into H with kernel K, we examine the quotient group G/K. We start with the most recent example.

```
Q = QuotientGroup[Dihedral[3], K]
```

The quotient group has just two elements, so it clearly must be isomorphic to $\mathbb{Z}_2$, which happens to be the image of τ. Let's formally establish this fact by creating an isomorphism between $G/\texttt{Kernel}[\tau]$ and $\texttt{Image}[\tau]$.

```
FormMorphoidSetup[Q, Image[τ]];
```

We want to map the identity of the domain to the identity of the codomain, so we define the `Morphoid` by position using the list $\{1, 2\}$ to indicate that the first element of Q goes to the first element of $\texttt{Image}[\tau]$ and the second to the second.

```
θ = FormMorphoid[{1, 2}, Q, Image[τ]]
```

```
IsomorphismQ[θ]
```

The isomorphism that we've just created is not coincidental. It follows from the following theorem.

Fundamental Homomorphism Theorem: Let f be a group homomorphism from G into H with kernel K and image $f(G)$. Then $f(G)$ is isomorphic to G/K, an isomorphism between these two groups being $\psi: G/K \to f(G)$ defined by $\psi(aK) = f(a)$.

In the example, notice how the coset $\texttt{Ref NS}$ is mapped to $\tau[\texttt{Ref}] = 1$, while $\texttt{NS} = 1 \texttt{ NS}$ is mapped to $\tau[1] = 0$.

Virtually every abstract algebra text has a proof of this theorem or some slight variation of the theorem. For example, one popular text starts with the assumption that f is onto (and so $f(G) = H$). The proof is still essentially the same. We close this section by illustrating the theorem with two more examples.

■ 13.7.1 Example 2

Consider the direct product $\mathbb{Z}_5 \times \mathbb{Z}_5$ and map each pair (a, b) into $4a + 3b$ (mod 5) in $\mathbb{Z}_5$.

```
G = DirectProduct[Z[5], Z[5]]
```

```
α = FormMorphoid[Mod[Plus @@ ({4, 3}.#1), 5] &, G, Z[5]]
```

It may not be obvious whether α is a homomorphism, so we verify that it is.

```
MorphismQ[α]
```

We compute the kernel and image.

```
K = Kernel[α]
```

```
Image[α]
```

Notice that α is onto, so we can establish an isomorphism between G/K and $\mathbb{Z}_5$.

```
Q2 = QuotientGroup[G, K]
```

```
γ = FormMorphoid[α[First[#1]] &, Q2, Image[α]]
```

```
IsomorphismQ[γ]
```

Q12. Let G = DirectProduct[Z[4], Z[4]] and K = {{0,0}, {0,2}, {2,0}, {2,2}}. To what group is G/K isomorphic? To prove your answer, find a homomorphism from G into some other group that has K as its kernel.

▪ 13.7.2 Example 3

In this example, we map the divisors of 42 into the divisors of 30 with the function GCD[#, 30]&. We form the Morphoid between the MixedDivisors Groupoids based on these sets.

```
? MixedDivisors
```

```
η = FormMorphoid[GCD[#1, 30] &,
    MixedDivisors[42], MixedDivisors[30]]
```

The kernel will be the divisors of 42 that are relatively prime to 30.

```
K = Kernel[η]
```

Not every divisor of 30 is in the range, so η is not onto.

```
Image[η]
```

The Fundamental Homomorphism Theorem tells us that the image is isomorphic to the following quotient group.

```
Q3 = QuotientGroup[MixedDivisors[42], K]
```

The isomorphism we define maps each of these cosets to the image under η of its first element.

```
θ = FormMorphoid[η[First[#1]] &, Q3, Image[η]]
```

```
IsomorphismQ[θ]
```

Again, our result is not a surprise.

The Fundamental Homomorphism Theorem is occasionally called the First Homomorphism Theorem. You might wonder about a Second Homomorphism

Theorem. There is one, and it is often called the Diamond Homomorphism Theorem. There is also a Third (Zassenhauss' Theorem). These theorems are stated and proved in many introductory abstract algebra books.

> **Q13**. A *square-free integer* is an integer that has no square divisors other than 1. For example, 42 and 30 are square-free, while 50 is not. Mixed-Divisors[m] is a group if and only if m is square-free. As a general rule, if m and n are both square-free integers and we map MixedDivisors[m] into MixedDivisors[n] with the function GCD[#, n]&, what will be the kernel and range of this morphism?

▦ 13.8 The alternating group—parity as a morphism

We have seen that every permutation is either odd or even. Now we will think of this classification as a Morphoid mapping S_n into the group {1, −1} with multiplication, which we call IntegerUnits. Here we will work with $n = 4$.

> **sign = FormMorphoid[Parity, Symmetric[4], IntegerUnits]**

This happens to be a significant morphism for any n.

> **MorphismQ[sign]**

The kernel is quite significant here.

> **Kernel[sign]**

> **Q14**. Describe the elements of the kernel in terms of their parity.

The kernel, which happens to be a normal subgroup of S_4, is called the *alternating group* of degree 4, or A_4. This group can be accessed directly without morphisms using Alternating.

> **?Alternating**

> **Alternating[4]**

There are many interesting properties of the alternating groups that you can read about in your text.

Group Lab 14

Rotational Groups of Regular Polyhedra

14.1 Prerequisites

To complete this lab, you should know how a group can be generated from a set of elements and a binary operation. You should also be familiar with Euler angles (see the Rotations Lab on the CD for a review) and group actions.

14.2 Goals

The goal of this lab is to learn how to generate the rotational groups of polyhedra.

14.3 The rotational group of the tetrahedron

14.3.1 Statement of the problem, first rotation

First let's read in the packages and definitions needed for this lab.

```
Needs["Graphics`Polyhedra`"];
Needs["AbstractAlgebra`Master`"];
SwitchStructureTo[Group];
Needs["Geometry`Rotations`"];
Needs["Graphics`Shapes`"];
Needs["NumberTheory`Recognize`"];
```

Consider the tetrahedron.

```
object = Tetrahedron[];
object // disp
```

We examine certain rotation matrices that act on graphics objects like the `Tetrahedron`. The function `ActionOn` is used extensively in this lab.

```
? ActionOn
```

We want to consider the rotations that make up the so-called rotational group of the tetrahedron, specifically the ones that rotate a tetrahedron so that it occupies the same space as the original object. We can use the function `Compare` to visually confirm whether a rotation is in the rotation group.

```
? Compare
```

Here is an example of a rotation we will not want to consider, because the matrix clearly does not return the object to its original position.

$$\texttt{Compare}\big[\texttt{object, RotationMatrix3D}\big[\frac{\pi}{6},\ \frac{\pi}{4},\ \frac{\pi}{8}\big]\big];$$

One example of an element in the rotational group of this tetrahedron is `RotationMatrix3D[2Pi/3, 0, 0]`. The specific Euler angles that describe this matrix depend on the position of the tetrahedron. One of the faces of our object lies on a plane that is parallel to the *xy*–plane, so a $\frac{2\pi}{3}$ rotation about the *z*-axis is in the rotation group. Also, it is important to note that the tetrahedron needs to be centered about the origin. All polyhedra in the standard *Mathematica* package `Polyhedra.m` are centered about the origin.

```
Clear[r]
```

$$\texttt{r}_1 = \texttt{RotationMatrix3D}\big[\frac{2\,\pi}{3},\ 0,\ 0\big];$$

```
MatrixForm[r₁]

Compare[object, r₁];
```

■ 14.3.2 Generation of more rotations

It is not too difficult to identify r_1 as a member of the rotational group, but how do we identify more complicated rotations? One way is by multiplying rotations. Given any set of rotations, we can generate a group.

```
Clear[G]
G₁ = GenerateGroupoid[{r₁}, Simplify[#1.#2] &]
```

This is a cyclic subgroup of order three.

```
Order[G₁]
```

> **Q1**. Is this the whole rotation group? Based on the fact that the faces of a tetrahedron are four identical equilateral triangles, how many elements would you expect to find in its rotational group?

■ 14.3.3 Procedure for finding more complicated rotations

Although it may be clear that there are more elements, it is probably not obvious what the other rotation matrices are. At this point, we outline a systematic process to find these matrices.

Let's look at a labeled wire-frame picture of the object.

```
? wireFrame
```

```
wireFrame[Tetrahedron];
```

The first rotation that we identified, r_1, maps the (ordered) face {1, 3, 4} into the face {1, 2, 3} and, by applying it again, into {1, 4, 2}. Now suppose we want to map, for example, {1, 3, 4} into {4, 3, 2}. Let R be the unknown matrix.

```
R = Array[r, {3, 3}]
```

Before setting up a system of equations to solve for the rotation matrix, we need to take the vertex coordinates given in the *Mathematica* `Polyhedra` package and make them *exact*. For this we use the function `Exact`. This function is not very sophisticated, but it works for the kinds of numbers we are using.

```
? Exact
```

Here are the vertices given in the package.

```
Vertices[Tetrahedron]
```

The following are exact values.

```
Xvertices = Map[Exact, Vertices[Tetrahedron], {2}]
```

Although a system can be generated more efficiently, here we write it out a bit more descriptively.

```
sys =
{R.Xvertices[[1]] == Xvertices[[4]] ,
R.Xvertices[[3]] == Xvertices[[3]] ,
R.Xvertices[[4]] == Xvertices[[2]] }
```

Now we solve the system, use the rules to get values for R, pick out the one and only solution, and finally, simplify the solution.

```
r₂ = Simplify[First[R /. Solve[sys, Flatten[R]]]]
```

Now the moment of truth: Is r_2 in the rotational group?

```
Compare[object, r₂];
```

This is the rotation matrix that we were looking for. It would be nice to know the values of ϕ, θ and ψ that produce r_2.

Q2. Determining Euler Angles from a Rotation Matrix

Determine the values of ϕ, Θ, and ψ for which `RotationMatrix3D[`ϕ`,` Θ`,` ψ`]` is equal to r_2. Hint: To equate two matrices, A and B, and convert to a list of equations, you can use code something like `(A == B)//-Thread[#]&//Map[Thread[#]&,#]&//Flatten`.

Now that we have two distinct rotations of the tetrahedron, we might be able to generate a larger group.

```
G₂ = GenerateGroupoid[{r₁, r₂}, Simplify[#1.#2] &];
Order[G₂]
```

■ 14.3.4 Verification that we have the complete rotational group

This should be the whole group for which we are searching. To demonstrate how the group generates all the rotations, we mark one face and a vertex of the tetrahedron.

```
markedFace = object[[2]] //
    {#, RGBColor[0.022, 0.688, 0.717], Thickness[0.01],
      Line[{Apply[Plus, object[[2, 1]]] / 3, 1.1 #[[1, 1]]}],
      RGBColor[0.701, 0.038, 0.038],
      PointSize[0.03], Point[1.1 #[[1, 1]]]} & // disp;
```

First we look at the individual effects of each group element on this face. You may want to enlarge the graphic you get here.

```
Show[GraphicsArray[(Partition[#1, 4] &) [
    (ActionOn[markedFace, #1] &) /@First[G₂]]]];
```

Taken together, we get a single tetrahedron with three marks on each face.

```
Show[(ActionOn[markedFace, #1] &) /@First[G₂]];
```

14.4 Further exercises

Q3. Rotational Group of the Cube

    a. How many elements would you expect to find in the group of rotations of the cube?

    b. Generate the group of rotations of Cube[] that is contained in the Polyhedra package.

```
wireFrame[Cube];
```

Q4. Rotational Group of the Icosahedron

    a. How many elements would you expect to find in the group of rotations of the icosahedron?

    b. Generate the group of rotations of Icosahedron[] that is contained in the Polyhedra package.

```
wireFrame[Icosahedron];
```

Q5. Rotational Groups of the Octahedron and Dodecahedron

The octahedron is the dual of the cube. If each face of the cube is replaced with a point at its center and points that are derived from adjacent faces are connected, then the wire frame of an octahedron appears. Explain why the rotational group of the octahedron should be isomorphic to that of the cube. The icosahedron and dodecahedron pair up in the same way.

Q6. The cuboctahedron can be formed by taking a cube and slicing its eight corners off so that the exposed triangular faces just barely meet at one point. It can also be created in the same way starting with an octahedron, thus its name. Determine the number of elements in the rotational group of the cuboctahedron. Is it isomorphic to a group you've already seen?

```
AppendTo[Polyhedra, Cuboctahedron];
Cuboctahedron[Graphics`Polyhedra`Private`opts___] :=
 Polyhedron[Cuboctahedron,
   Graphics`Polyhedra`Private`opts][[1]]
Vertices[Cuboctahedron] ^= Map[N,
```

$$\left\{\left\{0, -\frac{1}{\sqrt{2}}, -\frac{1}{\sqrt{2}}\right\}, \left\{0, -\frac{1}{\sqrt{2}}, \frac{1}{\sqrt{2}}\right\}, \left\{0, \frac{1}{\sqrt{2}}, -\frac{1}{\sqrt{2}}\right\},\right.$$

$$\left\{0, \frac{1}{\sqrt{2}}, \frac{1}{\sqrt{2}}\right\}, \left\{-\frac{1}{\sqrt{2}}, 0, -\frac{1}{\sqrt{2}}\right\}, \left\{-\frac{1}{\sqrt{2}}, 0, \frac{1}{\sqrt{2}}\right\},$$

$$\left\{-\frac{1}{\sqrt{2}}, -\frac{1}{\sqrt{2}}, 0\right\}, \left\{-\frac{1}{\sqrt{2}}, \frac{1}{\sqrt{2}}, 0\right\}, \left\{\frac{1}{\sqrt{2}}, 0, -\frac{1}{\sqrt{2}}\right\},$$

$$\left.\left\{\frac{1}{\sqrt{2}}, 0, \frac{1}{\sqrt{2}}\right\}, \left\{\frac{1}{\sqrt{2}}, -\frac{1}{\sqrt{2}}, 0\right\}, \left\{\frac{1}{\sqrt{2}}, \frac{1}{\sqrt{2}}, 0\right\}\right\}, \{2\}\right];$$

```
Faces[Cuboctahedron] ^= {{2, 6, 7}, {2, 6, 4, 10},
   {2, 10, 11}, {2, 11, 1, 7}, {11, 10, 12, 9}, {6, 8, 5, 7},
   {4, 6, 8}, {3, 5, 8}, {1, 5, 3, 9}, {4, 8, 3, 12},
   {4, 10, 12}, {1, 11, 9}, {1, 5, 7}, {3, 9, 12}};

wireFrame[Cuboctahedron];
```

Q7. Why should this method of finding rotation matrices work?

Q8. Let T be a triangle in three-dimensional real space whose vertices are linearly independent. Prove that if R is a three-by-three matrix with the property that the triangle $R.T$ (i.e., Dot[R,T]) is congruent to T, then R is a rotation matrix.

Part II

Ring Labs

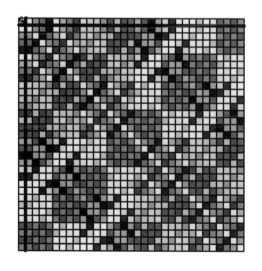 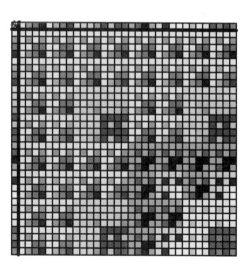

Introduction to Rings and Ringoids

1.1 Prerequisites

There are no prerequisites for this lab, although a brief introduction to the terminology related to rings might be beneficial.

1.2 Goals

This lab is intended to introduce both the mathematical concept of the structure of a ring and the corresponding *Mathematica* structure `Ringoid`. Various properties that a ring must and can have are introduced, and the *Mathematica* commands to explore these properties are illustrated.

1.3 Getting started? Begin here

In most of the labs in *Exploring Abstract Algebra with Mathematica*, we need first to read in extra *Mathematica* packages necessary to provide the functionality within the lab. This is done using a `Needs` statement. (Usually, just reading in the `AbstractAlgebra`Master`` package is done.) On opening this notebook you were probably asked, "Do you want to automatically evaluate all the initialization cells in the Notebook?" If you answered affirmatively, then you do not need to evaluate the following cell, but it doesn't hurt to do so. If you answered negatively, then you need to evaluate this cell.

```
Needs["AbstractAlgebra`Master`"];
SwitchStructureTo[Ring];
```

1.4 Ringoids and rings

Consider any set and two operations on that set. We will call this structure a `Ringoid`. An example of a `Ringoid` is the following.

```
R = FormRingoid[{0, 1, 2}, Mod[#1 + #2, 3] &, Mod[#1 #2, 3] &]
```

We always refer to the first operation of a `Ringoid` as *addition* and the second as *multiplication*. However, there is no restriction on either of the operations. (In other words, there is no implication that the addition could not *act like* multiplication and the multiplication *act like* some other type of operation.) One can access these operations as follows.

```
Addition[R]
Multiplication[R]
```

To compute the sum of 2 and 2, as well as the product of 2 and 2, one does as follows.

```
Addition[R][2, 2]
Multiplication[R][2, 2]
```

> **Q1.** Will it always be the case that the addition and multiplication return the same value? Why or why not?

The set of elements in a `Ringoid` is obtained using either of the following methods.

```
Elements[R]
Domain[R]
```

The operations on *R* are mod 3 addition and mod 3 multiplication, with which you may be well acquainted. The operation tables for any `Ringoid` may be displayed with the function `CayleyTables`.

```
CayleyTables[R, Mode → Visual]
```

The output of `CayleyTables`, using the `Visual` mode, appears in two forms. There are two graphical objects that resemble tables that appear in an abstract algebra text and a pair of *Mathematica* lists representing the tables. If you want to look at just the graphical tables, you need to end the input cell with a semicolon

(;) so that the list of lists is not displayed. We often do this from now on. On the other hand, if you are interested only in the output, you may want to drop the `Mode → Visual` argument.

We consider several types of `Ringoids`. Note that the `Ringoid` we have been calling R is simply the set of integers mod 3 with the appropriate operations. It can also be obtained by `Z[3]`.

> **Z[3]**

For every integer n, $n \geq 1$, $\mathbb{Z}_n$ is the ring with domain $\{0, 1, ..., n - 1\}$ and operations mod n addition and mod n multiplication. If you did some of the group labs, you may recall that we also use $\mathbb{Z}_n$ for the group consisting of the same set but with just addition mod n. By default, when we work with rings, we intend `Z[n]` to refer to the ring $\mathbb{Z}_n$. We can always obtain the group as follows.

> **G = Z[3, Structure → Group]**

We can change the default understanding of $\mathbb{Z}_n$ to be the group or ring by specifying the options for $\mathbb{Z}_n$. For example, when reading in the *Mathematica* code at startup, the equivalent of the following statement was evaluated.

> **SetOptions[Z, Structure → Ring]**

We could change `Ring` to `Group` and then $\mathbb{Z}_n$ would thereafter be understood to be the group instead of the ring. (Note: The `SwitchStructureTo` function may also be useful.) We use this family of rings extensively in this lab and subsequent labs. Other rings are introduced as well.

■ 1.5 Properties of rings

Let's consider some of the properties of a ring. Execute the following cell to be given a random $\mathbb{Z}_n$.

> **n = Random[Integer, {8, 19}]**
> **T = Z[n]**

■ 1.5.1 Additive properties

The first few properties pertain only to the addition. To test these properties, we form the additive `Groupoid` consisting of the elements and the addition of the `Ringoid`.

> **? AdditiveGroupoid**

```
AG = AdditiveGroupoid[T]
```

■ Addition must be closed

Recall that a set *H* is *closed* under the operation $*$ if $g*h$ is in *H* for all *g* and *h* in *H*. We can test for this property with the function `ClosedQ`. We can check either the additive `Groupoid` or the `Ringoid` with a restriction on the operation. (Almost all the ring properties are related to group properties of either the additive or multiplicative `Groupoid`. In each case, we can check the particular `Groupoid` or check the `Ringoid` by specifying a value for the `Operation` option.)

```
ClosedQ[AG]
ClosedQ[T, Operation → Addition]

?Operation
```

■ Addition must be associative

Recall that addition is *associative* if and only if $a + (b + c) = (a + b) + c$ for all values of *a*, *b*, and *c* in the domain. We can test for this property with the function `AssociativeQ`.

```
AssociativeQ[T, Operation → Addition]
```

For larger `Ringoids`, this test is time-consuming. Use of the function `Random-AssociativeQ` is advised for these `Ringoids`. This function tests random triples (*a*, *b*, *c*) for the associative property. If any triple violates the condition, the addition is known to be not associative. If no violation occurs, it is likely that the addition is associative.

■ Addition must be commutative

Addition is *commutative* if and only if $a + b = b + a$ for all values of *a* and *b* in the domain. You can test for this property with the function `AbelianQ` or `CommutativeQ`, but the former works only with `Groupoids`. (Thus we can use `AbelianQ` with the additive `Groupoid` or use `CommutativeQ` on the `Ringoid`.)

```
AbelianQ[AG]
CommutativeQ[T, Operation → Addition]
```

■ Existence of a "zero"

The term *zero* is used for the additive identity of a ring. That is, if *z* is the zero, then

$$z + r = r + z = r$$

for all r in the ring. To ask whether a zero exists, use the following.

> `HasIdentityQ[T, Operation → Addition]`

This test can be performed more directly as follows.

> `HasZeroQ[T]`

When a `Ringoid` has a zero, it can be called upon by name.

> `Zero[T]`

> **Q2.** What is the zero of `Ringoid[{0, 1, 2, 3}, Mod[#1 + #2, 4] & , Mod[#1*#2, 4] & ]`?

> **Q3.** What is the zero of `Ringoid[{0, 1, 2, 3}, Mod[#1*#2, 4] & , Mod[#1 + #2, 4] & ]`?

▪ Additive inverses

Assuming a `Ringoid` has a zero z, we say r is the additive inverse of s if $s + r = r + s = z$. All elements of a ring must have an additive inverse. The additive inverse of s is referred to as its *negation*. We can ask whether all elements have an additive inverse.

> `HasInversesQ[T, Operation → Addition]`

We can also ask for a list that displays how the elements of the `Ringoid` pair up as inverses. As long as the addition is associative, it can be proven that no element can have more than one negation. You will find that in some cases an element is its own negation.

> `Inverses[T, Operation → Addition]`

> **Q4.** How many elements in T are their own negation? (What is your ring T?)

The negation of an individual `Ringoid` element can be determined using `NegationOf`.

> `NegationOf[T, 1]`

> **Q5.** Below, form another random $\mathbb{Z}_n$, named T2. What ring did you get? For each element, determine its negation without *Mathematica*. Use either `Inverses` or `NegationOf` to check your work.

```
n = Random[Integer, {5, 12}]
T2 = Z[n]
```

■ 1.5.2 Multiplicative properties

We use MG for the multiplicative Groupoid of the ring T.

```
? MultiplicativeGroupoid

MG = MultiplicativeGroupoid[T]
```

■ Multiplication must be closed

Not only must the addition be closed, the multiplication must be closed as well. As noted, we can either check the multiplicative Groupoid or the Ringoid and restrict the operation.

```
ClosedQ[MG]
ClosedQ[T, Operation → Multiplication]
```

■ Multiplication must be associative

Besides being closed, the multiplication must also be associative.

```
AssociativeQ[T, Operation → Multiplication]
```

■ 1.5.3 Distributive property

There is one property that considers how the addition and multiplication interact: multiplication must be *distributive* over addition in order for a Ringoid to be a ring. That is, we must have

$$a(b+c) = ab + ac \qquad \text{and} \qquad (b+c)a = ba + ca$$

for all values of a, b, and c in the domain of the Ringoid. The function DistributiveQ tests for this property.

```
DistributiveQ[T]
```

A function called RandomDistributiveQ, which is similar to RandomAssociativeQ, is also available.

Ring Lab 2

Introduction to Rings, Part 2

▓ 2.1 Prerequisites

You should complete Ring Lab 1 before attempting this lab.

▓ 2.2 Goals

The goals of this lab are familiarization with several important types of rings and awareness of how to work with rings in the context of the `AbstractAlgebra` packages.

▓ 2.3 Units and zero divisors

■ 2.3.1 Units of a ring

First we read in the code that provides the functionality needed for this lab.

```
Needs["AbstractAlgebra`Master`"];
SwitchStructureTo[Ring];
```

In a ring with unity (i.e., multiplicative identity), an element that has a multiplicative inverse in the ring is called a *unit*. The percentage of elements that are units in a ring can vary quite a bit. The unity of a ring is always a unit, and the zero of a ring is never a unit. If all the nonzero elements of a commutative ring with unity are units, the ring is called a *field*. (We discuss fields at length in Ring Labs 10,

11, and 13.) It is also possible for a ring to have no units other than the unity. We now consider the set of units in a variety of rings.

▪ Units of $\mathbb{Z}_8$

Consider the ring $\mathbb{Z}_8$. What do the Cayley tables tell us?

```
CayleyTables[Z[8], Mode → Visual];
```

> **Q1**. What is the unity of the ring $\mathbb{Z}_8$? What are the units of $\mathbb{Z}_8$? Explain how you obtained your answers.

▪ Boolean rings

A *Boolean ring* consists of the set of all subsets of a given set, combined with the operations symmetric difference for the addition, and intersection for the multiplication. Let's consider an example.

```
R = BooleanRing[{Curly, Moe, Larry}]
```

This is the collection of all subsets of the set {Curly, Moe, Larry}. Note the cardinality of the set.

> **Q2**. Suppose you replaced {Curly, Moe, Larry} with the names of the students in your class. (This would increase the average intelligence of the set by a huge factor.) How many elements would the new Ringoid have? Justify your answer.

Note that this Ringoid has a different approach to addition and multiplication. The following confirms the description.

```
Addition[R]
Multiplication[R]
```

The multiplication is simply the intersection of two subsets. The addition, as indicated, is the symmetric difference of two subsets. In case you need a quick review of the symmetric difference, consider the following examples.

```
setA = {2, 3, 4}
setB = {1, 3, 4, 5}
setC = {3, 4}

Addition[R][setA, setB]
Addition[R][setA, setC]
```

You should note that the symmetric difference of two sets returns the union of the two sets with the elements in the intersection removed.

Consider the Cayley tables of this ring. After the tables are produced, enlarge them (until you can see them more clearly) by selecting the graphic and then stretching from one of the corners.

```
CayleyTables[BooleanRing[{Curly, Moe, Larry}], Mode → Visual];
```

Q3. What is the zero of this ring? Justify your answer.

Now we focus on the Boolean ring of subsets of {1, 2} and its multiplication table.

```
R1 = BooleanRing[2]
```

```
MultiplicationTable[R1, Mode → Visual];
```

Q4. What is the zero for this ring? Is there a unity for this ring? If so, what is it (and explain how you obtained it). If not, explain why not.

Q5. Determine all the units for this ring.

Q6. Generalize by determining all the units for the ring `BooleanRing[n]`.

▪ Gaussian integers mod *n*

The Gaussian integers, $\mathbb{Z}[i]$, are defined as the ring whose elements are $\{a + b\,i \mid a, b \in \mathbb{Z}\}$ and whose operations are ordinary (complex) addition and multiplication. Note that this is an infinite subring of the complex numbers.

Q7. Prove the last statement.

We can consider a finite variation of $\mathbb{Z}[i]$ by reducing the elements modulo some positive integer. Try, for example, 3.

```
R2 = Z[3, I, Structure → Ring]
```

What can we see in the tables? (Again, enlarge the graphic so that it is more visible.)

```
CayleyTables[R2, Mode → Visual];
```

Q8. What is the zero of this ring?

> **Q9**. Is there a unity for this ring? If so, what is it (and explain how you obtained it). If not, explain why not.

> **Q10**. Determine all the units for this ring.

■ Units of a ring

There are three basic functions related to units that can be used: `UnitQ`, `Units`, and `MultiplicativeInverse`.

```
? UnitQ
```

```
? Units
```

```
? MultiplicativeInverse
```

You can test your answer to the last question.

```
Units[R2]
```

■ Trivial ring

Consider the "trivial ring," for which all products are zero.

```
? TrivialZR
```

```
CayleyTables[TrivialZR[5], Mode → Visual];
```

```
UnitQ[TrivialZR[5], 3]
```

Since the "trivial ring" has no unity, appropriate messages are given. Observe that 3 does not have a multiplicative inverse.

■ 2.3.2 Zero divisors of a ring

In solving a quadratic equation such as $x^2 - 3x - 10 = 0$, we routinely factor the left-hand side to $(x - 5)(x + 2)$ and then solve $x - 5 = 0$ and $x + 2 = 0$. We are accustomed to assuming that for a product to be zero, one of the factors must be zero. This is an assumption that cannot be made in the general theory of rings. Consider the ring of integers modulo 15, $\mathbb{Z}_{15}$. As can be seen from the multiplication table, there are many instances where products of nonzero elements of $\mathbb{Z}_{15}$ are zero.

```
CayleyTables[Z[15],
     Mode → Visual, Operation → Multiplication];
```

A *zero divisor* is a nonzero ring element that can produce a product of zero on multiplication with another nonzero element. For example, in $\mathbb{Z}_{15}$, 10 is a zero divisor because its product with 3 is zero.

Q11. Determine all the zero divisors for $\mathbb{Z}_{15}$.

Q12. Determine all the zero divisors for $\mathbb{Z}_{11}$. (If you wish, change the 15 to 11 in the `CayleyTables` command.)

Recall the two rings with which we worked earlier.

```
R1 = BooleanRing[2]
R2 = Z[3, I, Structure → Ring]
```

Q13. Determine all the zero divisors of both *R1* and *R2*.

■ 2.3.3 Working with zero divisors

To work with zero divisors, we can use either of the following functions.

```
? ZeroDivisors
```

```
? ZeroDivisorQ
```

Here is how we might test our answers to questions 11 and 12.

```
ZeroDivisors[Z[15]]
```

```
ZeroDivisorQ[Z[15], 8]
ZeroDivisorQ[Z[15], 9]
```

The number of zero divisors can vary substantially from one `Ringoid` to another.

```
ZeroDivisors[Z[11]]
```

```
ZeroDivisors[TrivialZR[11]]
```

You may want to check your answer to question 13.

```
ZeroDivisors[R1]
ZeroDivisors[R2]
```

■ 2.4 Integral domains

An integral domain is a commutative ring with unity with no zero divisors. The ring of integers is a prototypical example.

> **Q14**. Recall the rings used in this lab thus far. Which are integral domains and which are not? Come up with at least one of each.

■ 2.5 Fields

A field is a commutative ring with unity with the property that every nonzero element is a unit.

> **Q15**. Recall the rings used in this lab thus far. Which are fields and which are not? Come up with at least one of each.

■ 2.6 Additional exercises

> **Q16**. What are the units of $Z[20, 5]$? What are the zero divisors?

```
CayleyTables[Z[20, 5], Mode → Visual];
```

> **Q17**. Using the functions provided thus far, find a `Ringoid` that has a unity and this element is the only nonzero element that is not a zero divisor.

> **Q18**. Using the functions provided thus far, find a `Ringoid` that has a unity and the only nonunit is the zero. Is this ring a field?

> **Q19**. True or false: In a commutative ring with unity, the intersection of units and zero divisors is always empty. Prove that your answer is correct.

To test if a `Ringoid` is an integral domain or field, appropriate functions can be used (besides testing the parts of the definition).

```
IntegralDomainQ[Z[7]]
IntegralDomainQ[BooleanRing[2]]
```

```
FieldQ[Z[7]]
FieldQ[Z[15]]
FieldQ[Z[20, 4]]

R = Z[30, 6]

FieldQ[R]
```

Ring Lab 3

An Ideal Part of Rings

3.1 Prerequisites

Before attempting this lab, you should have completed Ring Lab 1. You should also be familiar with cosets of normal subgroups.

3.2 Goals

The goal of this lab is to develop the concept of an *ideal* through examples, leading to discovery of some of the properties of ideals. Quotient rings are also introduced through the examples.

3.3 What is the ideal part of a ring?

We begin by reading in a package that provides the functionality needed for the lab.

```
Needs["AbstractAlgebra`Master`"];
SwitchStructureTo[Ring];
```

Let's consider a subset of the integers, $\mathbb{Z}$.

```
H = Table[i, {i, -40, 40, 5}]
```

Since we are working on a finite machine, we can only show finite subsets of $\mathbb{Z}$. Imagine, however, that H extends infinitely (in both directions) with the same pattern.

> **Q1**. Once H is extended as stated, is H a ring? Is it a subring of $\mathbb{Z}$? How do you verify this?

Now consider multiplying every element in the (extended) set H by an element in $\mathbb{Z}$, say 3.

```
M = 3 H
```

Of course, this only shows a finite part. Are these products in the (extended) set H?

```
M ∩ H

H
```

Certainly some of the elements are in the original subset H. Let's use *Mathematica* to enlarge H.

```
H = Table[i, {i, -150, 150, 5}]
SubsetQ[M, H]
```

By enlarging H (following the given pattern), we see that $M = 3H$ is a subset of H.

> **Q2**. M was determined by multiplying by 3 on the left of every element in H. We then found that M was a subset of (the enlarged) H. Is there anything special about 3? Could this work for other integers? Does the multiplication need to occur on the left? Justify your answers.

> **Q3**. Consider the M we found by multiplying H by 3. Actually, consider the extended M found by multiplying the extended H by 3. In other words, consider $3 \langle 5 \rangle$. Is this a ring? Is this a subring of $\mathbb{Z}$? Is this a subring of (the extended) H (i.e., $\langle 5 \rangle$)?

Let's try another subset of the integers.

```
P = Table[i, {i, -49, 50, 5}]
```

Q4. What happens if we multiply P on the left by 3? Call this result M (as before). What is the intersection of M and P? Adopt the notation that EX means the infinite, extended version of the finite X. (This notation is used later as well, so take note.) What can you say about the intersection of EM and EP? Try to prove your result.

Q5. What makes P different from H? How do they differ in their properties?

Consider another subset.

```
K = Table[i, {i, -48, 50, 3}]
```

Q6. Let M be the result of multiplying K by some integer q. Answer the kinds of questions that were asked regarding sets H and P in questions 3-5.

Hopefully, you made the observation that EP and EH are both cosets in the group $\mathbb{Z}/EH$, with $EP = 1 + \langle 5 \rangle$ and $EH = \langle 5 \rangle$. (For the moment, think of $\mathbb{Z}/EH$ as a group, not a ring.) Let's take a look at one more example.

```
R = Z[12]
els = Elements[R]
```

Now consider two subsets of the elements of R.

```
NR = {0, 3, 6, 9}
H = {0, 5, 7}
```

What happens if we multiply the set NR and the set H by an element r in R? Suppose we use $r = 5$.

```
5 NR
5 H
```

Whoops! We need to remember that the multiplication really has to be taking place mod 12.

```
Mod[5 NR, 12]
Mod[5 H, 12]
```

Now repeat for all elements r in R.

```
TableForm[Map[{#, Mod[# NR, 12]} &, els], TableDepth → 2,
  TableHeadings → {None, {"k", "k*{0,3,6,9}\n"}}]
(* working with NR *)

TableForm[Map[{#, Mod[# H, 12]} &, els], TableDepth → 2,
  TableHeadings → {None, {"k", "k*{0,3,6,9}\n"}}]
(* working with H *)
```

Q7. Consider the sets *NR* and *H* and let *r* be any element in *R*. What can you say about the sets *r NR* and *r H*? What difference between *NR* and *H* causes the results shown?

■ 3.3.1 Definition of an ideal

Given a ring *R* and a subring *J* of the ring *R*, we call *J* a (two-sided) *ideal* of *R* if for every element *r* in *R* and for every element *x* in *J* we have both *r x* and *x r* in *J*.

Q8. Describe the examples of ideals we have seen thus far.

■ 3.3.2 Another ring

Consider the ring of two-by-two matrices with entries from $\mathbb{Z}_2$.

```
M = Mat[Z[2], {2, 2}]
```

You can count the elements; there are 16.

```
Elements[M]
Length[Elements[M]]
```

As defined, although *M* is a ring, it is not a `Ringoid` in *Mathematica*. Let's make it one.

```
R = ToRingoid[Mat[Z[2], 2]]
```

Consider two subsets of *R*.

```
Map[MatrixForm, H = {{{1, 0}, {0, 1}},
   {{0, 0}, {0, 0}}, {{1, 0}, {0, 0}}, {{0, 0}, {0, 1}}}]

Map[MatrixForm,
  K = {{{1, 0}, {0, 1}}, {{0, 0}, {0, 0}}, {{1, 0}, {0, 0}},
     {{0, 0}, {0, 1}}, {{0, 0}, {1, 0}}, {{1, 0}, {1, 0}},
     {{0, 0}, {1, 1}}, {{1, 0}, {1, 1}}}]
```

You can verify by hand, or check the following, that these are indeed subrings of *R*.

```
SubringQ[H, R]

SubringQ[K, R]
```

Are H and K ideals? We need to pick an element in R and multiply it by an element in H (or K) and see if we land back in H (or K). For example, let M_1 be the following matrix in R.

```
Clear[M]
(M₁ = {{1, 1}, {0, 0}}) // MatrixForm
```

What happens if we try multiplying as we did earlier?

```
M₁ H
```

> **Q9**. This is a *Mathematica* question, not an algebraic one—answer it if you can. What went wrong?

Recall that the multiplication of R can be obtained as follows.

```
op = Multiplication[R]; (* no output is given *)
```

We can `Map` the operation `op`, with M_1 as one of the operands, across the whole set H.

```
Map[MatrixForm, Map[op[M₁, #1] &, H]]
```

Is it true that $M_1 * N_1$ is in H for every element N_1 in H? If not, what does this say about H being an ideal? If so, what does this say about H being an ideal?

> **Q10**. Is H an ideal? What about K?

Note that both H and K are normal subgroups of the additive group of R.

```
NormalQ[H, AdditiveGroupoid[R]]
NormalQ[K, AdditiveGroupoid[R]] (* may take some time *)
```

3.4 Ideals factor into other ring properties

Consider the following ring R and subset B.

```
R = Z[15]
B = {0, 3, 6, 9, 12}
```

Observe that B is an ideal of R.

```
IdealQ[B, R]
```

We know that B is a normal subgroup of R when we view R as an additive group. Therefore, the quotient group R/B makes sense and we can talk about the cosets as elements. Each coset has the form $x + B$, where x is an element of R. We already know how to add cosets: $(x + B) + (y + B) = x + y + B$. Can we form a product with these cosets? How about multiplying $x + B$ times $y + B$ to yield $x\,y + B$? We need to verify that this is a valid, well-defined operation, which we leave to your text and/or classroom. With this, we can form the *quotient ring* (also known as the *factor ring*) R/B.

```
QR = QuotientRing[R, B]
```

Here is an example of how the multiplication works.

```
Multiplication[QR][2 + NS, 2 + NS]
```

Consider the Cayley tables of this quotient ring.

```
CayleyTables[QR, Mode → Visual]
```

Q11. To what ring is this quotient ring isomorphic?

Ring Lab 4

What Does $\mathbb{Z}[i]/\langle a + b\,i\rangle$ Look Like?

4.1 Prerequisites

Prior to working on this lab, you should be familiar with the term *ideal* through discussions in class or from Ring Lab 3. You should also be familiar with an integral domain, a field, and the characteristic of a ring.

4.2 Goals

The goal of this lab is to explore the quotient structure of the Gaussian integers modulo an ideal generated by an arbitrary Gaussian integer.

4.3 Example 1

To work on this lab, we load the necessary packages.

```
Needs["AbstractAlgebra`Master`"];
SwitchStructureTo[Ring];
```

Consider the problem of investigating the ideal $J = \langle 2 + 2\,i\rangle$ in the Gaussian integers, $\mathbb{Z}[i]$. In particular, what does the quotient ring $\mathbb{Z}[i]/\langle 2 + 2\,i\rangle$ look like?

First recall that the set of Gaussian integers is the set $\{a + b\,i \mid a, b \in \mathbb{Z}\}$. This forms a lattice in the plane.

```
gr = IntegerLatticeGrid[
    {-6, 6}, {-6, 6}, AspectRatio → Automatic];
```

Now consider the ideal generated by $z = 2 + 2\,i$.

```
z = 2 + 2 I
```

What elements are in this ideal? We know, by definition, that all elements of the form $r\,z$ (where r is in $\mathbb{Z}[i]$) are in this ideal. Suppose we consider r to be an (ordinary) integer and multiply it by z. We might as well try a number of integers at a time.

```
someMultipliers = Table[k, {k, -3, 3}]
```

```
idealList = someMultipliers * z
```

Let's plot them on the lattice. To use `ListPlot`, we need to convert the complex numbers to ordered pairs.

```
plotOfIdeals = ListPlot[Map[ComplexToPoint, idealList],
    PlotStyle → {PointSize[0.030], RGBColor[1, 0, 0]},
    DisplayFunction → Identity];
Show[{gr, plotOfIdeals}];
```

What other multiples should we consider? We could multiply by multiples of i.

```
addToIdealList = Table[k I, {k, -3, 3}] z
```

```
idealList = Join[idealList, addToIdealList]
```

Now look at the set of multiples we have so far.

```
plotOfIdeals = ListPlot[Map[ComplexToPoint, idealList],
    PlotStyle → {PointSize[0.030], RGBColor[1, 0, 0]},
    AspectRatio → Automatic, DisplayFunction → Identity];
Show[{gr, plotOfIdeals}];
```

To make sense of this picture, recall that multiplying a complex number by i has the effect of increasing the argument of the complex number (the angle between it and the positive x-axis) by $90°$. This explains why the original diagonal is rotated to give us the preceding pattern.

There are still quite a few elements in the ideal that reside in this rectangle that are not being shown. Perhaps we should multiply all the complex numbers in this grid by z and then plot all the products that are still within the grid.

```
g = 6;
pts = Flatten[Table[{i, j}, {i, -g, g}, {j, -g, g}], 1];
cpts = Map[Apply[Complex, #] &, pts];
idealList = Intersection[cpts, z * cpts]
```

Here is the new plot.

```
plotOfIdeals = ListPlot[Map[ComplexToPoint, idealList],
    PlotStyle → {PointSize[0.030], RGBColor[1, 0, 0]},
    AspectRatio → Automatic, DisplayFunction → Identity];
Show[{gr, plotOfIdeals}];
```

Q1. We now see all the elements in $J = \langle 2 + 2i \rangle$ that are within the window shown. How many elements are there in $\mathbb{Z}[i]/J$? In other words, how many other cosets do you think there are (and then add 1 to account for J)? Note that in the code, the variable `pts` records the lattice points and the variable `ideal-List` records the members of J, so perhaps they may be useful in helping to answer this question. What do you think is the size of $\mathbb{Z}[i]/J$? Why?

Let's refer to the list of points in `idealList` by a different variable so that we can more efficiently do the bookkeeping as we try finding the other cosets. Suppose we use `J[0]`.

```
Clear[J]
J[0] = idealList
```

Now we need to find the other cosets. We will be *adding* elements to the ideal, as opposed to multiplying, to generate the other cosets. Since we have defined J to be `J[0]`, and we know all cosets are of the form $r + J$, finding another coset is not too difficult. We just need to pick another ring element and add it to the elements in `J[0]`. Which ring element should we add? Since we clearly want to pick one that is not already in J, we may need to look at our picture again.

```
Show[{gr, plotOfIdeals}];
```

How about using 1?

```
J[1] = 1 + J[0]
(* This conveniently adds 1 to every
   element in J[0]. This works well since we are
   dealing with complex numbers. Equivalently,
  one could use J[1] = Map[(# + 1)&, J[0]]. *)
```

This returns all the elements of `J[0]` with 1 added to them. In other words, we are building (part of) another coset. Here is a graph showing what we have so far.

```
cosetPlot[2] = ListPlot[ComplexToPoint /@ J[1],
    PlotStyle → {PointSize[0.030], Hue[ 1/9 ]},
    AspectRatio → Automatic, DisplayFunction → Identity];
Show[{gr, plotOfIdeals, cosetPlot[2]}];
```

Q2. How does this graph reflect that `J[1]` = `1` + `J[0]`?

Q3. Do you want to modify your guess regarding how many cosets there will be? What is your current response?

Let's pick another element to help find another coset. Additionally, let's begin to automate our process by using the variable `index` to keep track of the count of the cosets. Suppose we choose 2 for our new element.

```
newElement = 2
index = 2
J[index] = newElement + J[0]
(* now calculate the new J[index] by adding newElement *)
```

We have now calculated our third coset, `J[2]`. (Recall that the first was `J[0]` and the second was `J[1]`.) Now let's show a graph of the current situation.

```
cosetPlot[index + 1] = ListPlot[ComplexToPoint /@ J[index],
    PlotStyle → {PointSize[0.030], Hue[ index/9 ]},
    AspectRatio → Automatic, DisplayFunction → Identity];
Show[Join[{gr, plotOfIdeals},
    Table[cosetPlot[k + 1], {k, 1, index}]]];
```

Q4. Do you want to modify your guess regarding how many cosets there will be? What is your current response?

Q5. By choosing a value for `newElement`, step through the following code until you have found all the cosets. A suggestion for choosing `newElement`: Pick an uncolored point that is "closest" to the *x*-axis (where you measure the distance by the absolute value of the angle from the *x*-axis to the point) but also within a circle of radius $|2 + 2\,i| = 2\sqrt{2}$. Keep track of your elements.

```
newElement =

index += 1
J[index] = newElement + J[0]
cosetPlot[index + 1] = ListPlot[ComplexToPoint /@ J[index],

    PlotStyle → {PointSize[0.030], Hue[ index/9 ]},

    AspectRatio → Automatic, DisplayFunction → Identity];
Show[Join[{gr, plotOfIdeals},
    Table[cosetPlot[k + 1], {k, 1, index}]]];
```

Q6. How many cosets are there? List them.

■ 4.3.1 What ring is it?

Now that we have found the cosets, let's explore what ring this quotient $\mathbb{Z}[i]/\langle 2+2i\rangle$ might be. Since each coset can have one of many different names (or representatives, in the sense that $2+\langle 5\rangle$ and $7+\langle 5\rangle$ are the same cosets in $\mathbb{Z}/\langle 5\rangle$ but use different representatives), we should first agree on what names to use so that we can use the same language. (Technically, this is not necessary, but it is convenient.) Therefore, here is one choice.

```
cosetReps = {0, 1, 2, 2 - I, 2 + I, 1 - I, 1 + I, 1 - 2 I}

cosetRepsPlot = ListPlot[ComplexToPoint /@ cosetReps,
    PlotStyle → {PointSize[0.030], Hue[0]},
    AspectRatio → Automatic, DisplayFunction → Identity];
Show[{gr, cosetRepsPlot,
    Graphics[Circle[{0, 0}, Abs[2 + 2 I]]]}];
```

Here is another choice, with a different visualization. We use this choice in further analysis of the quotient ring.

```
QuotientRing[Z[I], 2 + 2 I,
    Mode → Visual, Form → Representatives]
```

The representatives are the elements listed in this `Ringoid`. Additionally, these elements can be found within and on the bold square. In general, the corners of the square are different representatives of the same coset. Also, any element that appears on a side of a square has a second representative of the same coset on the opposite side of the square.

Q7. There are more Gaussian integers within or on this bold square than there are coset representations in this quotient ring. Which ones are not included in the quotient ring? How are these "extras" related to those in the quotient ring?

Q8. Give an explanation for the role the collection of squares is playing. In particular, address how the other squares relate to the bold one.

Since we have found all the cosets, we should consider how to add and multiply them. We know that if we add the cosets $4 + 5i + J$ and $8 + 9i + J$, we get $12 + 14i + J$. Similarly, the product is $-13 + 76i + J$. (Double-check each of these calculations.) We would, however, like to have them represented by elements of the form $z + J$, where z is a representative in our quotient ring. First, let's look at another version of the quotient ring, where the elements are given in the form $z + J$.

```
QR = QuotientRing[Z[I], 2 + 2 I]
```

Now we can look at the sum and product.

```
Addition[QR][4 + 5 I + J, 8 + 9 I + J]
Multiplication[QR][4 + 5 I + J, 8 + 9 I + J]
```

Q9. From these computations, we see that $(4 + 5i + J) + (8 + 9i + J) = 2 + J$ and $(4 + 5i + J) * (8 + 9i + J) = 3 + J$. Why is this the case? Look again at the plot given in the visualization of the quotient ring.

    a. First, find the representatives in the bold square for both $4 + 5i + J$ and $8 + 9i + J$. (The second one may take a little ingenuity with this graph, but you can do it.)

    b. Now add and multiply (by hand) the two representatives. What are these results (before reducing mod $\langle 2 + 2i \rangle$)?

    c. Finally, reduce the preceding results mod $\langle 2 + 2i \rangle$ so that you can find the corresponding representative in the bold square. Are you convinced of these results?

By looking at the Cayley table, we should be able to answer some questions regarding this ring. (You may wish to enlarge the graphics so you can view them better.)

```
CayleyTables[QR, Mode → Visual];
```

Q10. Is QR a field? Why or why not? Is $J = \langle 2 + 2\,i \rangle$ a maximal ideal? Why or why not?

Q11. Is QR an integral domain? Why or why not? Is $J = \langle 2 + 2\,i \rangle$ a prime ideal? Why or why not?

Q12. What is the characteristic of QR?

Q13. Since the additive group of this ring has order eight, what are the candidates for a group that is isomorphic to this? In other words, what are the groups of order eight?

Q14. Which one is this? What group is isomorphic to the additive group of this ring?

4.4 Example 2

Suppose we now let $J = \langle 3 + i \rangle$ and consider the quotient ring $\mathbb{Z}[i]/J$. As before, let's view this visually.

```
QuotientRing[Z[I], 3 + I, Mode → Visual]
```

Try the following to see how this graphic image is built up in stages. (Note: The following cell uses a lot of memory; you may wish to delete previous graphics cells first.)

```
QuotientRing[Z[I], 3 + I, Mode → Visual, Staged → All];
```

You can now double-click on any one of the cells and animate the graphics. The speed can be controlled by typing a number from 1 to 9 (1 slow, 9 fast). If you want to see just one stage at a time, the following is an alternative. (If you have limited memory, you may wish to first clear the graphics just generated.)

```
QuotientRing[Z[I], 3 + I, Mode → Visual, Staged → True];
```

Now use `NextStage` to advance (or `PreviousStage`, if you wish to reverse).

```
NextStage[QuotientRing]
```

Q15. How many cosets are there?

Q16. What can you say about this quotient ring? What is the size? Is it a field? Is it an integral domain?

The following are some Gaussian integers. You may recall that the absolute value of a complex number $a + bi$ is the length of the vector starting at the origin and terminating at the point (a, b), having the value $\sqrt{a^2 + b^2}$.

```
someTestCases = {2 + I, 2, 3 + I, 3 + 2 I, 2 I, 4 - 3 I, 7 - I}
```

```
TableForm[({#1, Abs[#1], Size[QuotientRing[Z[I], #1,
        Form → Representatives]]} &) /@ someTestCases,
  TableHeadings → {None, {"z", "|z|", "|Z[i]/<z>|\n"}}]
```

The output consists of a complex number z, the absolute value of z, followed by the size of the quotient ring formed by $\mathbb{Z}[i]/\langle z \rangle$ (i.e., the number of cosets).

Q17. Can you make a conjecture regarding the sizes measured?

Q18. Try to prove your conjecture.

Q19. Which of the quotient rings above do you think are fields? Integral domains?

You may be interested in trying the following.

```
Map[{#, PrimeQ[#, GaussianIntegers → True]} &,
  someTestCases]
```

```
Map[QuotientRing[Z[I], #1,
      Form → Representatives, Mode → Visual] &, someTestCases];
(* This will likely take some time — good
    opportunity for a break *)
```

Ring Lab 5

Ring Homomorphisms

▦ 5.1 Prerequisites

Before you start this lab, you should be familiar with Ringoids and the ideas found in Ring Labs 1 and 2, as well as normal subgroups and ideals.

▦ 5.2 Goals

This lab explores the notion of a ring homomorphism. First we define one, and then we see how one can be constructed.

This lab is designed to be independent of the group labs on isomorphisms and homomorphisms. If you have done them, you can skip the first section of this lab, except to evaluate the inputs that define Morphoids f, g, and w.

▦ 5.3 Morphoids on rings

A Morphoid is a *Mathematica* object that consists of either a function or a set of rules followed by two structured sets (Ringoids or Groupoids). The function or rules serve to map each element in the first structured set into the second one. The only principle governing construction of a Morphoid is that an actual (mathematical) function must be defined by the rules or function. To continue, let's read in the *Mathematica* functions we will need for this lab.

```
Needs["AbstractAlgebra`Master`"];
SwitchStructureTo[Ring];
```

The way to create a `Morphoid` is with the function `FormMorphoid`. For example, to define the `Morphoid` from the ring $\mathbb{Z}_{15}$ into the ring $\mathbb{Z}_{30}$, where the image of each element x in $\mathbb{Z}_{15}$ is $2\,x$, we evaluate the following.

```
f = FormMorphoid[2 # &, Z[15], Z[30]]
```

If you try to create a `Morphoid` with an invalid function, it will fail to be created.

```
FormMorphoid[2 # &, Z[15], Z[20]]
```

One family of built-in `Morphoids` is `ZMap[m, n]`, where m and n are positive integers.

```
? ZMap
```

```
g = ZMap[15, 5, Structure → Ring]
```

You can get a graphical representation of a `Morphoid` with the function `Visual izeMorphoid` (or by using the `Visual` mode of `FormMorphoid`).

```
? VisualizeMorphoid
```

```
VisualizeMorphoid[g, ColorCodomain → Automatic];
```

Any pair of `Ringoids` can potentially have a `Morphoid` connecting them. Here is an example that is defined with a list of rules instead of a function.

```
w = FormMorphoid[
     {{} → 0, {1} → 0, {2} → 0, {1, 2} → 1}, BooleanRing[2], Z[2]]
```

You may have noticed that depending on how you create a `Morphoid`, the first argument that appears in the output can vary. Ideally, what appears is the simplest description of the function that defines the `Morphoid`. As you will see later, the single rule $1 \rightarrow 1$ defines the `Morphoid` g, so the rest of the rules that define g do not need to be displayed.

▣ 5.4 Ring homomorphisms

A *homomorphism* between rings R and T is a function that preserves the operations between the two rings. That is, if f is a homomorphism, then for all values of x and y selected from R,

$$f(x + y) = f(x) + f(y) \quad \text{and} \quad f(x * y) = f(x) * f(y).$$

(Note that the operations + and ∗ on the left-hand side of each equation occur within *R*, while those on the right-hand side occur within *T*.) If *f* is also one-to-one and onto, then it is an *isomorphism*. The existence of an isomorphism between two rings establishes them as being "equal" in an algebraic sense.

To review the notion of operation preserving, we start by looking at the first two `Morphoids` (*f* and *g*), since their domains are the same. For both of them we use the pair of elements {11, 7} and use the function `PreservesQ` to see if the conditions above are satisfied when *x* = 11 and *y* = 17.

> `PreservesQ[f, {11, 7}]`

> `PreservesQ[g, {11, 7}]`

To see why *f* does not preserve operations, we can use the `Visual` mode.

> `PreservesQ[f, {11, 7}, Mode → Visual];`

From this graphical representation of the tests, we see that the addition condition is satisfied but the multiplication condition fails: $f(11 * 7) = 4$ but $f(11) * f(7) = 8$. Therefore, *f* is not a ring homomorphism. Although *g* did preserve the operations for the pair {11, 7}, we cannot conclude that *g* is a homomorphism, since the conditions were checked for only one of the $15^2 = 225$ pairs of values that need to be checked. `MorphismQ` will check them all at once.

> `MorphismQ[g, Mode → Visual]`

Q1. The third `Morphoid` that was defined in the first section of this lab is also not a homomorphism. Explain why this is the case with a specific example that shows it is not.

5.5 The kernel and image

5.5.1 The kernel

Since every ring "contains" the additive group consisting of the domain and the addition operation, a ring homomorphism can also be viewed as a group homomorphism. This observation extends to the more general structures of a `Morphoid` on a `Ringoid` and a `Morphoid` on a `Groupoid`. If you have already studied group homomorphisms, you are aware of the concept of the kernel. In the group setting, the kernel of a homomorphism is the set of elements in the domain that map onto the identity of the codomain. With rings, the additive identity is the zero of the ring. Therefore it is natural to make the following definition: The *kernel* of a ring homomorphism is the set of elements in the domain that map onto the zero of the codomain.

Our first true ring homomorphism was called *g*. What is its kernel? The `Visual` mode of `Kernel` should clearly illustrate the elements in the kernel.

```
Kernel[g, Mode → Visual]
```

> **Q2.** First review the definition of an ideal. Without making any assumptions about the kernel (except its definition), answer the following questions.
>
> a. If *h* is a ring homomorphism and *r* and *s* are in $K = $ `Kernel`[*h*], explain why $r - s$ must also be in *K*.
>
> b. If *h* is a ring homomorphism and *r* is in $K = $ `Kernel`[*h*] and *s* is in the domain of *h*, explain why *r s* and *s r* must also be in *K*.
>
> c. Properties from (a) and (b) imply what about the kernel?

`Kernel` returns a result that is a `Ringoid`. It is a well-defined *Mathematica* function, provided the codomain has a zero.

```
f
Kernel[f]
```

```
w
Kernel[w]
```

> **Q3.** What conclusions, if any, can be drawn from the observations made in question 2 and the values of `Kernel[f]` and `Kernel[w]`? Assume that you do not know whether *f* or *w* is a homomorphism.

■ 5.5.2 The image

The image of a ring homomorphism is the range of the function, but it inherits the operation of the codomain and is also a `Ringoid`. For our first homomorphism, *g*, the image is the whole codomain, which is not uncommon.

```
g
Image[g]
```

Although a `Morphoid` that is not a homomorphism can have an image, and that image may also be a subring (as we see with *f* that we defined above), there is no guarantee that it will be a subring.

```
Image[f]
ClosedQ[%]
```

> **Q4**. Find an example of a `Morphoid` of rings for which the image is not a subring of the codomain.

📓 5.6 The kernel is an ideal

It was observed at the conclusion of question 2 that the kernel of a ring homomorphism is an ideal of the domain. This closely parallels the situation with groups. The kernel of a group homomorphism is a normal subgroup. Ideals and normal subgroups are precisely the subsets of their respective systems that allow quotient structures of cosets. We illustrate this situation in the ring case here.

■ 5.6.1 Example 1

The first true ring homomorphism was called g.

```
g
Kernel[g]
```

We can create the quotient ring with `QuotientRing`.

```
QR = QuotientRing[Z[15], Kernel[g]]
```

The theory that has led up to this point assures us that QR is a ring.

> **Q5**. To what common ring is QR isomorphic?

Your response to this question should be implied by the following theorem. (The proof appears in nearly every introductory abstract algebra text.)

■ 5.6.2 First Isomorphism Theorem (for rings)

Theorem Let $f : R \to S$ be a surjective (onto) homomorphism of rings with kernel K. Then the quotient ring R/K is isomorphic to the image of f, with the isomorphism $g : R/K \to f(R)$ defined by $g(a + K) = f(a)$.

■ 5.6.3 Example 2

In the second example, we define and work with a `Morphoid` based on a direct product.

```
R = DirectProduct[Z[3], Z[5]]
```

We define a `Morphoid` from $\mathbb{Z}_{30}$ into R.

```
β = FormMorphoid[{Mod[#1, 3], Mod[#1, 5]} &, Z[30], R]
```

```
MorphismQ[β]
```

```
K = Kernel[β]
```

> **Q6**. Based on the size of the kernel, how many elements should we expect to have in the quotient ring $\mathbb{Z}_{30}/K$?

We can generate the quotient ring to verify a conjecture from the last question.

```
Q2 = QuotientRing[Z[30], K]
```

> **Q7**. Based on the First Isomorphism Theorem, to what ring is the quotient ring isomorphic?

5.7 One-rule `Morphoids`

It was noted earlier that `ZMap[15, 5]` and similar expressions form `Morphoids` with a single rule displayed. Internally, the `Morphoid` is actually defined with a function, but the rule may be somewhat shorter, so for simplicity we display the rule.

```
h = ZMap[4, 2]
```

> **Q8**. a. Assuming that h is a morphism, which is true, explain how and why the rule $1 \rightarrow 1$ determines rules for the other elements in $\mathbb{Z}_4$. (What are the other rules?)
>
> b. Explain what goes wrong with `ZMap[3, 2]` and why $1 \rightarrow 1$ cannot define a morphism.

5.8 The Chinese Remainder Theorem

The image of the homomorphism β, considered in section 5.6, was the direct product of $\mathbb{Z}_3$ with $\mathbb{Z}_5$. This direct product happens to be isomorphic to $\mathbb{Z}_{15}$. This follows from the Chinese Remainder Theorem.

Chinese Remainder Theorem. If $m_1, m_2, \ldots, m_r$ are positive integers such that no two have a common divisor greater than one, then the ring $\mathbb{Z}[m_1 m_2 \ldots m_r]$ is isomorphic to the direct product $\mathbb{Z}[m_1] \times \mathbb{Z}[m_2] \times \cdots \times \mathbb{Z}[m_r]$.

We do not provide a proof of this theorem since it appears in many standard texts. We do, however, illustrate one approach from a computational point of view.

An isomorphism from $\mathbb{Z}_{15}$ into $\mathbb{Z}_3 \times \mathbb{Z}_5$ is easy to construct. We map each element of $\mathbb{Z}_{15}$ into the pair of remainders on division by 3 and 5 respectively.

```
θ = FormMorphoid[{Mod[#1, 3], Mod[#1, 5]} &, Z[15], R]
```

```
IsomorphismQ[θ]
```

For this case, the matter is settled. In the general situation, if we were to define a similar function from $\mathbb{Z}[m_1 m_2 \ldots m_r]$ into the direct product, it is not totally clear that the function would be one-to-one and onto. This can be proven, however, and there is an algorithm that determines the inverse of the function. The inverse, which is not as simple to compute, is available as a function called ChineseRemainderTheorem in the NumberTheory`NumberTheoryFunctions package. Let's load this package and learn about this function.

```
Needs["NumberTheory`NumberTheoryFunctions`"];
```

```
? ChineseRemainderTheorem
```

In the example, we might want to know what the inverse image of $\{2, 3\}$ is, that is, for what integer n is $\theta(n)$ equal to $\{2, 3\}$. We use $\{2, 3\}$ for the first argument and $\{3, 5\}$ for the second, since 3 and 5 are the two moduli used to form the codomain of θ.

```
ChineseRemainderTheorem[{2, 3}, {3, 5}]
```

It turns out that 8 is the smallest integer congruent to 2 mod 3 and also congruent to 3 mod 5. We can easily verify this result.

```
θ[8]
```

The inverse function can be used to create an inverse isomorphism

```
γ =
 FormMorphoid[ChineseRemainderTheorem[#1, {3, 5}] &, R, Z[15]]
```

```
IsomorphismQ[γ]
```

Of course, we should also verify that γ is really the inverse of θ.

```
MorphoidComposition[θ, γ]
```

`MorphoidComposition[γ, θ]`

> **Q9**. Explain why the last two results verify that these two functions are inverses.

The power of the `ChineseRemainderTheorem` function may not be clear, since our original function was based on a relatively small set. Here is a question for which the Chinese Remainder Theorem can be employed.

> **Q10**. Suppose that Franklin Street Clothing is having a sale. You give x dollars to person A, who then buys as many $19 shirts as possible with the x dollars and then has $3 left over. You also give x dollars to person B, who then buys as many $29 sweatshirts as possible with the x dollars and then has $26 left over. What is the least positive number of dollars that x could be? Explain how the answer to this problem can be obtained using isomorphisms.

We finish up the lab with a couple of follow-up questions.

> **Q11**. Does the Chinese Remainder Theorem tell us that $\mathbb{Z}_{12}$ is isomorphic to the direct product of $\mathbb{Z}_2$ and $\mathbb{Z}_6$? Explain your answer.

> **Q12**. Explain, from what we have done in this lab, why the rings $\mathbb{Z}_{30}/\{0, 15\}$ and $\mathbb{Z}_{15}$ are isomorphic.

Ring Lab 6

Polynomial Rings

6.1 Prerequisites

To work on this lab, you need only a cursory familiarity with `Ringoids`, mostly $\mathbb{Z}_n$.

6.2 Goals

The goal of this lab is discovering some of the basic properties of polynomial algebra over a ring, through division and the `GCD` function. Factorization is discussed in detail in Ring Lab 7.

6.3 Introduction to polynomials

To get started, let's read in the package that provides the functionality needed for this lab.

```
Needs["AbstractAlgebra`Master`"];
SwitchStructureTo[Ring];
```

Consider the polynomial $p = x^2 + 3x - 4$. This type of expression has been familiar to you for many years. You should know how to factor it and how to graph $y = p$.

> **Q1**. How do you factor this polynomial? What does the graph of y look like? Try also to get *Mathematica* to do both of these tasks.

We implicitly consider viewing the polynomial p as having coefficients coming from the integers (or possibly real numbers). Do we need to confine our coefficients to the integers, rationals, reals or complex numbers? No.

Let R be any commutative ring. We define the *ring of polynomials over R in the indeterminate x* as the set of all (formal) symbols of the form

$$a_n x^n + a_{n-1} x^{n-1} + \cdots + a_2 x^2 + a_1 x + a_0,$$

where the coefficients a_i are from the ring R and n is a nonnegative integer. We denote this ring of polynomials by $R[x]$. In *Mathematica*, we can create a ring of polynomials as follows.

```
P = PolynomialsOver[Z[7]]
```

The ring P is the set of all polynomials whose coefficients come from $\mathbb{Z}_7$.

> **Q2**. Why don't we list all the elements? How many are there?

The first thing we need to think about when working with polynomials is to be able to identify two elements as different or the same. Suppose we have a polynomial $f(x)$ of the form

$$a_n x^n + a_{n-1} x^{n-1} + \cdots + a_2 x^2 + a_1 x + a_0.$$

In this case, if a_n is not zero, we say $f(x)$ has *degree n* and we call the coefficient a_n the *leading coefficient*. We say that the two polynomials $f(x)$ and

$$g(x) = b_m x^m + b_{m-1} x^{m-1} + \cdots + b_2 x^2 + b_1 x + b_0$$

are equal if $n = m$ (i.e., they have the same degree) and if $a_i = b_i$ for all $i \leq n$. Note that what is important in a polynomial is only the list of coefficients used; the variable (and its powers) act merely as place holders, indicating the position of the coefficients. Thus, in *Mathematica*, one way we create a polynomial is to simply give the function `Poly` a sequence of coefficients, prefixed by the ring from which the coefficients come. For example,

```
Clear[x]
p = Poly[Z[7], 6, 1, 0, 2]
```

yields the polynomial $2x^3 + x + 6$. In this case, the indeterminate is denoted by x. You can specify some other indeterminate.

```
Clear[y]
p2 = Poly[Z[7], 6, 1, 0, 2, Indeterminate → y]
```

You can also enter a polynomial directly:

```
p3 = Poly[Z[7], 2 x³ + x + 6]
```

It is important to remember to specify the ring from which the coefficients come. If this is forgotten, the expression will be returned as entered (which is a standard *Mathematica* means for communicating that the input is unsuitable).

```
p4 = Poly[2 x³ + x + 6]
```

As usual, asking for information about a function is often useful.

```
? Poly
```

In particular, note that if you are accustomed to entering polynomials (as well as viewing them) the way *Mathematica* returns them, you will feel at home with the default setting of `PowersIncrease`. On the other hand, if you prefer to have the powers of the polynomial increase from right to left for both input and output, you may wish to use `SetOptions` and change the default on `PowersIncrease`. Note how this option works.

```
Poly[Z[5], 1, 2, 3]
Poly[Z[5], 1, 2, 3, PowersIncrease → RightToLeft]
```

Note that both the input method (when just giving a sequence of coefficients) and the output display is governed by this option. Based on the method you prefer, evaluate one of the following two cells.

```
SetOptions[Poly, PowersIncrease → LeftToRight]
(* output will be similar to 1 + 2 x + 3 x² *)

SetOptions[Poly, PowersIncrease → RightToLeft]
(* output will be similar to 3 x² + 2 x + 1 *)
```

Consider the following polynomial *r*.

```
r = Poly[Z[7], 7 x⁴ - 4 x + 4 x² + 9]
```

When the base ring is $\mathbb{Z}_n$, polynomials can be entered with coefficients from the integers (positive or negative), and they will be reduced mod *n*. (This property can be turned off by adding the option `FlexibleEntering → False`.)

Finally, recall the Boolean ring over $\{a, b\}$ and consider the following polynomials.

```
Clear[a, b]
R3 = BooleanRing[{a, b}]

s1 = Poly[R3, {b}, {a, b}, {}, {b}]
```

```
s2 = Poly[R3, {c}, {a, b, c}, {y}, {b}]
```

When a polynomial is constructed, the coefficients are checked for membership against the base ring; an error message is given if the polynomial is ill-formed.

Q3. Construct a well-formed and an ill-formed third-degree polynomial over $\mathbb{Z}_{10}$.

Let's pick a couple of random polynomials of degree 3. (Recall that P is our extension ring of polynomials over $\mathbb{Z}_7$.)

```
a = RandomElement[P, 3]
b = RandomElement[P, 3]
```

Q4. List your polynomials. You should have a reasonable guess or idea how to add these two polynomials. Do so (by hand). Also, determine their product.

In the ring of polynomials, there is a built-in addition and multiplication. Let's define some *Mathematica* aliases for these functions in our ring of polynomials over $\mathbb{Z}_7$.

```
add = Addition[P];
mult = Multiplication[P];
(* no output is given *)
```

Now let's add and multiply a and b (and note that this should check your answer to question 4).

```
add[a, b]
mult[a, b]
```

Conventional notation works as well.

```
a + b
a b
```

Q5. What is the degree of the product of polynomials a and b?

The `Degree` function determines the degree of a polynomial.

```
Degree[Poly[Z[8], 4 x^2 + 7 x - 2]]
```

```
Degree[a b]
```

Q6. What are the constant polynomials in P? (These are the polynomials without the indeterminate x.)

Q7. Does this ring of polynomials over $\mathbb{Z}_7$ have a zero? What about a unity? If you answered yes to either, what is it?

Q8. Can we add two polynomials of degree k and obtain a sum of a lower degree? Try the following code once or twice (or more times, if you feel it is necessary) and consider the results. If obtaining a sum with lower degree cannot happen, explain why not; if it can, cite circumstances under which it can.

```
Print["Let d(p) be the degree of the polynomial p:"];
(TableForm[#1, TableHeadings → {None, {"polynomial a",
        "polynomial b", "{d(a), d(b)}", "d(a + b)\n"}},
    TableSpacing → {0, 3}, TableDepth → 2] &)[
 Table[a = RandomElement[P, 2]; b = RandomElement[P, 2];
   {a, b, {Degree[a], Degree[b]}, Degree[a + b]}, {25}]]
```

Q9. What happens when we multiply a polynomial of degree 3 with one of degree 2? What is the degree of the product? Try the code below once or twice (or more times, if you feel it necessary) and consider the results. Explain your conclusion.

```
(TableForm[#1, TableHeadings →
        {None, {"polynomial a", "polynomial b", "d(a * b)\n"}},
     TableSpacing → {0, 3}] &)[
 Table[a = RandomElement[P, 3]; b = RandomElement[P, 2];
   {a, b, Degree[P, Multiplication[P][a, b]]}, {20}]]
```

Since P is the ring of all polynomials over $\mathbb{Z}_7$, we can ask about the `Zero` and `Unity` of this ring.

```
{z, u} = {Zero[P], Unity[P]}
```

Of course, we need to be careful about how we interpret what we see here. If we ask if z is equal to the number 0 and if u is equal to the number 1, we get the following.

```
z === 0
u === 1
```

Q10. Explain these results.

Let's shift gears for a moment and consider polynomials over $\mathbb{Z}_6$ and look at two polynomials and their product in this context.

```
Clear[P]
P₂ = PolynomialsOver[Z[6]]
a = Poly[Z[6], 3 x³ + x² + x + 1]
b = Poly[Z[6], 2 x² + 5]
a b
```

> **Q11**. What can you surmise from this example? Look carefully at the output. Give another example of a similar occurrence in the ring of polynomials over $\mathbb{Z}_6$.

6.4 Divide and conquer

The main reason arithmetic with polynomials is interesting (and so similar to integer arithmetic) is because of the division property. The division property is often referred to as the Division Algorithm in algebra texts.

Let's do some quick reviewing of familiar territory before looking at general polynomial rings. When we divide 159 by 13, we seek to find the number of times we can multiply 13 and still remain less than 159. Or in the vernacular, "How many times does 13 go into 159?" In this case, 12 times 13 equals 156, leaving a remainder of 3.

```
159 == 13 12 + 3
```

Note that we have written 159 as the product of an integer times 13 plus a remainder whose value is less than 13.

Now consider the following polynomials (over the integers).

```
a = x³ + 5 x² - 3 x + 8
b = x - 7
```

What do we get when we divide a by b? In other words, can we write $a = b\,q + r$ for some polynomials q and r?

> **Q12**. You should know how to do this by hand. Do so and report your results.

Mathematica has some built-in functions to find the quotient and remainder when considering polynomials over the integers (as well as some other standard rings).

```
q = PolynomialQuotient[a, b, x]
r = PolynomialRemainder[a, b, x]
```

Do these work as we suppose?

```
a == b q + r
```

We need to do some coaxing.

```
a == Expand[b q + r]
```

To consider equivalent functionality over arbitrary rings, we need to use the functions built into the packages that were read in at the beginning. In this scenario, we use the function `PolynomialDivision`. Here, we define polynomials a and b in the ring of polynomials over $\mathbb{Z}_7$.

```
a = Poly[Z[7], x⁴ + 5 x² + 3 x + 4]
b = Poly[Z[7], x² + 3 x + 2]
```

Try your hand, using paper and pencil, at determining the quotient and remainder of a divided by b. Confirm your results with the following cell.

```
{q, r} = PolynomialDivision[a, b]
```

We can test this result by multiplying the quotient q times b, then adding the remainder r.

```
a == b * q + r
```

Often it's necessary to know only the remainder or the quotient. The built-in functions already used also have extensions in these packages.

```
? PolynomialQuotient
```

```
? PolynomialRemainder
```

```
PolynomialQuotient[a, b]
PolynomialRemainder[a, b]
```

Let's consider another example by changing b to a linear polynomial

```
a = Poly[Z[7], 4 x⁴ + 3 x³ + 5 x² + 1]
b = Poly[Z[7], x - 3]
```

Note that $x - 3$ is the same polynomial as $x + 4$ in the ring of polynomials over $\mathbb{Z}_7$.

> **Q13**. Use long division to determine the quotient and remainder. Now evaluate the polynomial a at the value 3.

To evaluate any polynomial, simply substitute in the value and compute in the ring. In this case, we reduce mod 7 as we go. Try it. We can also use the following function.

```
PolynomialEvaluation[a, 3]
```

Compare this to the following.

```
PolynomialRemainder[a, b]
```

Why did we divide by $x - 3$ and evaluate at 3? See if you can understand why with a couple more examples.

Let's change b to a different linear polynomial. We divide and examine the remainder.

```
b = Poly[Z[7], x - 2]
PolynomialRemainder[a, b]
```

Also, evaluate a when x is given the value 2.

```
PolynomialEvaluation[a, 2]
```

And again, one more time, over a different ring.

```
a = Poly[Z[13], 4 x⁴ + 3 x³ + 5 x² + 1]
b = Poly[Z[13], x + 5]
PolynomialEvaluation[a, 8]
PolynomialRemainder[a, b]
```

Q14. Can you make a conjecture based on these examples? You may wish to try some other linear polynomials. Note that you need to determine where to evaluate the polynomial, and this depends on the linear polynomial. What is this relationship?

Q15. Suppose that $a = f(x)$ and $b = x - k$. Based on your answer to question 14, what can you say about the remainder r when we have $f(k) = 0$? (Consider the following example, if you wish.)

```
a = Poly[Z[7], x⁴ + 2 x³ + 2 x² + 1]
b = Poly[Z[7], x - 3]
PolynomialEvaluation[a, 3]
PolynomialRemainder[a, b]
```

Q16. In all the examples we have encountered, what is the relationship between the degree of the remainder r and the degree of the divisor b?

Now let's fix a polynomial and consider the zeros of the polynomial over various rings $\mathbb{Z}_n$. By a zero, we mean a value such that the polynomial evaluates to the zero of the base ring (which, in $\mathbb{Z}_n$, is 0). For example, consider the polynomial $x^2 + 3x + 2$.

```
p = x² + 3 x + 2
```

Consider finding the zeros for this polynomial over $\mathbb{Z}_{15}$, $\mathbb{Z}_{17}$ and $\mathbb{Z}_{37}$.

```
Zeros[Poly[Z[15], p]]
Zeros[Poly[Z[17], p]]
Zeros[Poly[Z[37], p]]
```

Suppose we consider all $\mathbb{Z}_k$ for $k = 2$ to $k = 11$.

```
TableForm[
  Table[{k, Zeros[Poly[Z[k], x^2 + 3 x + 2]]}, {k, 2, 11}],
  TableHeadings → {None, {"k", "zeros over \!\(\*SubscriptBox[\(\[DoubleStruckCapitalZ]\), \(k\)]\)\n"}},
  TableDepth → 2, TableSpacing → {0, 3}]
```

> **Q17.** While several indices k resulted in four zeros, most resulted in two. For those that yielded only two, is there anything significant about the two zeros? What is the explanation for this?

Note that finding zeros is a special case of solving an equation set equal to zero.

```
Solve[Poly[Z[11], x^2 + 3 x + 2] == 0]
```

Of course, we can also solve equations involving other constants.

```
solns = Solve[Poly[Z[11], x^2 + 3 x + 2] == 1]
```

We can also verify the solutions. (Recall that $x + 3$ /. $\{x \to 5\}$ results in 8.)

```
x + 3 /. {x → 5}
```

```
Poly[Z[11], x^2 + 3 x + 2] /. solns
```

Let's try another polynomial.

```
TableForm[
  Table[{k, Zeros[Poly[Z[k], 3 x^3 + x^2 + 3 x + 2]]}, {k, 2, 11}],
  TableHeadings → {None, {"k", "zeros over Z_k\n"}},
  TableDepth → 2, TableSpacing → {0, 3}]
(* an empty position indicates no zeros *)
```

> **Q18.** By modifying this code, try another polynomial and/or other rings to see if you can determine a relationship between the (maximum) number of zeros and the index k of the ring $\mathbb{Z}_k$ under consideration.

■ 6.4.1 The Euclidean Algorithm

The greatest common divisor of two polynomials, *a* and *b*, *b* not zero, can be determined using the fact that if *r* is the result of `Polynomial-Remainder[a, b]`, then gcd(*a*, *b*) = gcd(*b*, *r*). This is an extension with polynomials of what you have already seen with integers. In other words, if *a* = 24 and *b* = 40, we know that gcd(*a*, *b*) = 8. Below we use `Mod`, which is the integer equivalent to `PolynomialRemainder`, to implement the Euclidean Algorithm. First define *a* and *b*.

```
a = 40
b = 24
```

Calculate the new values for *a* and *b*. Let the new *a* become the old *b* and the new *b* become the remainder on dividing *a* by *b* (`Mod[a, b]`). Now call these values *a* and *b* again.

```
newa = b
newb = Mod[a, b]
{a, b} = {newa, newb}
```

Repeat.

```
newa = b
newb = Mod[a, b]
{a, b} = {newa, newb}
```

Repeat.

```
newa = b
newb = Mod[a, b]
{a, b} = {newa, newb}
```

You are done when the remainder (`newb`) becomes 0; the gcd is then the last nonzero remainder (*b*).

Q19. Employ this algorithm by verifying that the gcd of 21 and 13 is 1.

Q20. Employ the Euclidean Algorithm for polynomials to determine the gcd of the polynomials $x^4 + 2 x^3 + 2 x^2 + 2$ and $x^4 + 3 x^3 + 3 x^2 + 3 x + 2$, both over $\mathbb{Z}_5$.

```
a = Poly[Z[5], x⁴ + 2 x³ + 2 x² + 2]
b = Poly[Z[5], x⁴ + 3 x³ + 3 x² + 3 x + 2]
```

You might find this cell handy.

```
newa = b
newb = PolynomialRemainder[a, b]
{a, b} = {newa, newb}
```

■ 6.4.2 Another approach

Algorithms are great things to instruct computers to do. The Euclidean Algorithm has been implemented in the function `PolynomialGCD`.

```
PolynomialGCD[a, b]
```

Ring Lab 7

Factoring and Irreducibility

▦ 7.1 Prerequisites

Before working on this lab, you should be familiar with polynomial arithmetic over integral domains. No previous labs need to be completed prior to attempting this lab.

▦ 7.2 Goals

The goal of this lab is to introduce some of the tools available for polynomial factorization over a variety of rings.

▦ 7.3 Introduction to factoring and irreducibility

To get started, we read in the package that provides some of the functionality needed for this lab.

```
Needs["AbstractAlgebra`Master`"];
SwitchStructureTo[Ring];
```

In this lab we consider how to factor polynomials. From high school math, you should already have a sense of what this means, but now we will make the definition more formal. For example, suppose you were asked if you could factor the following polynomial.

```
Clear[f, x]
f = x² - 3
```

Would you answer "No," since 3 is not a perfect square, so this is not a difference of two squares? Or would you answer "Yes," since one can factor this as $(x - \sqrt{3})(x + \sqrt{3})$? Actually, it is a poorly worded question. We will learn how we should ask it.

We consider polynomials from $D[x]$, where D is an integral domain. This means that the coefficients of the polynomials in the indeterminate x come from the integral domain D. Given any nonconstant polynomial $f(x)$ in $D[x]$ (i.e., degree($f(x)$) > 0), we call $f(x)$ *irreducible* if whenever we write $f(x) = g(x)h(x)$ (with $g(x)$ and $h(x)$ from $D[x]$) we have $g(x)$ or $h(x)$ a unit in $D[x]$. In other words, $f(x)$ cannot be factored except when one of the factors is a unit in $D[x]$. When a nonconstant polynomial is not irreducible, we call it *reducible over D*, or *factored over D*. Note that frequently the domain D is a field.

Thus, in our original question we should have asked if f was irreducible over $\mathbb{Q}$ (the rationals) or if f was irreducible over $\mathbb{R}$ (the reals) or over some other integral domain. In other words, we need to specify the ring from which our coefficients come.

Q1. Determine whether $x^2 - 3$ is irreducible or reducible over the integers, rationals, reals, and complex numbers.

Q2. Determine whether $2x^2 - 6$ is irreducible or reducible over the integers, rationals, reals, and complex numbers.

Q3. Determine whether $2x^2 + 6$ is irreducible or reducible over the integers, rationals, reals, and complex numbers.

▦ 7.4 Some techniques on testing the irreducibility of polynomials

▪ 7.4.1 Polynomials over $\mathbb{Z}_p$ —Theorem 1

Suppose we consider the polynomial $f(x) = 4x^3 + 3x^2 - 2x + 1$ over the field $\mathbb{Z}_p$ for some prime p. If $f(x)$ is reducible, then there exists $g(x)$ and $h(x)$ from $\mathbb{Z}_p[x]$ such that $f(x) = g(x)h(x)$ and neither $g(x)$ nor $h(x)$ is a unit. Consequently, these two polynomials must have degrees 1 and 2 (or 2 and 1), since the sum of their degrees must be 3. But then the linear factor, whose form is $ax + b$, has a zero, namely $-a^{-1} b$. Consequently, $f(x)$ also has a zero. Therefore, all we have to do is look for the zeros of f. This can be summarized in the following theorem; you will probably find the theorem in your text. (Note that this argument works whether the field is finite or infinite.)

Theorem 1: Let F be a field. A polynomial of degree 2 or 3 over F is reducible over F if and only if it has a zero in F.

Let's pursue the irreducibility of $f(x) = 4 x^3 + 3 x^2 - 2 x + 1$ over some finite field $\mathbb{Z}_p$. In this case, the task is particularly simple, since all we need to do is check to see if f has a zero over the field. Here is how we might proceed for $p = 3$. First we define f.

```
f = Poly[Z[3], 4 x^3 + 3 x^2 - 2 x + 1]
```

Next we map f over the entire domain of the field.

```
Map[PolynomialEvaluation[f, #1] &, {0, 1, 2}]
```

Thus we see that 1 is a zero for f. (Why 1? Look at the set over which we mapped f.) Consequently, by the theorem, we know that f is reducible over $\mathbb{Z}_3$.

> **Q4.** Since $f(x)$ is reducible, you should be able to find $g(x)$ and $h(x)$ such that $f(x) = g(x) h(x)$. Do so.

> **Q5.** Is $f(x)$ reducible over $\mathbb{Z}_7$?

> **Q6.** Consider $f(x) = 7 x^3 + 13 x^2 + 2 x + 7$. Starting with $p = 2$, find the first prime p for which $f(x)$ is reducible over $\mathbb{Z}_p$.

■ 7.4.2 Rational Root Theorem—$2 x^3 + 3 x^2 - 1$ over $\mathbb{Q}$

Consider the polynomial $f = 2 x^3 + 3 x^2 - 1$ over the rationals, $\mathbb{Q}$.

```
Clear[f, x]
f = 2 x^3 + 3 x^2 - 1
```

Since this is a cubic, we could use Theorem 1. However, since the rationals are an infinite field, we don't want to look for zeros by brute force (i.e., test every element to see if it is a zero), so we introduce another tool, designed specifically for finding rational zeros: the Rational Root Theorem.

Rational Root Theorem: Suppose $f(x) = a_n x^n + a_{n-1} x^{n-1} + \cdots + a_2 x^2 + a_1 x + a_0$ is a polynomial in $\mathbb{Z}[x]$ (with a_n not zero). If r and s are relatively prime and $f(\frac{r}{s}) = 0$, then $r \mid a_0$ and $s \mid a_n$.

Note what this says. If we have a rational zero of the form $\frac{r}{s}$, then the numerator r must divide the constant term of the polynomial and the denominator s must divide the leading coefficient. This reduces the search for a root from an infinite set to a finite number of possibilities. (Why does this lead to a finite list?)

For our polynomial $f = 2x^3 + 3x^2 - 1$, we are looking for integers r and s such that r divides -1 (thus ± 1) and s divides 2 (thus $\pm 1, \pm 2$). Hence, the list of *candidates* for rational roots is as follows. (Do you agree with this list?)

$$\texttt{possibleRationalRoots} = \left\{ \texttt{1, -1, } \frac{1}{2}\texttt{, } -\frac{1}{2} \right\}$$

Now we just need to test f at each of these and look for a zero.

$$\texttt{f /. \{x} \rightarrow \texttt{possibleRationalRoots\}}$$

In this case, both -1 and $1/2$ are zeros for f. We can now express f in factored form.

Q7. Since f is reducible, you should be able to find $g(x)$ and $h(x)$ such that $f = g(x)h(x)$. In fact, factor it completely into linear terms. (You should already know two of them.)

Q8. Given the polynomial $4x^5 + 3x^3 - 8x^2 + 7x - 6$, what are the *possible* rational roots to try? Do any of these work?

■ 7.4.3 Mod p Irreducibility Test—$x^4 - 2x^3 - 7x^2 - \frac{11}{3}x - \frac{4}{3}$ over $\mathbb{Q}$

First we define f to be a polynomial we wish to factor.

```
Clear[f, x]
```
$$\texttt{f = x}^4 - \texttt{2 x}^3 - \texttt{7 x}^2 - \frac{11}{3}\texttt{ x} - \frac{4}{3}$$

Since we need our polynomial to have integer coefficients (look at the formulation of the Rational Root Theorem), we multiply f by 3 to get a new polynomial over $\mathbb{Z}$.

```
Clear[g]
g = Expand[3 f]
```

If we find that $g(x)$ can be factored into a product $h(x)k(x)$, we will then have a factorization of $f(x) : 3^{-1}h(x)g(x)$. Note that $g(x)$ is a quartic, so it does not satisfy the hypotheses of Theorem 1. However, if we can find a linear factor, then the remaining factor will be cubic and we can complete the process with the cubic (if we are looking for a complete factorization, not just trying to determine reducibility). If we cannot find a linear factor, it still may be factorable into two quadratic polynomials. (Why?)

We will approach the problem first with the Rational Root Theorem and then introduce another technique.

■ Using the Rational Root Theorem

First we determine all *possible* rational roots (i.e., form the list of candidates).

```
possibleRationalRoots =
```
$$\left\{1, -1, \frac{1}{3}, -\frac{1}{3}, 2, -2, \frac{2}{3}, -\frac{2}{3}, 4, -4, \frac{4}{3}, -\frac{4}{3}\right\}$$
```
(* Do you agree with this list? *)
```

Now we evaluate g at each possible root.

```
g /. {x → possibleRationalRoots}
```

This indicates that 4 is a zero. Therefore, we need to find the quotient of g and $x - 4$. (Note that since 4 is a zero, $x - 4$ is a factor.)

```
g2 = PolynomialQuotient[g, x - 4, x]
```

This is our new polynomial to pursue. Continuing with the Rational Root Theorem, let us continue factoring it.

```
possibleRationalRoots =
```
$$\left\{1, -1, \frac{1}{3}, -\frac{1}{3}\right\}$$
```
(* Do you agree with this list? *)
```

```
g2 /. {x → possibleRationalRoots}
```

We see that none of these are zeros. Hence g is factored as $(x - 4)(3x^3 + 6x^2 + 3x + 1)$, so f is factored as $3^{-1}(x - 4)(3x^3 + 6x^2 + 3x + 1)$. Thus f is reducible.

Note: There is a function to determine the list of candidates, though usually it is fairly easy to create the list by hand.

```
RationalRootCandidates[g]
```

■ Using the Mod p Irreducibility Test

Even though we have seen that f is not irreducible (i.e., it is reducible), let us continue with a general approach to testing a polynomial using the Mod p Irreducibility Test.

Mod *p* Irreducibility Test: Let p be a prime and suppose that $f(x)$ is a polynomial over $\mathbb{Z}$ (the integers) with $\deg(f(x)) \geq 1$. Let $h(x)$ be the polynomial in $\mathbb{Z}_p[x]$ obtained from $f(x)$ in $\mathbb{Z}[x]$ by reducing all the coefficients of $f(x)$ modulo p. If the degree of $h(x)$ equals the degree of $f(x)$ and if $h(x)$ is irreducible over $\mathbb{Z}_p$, then $f(x)$ is irreducible over $\mathbb{Q}$ (the rationals).

We still need the g we defined above.

```
g
```

Now we wish to reduce this modulo some prime. (Although the built-in function `PolynomialMod` could be used here—specifically, `PolynomialMod[g, 2]`—we use another approach that is more general.) Try 2 and call this new polynomial *g2*.

```
g2 = Poly[Z[2], g]
```

It may be clear that this is reducible (since x is a factor), but for the record, this is how we can check for zeros.

```
Map[PolynomialEvaluation[g2, #1] &, Elements[Z[2]]]
```

We see that 0 is a zero and hence x is a factor. This does *not* give us any indication of a factor in $\mathbb{Q}[x]$, nor does it imply that g is reducible over $\mathbb{Q}[x]$ (though in this case we know it is reducible from our work with the Rational Root Theorem). For the sake of argument, let us try a few other primes.

```
g3 = Poly[Z[3], g]
```

Note that the theorem does not apply when reducing mod 3. (Why?)

```
g5 = Poly[Z[5], g]
Map[PolynomialEvaluation[g5, #] &, Elements[Z[5]]]
```

You can try any other prime modulus with the following.

```
p = Input["Enter a prime"];
If[PrimeQ[p],
  {"prime = " <> ToString[p] <> ":", gp = Poly[Z[p], g],
    If[Degree[gp] == 4, Map[PolynomialEvaluation[gp, #] &,
      Elements[Z[p]]], "does not apply"]},
  Print["You didn't enter a prime."]]
```

We see that every time we get at least one factor, and hence the polynomial is not irreducible over $\mathbb{Z}_p$. When using the Mod p Irreducibility Test, we either continue with more primes, hoping to strike gold, or we begin to think that it is reducible over $\mathbb{Q}$ and try another test (such as the Rational Root Test).

■ 7.4.4 How to handle quartics—$x^4 - 3x^2 + 2x + 1$ over $\mathbb{Q}$

First let's define our next polynomial.

```
h = x⁴ - 3 x² + 2 x + 1
```

Now let's look for a linear factor. We can try the Mod 2 Irreducibility Test.

```
h2 = Poly[Z[2], h]
Map[PolynomialEvaluation[h2, #] &, {0, 1}]
```

This shows that the polynomial reduced mod 2, *h2*, has no linear factors over $\mathbb{Z}_2$, but this does not make it irreducible. What about quadratic factors over $\mathbb{Z}_2$? Could one of these be a factor? What are the irreducible quadratics in $\mathbb{Z}_2[x]$? There are only 4 quadratics, so we can investigate them quite easily.

```
quads = Table[x² + a x + b, {a, 0, 1}, {b, 0, 1}] // Flatten //
    Map[Poly[Z[2], #] &, #] &
```

It's obvious that the first and third polynomials in this list have x as a factor, so we need to investigate only the second and fourth. First we try the second factor.

```
Map[PolynomialEvaluation[quads[[2]], #] &, {0, 1}]
```

The second factor has a zero, namely 1. (This is not surprising; in $\mathbb{Z}_2$ $x^2 + 1 = x^2 - 1 = (x-1)(x+1) = (x-1)(x-1)$.)

```
Map[PolynomialEvaluation[quads[[4]], #] &, {0, 1}]
```

From this, we can see that the only irreducible quadratic in $\mathbb{Z}_2[x]$ is the fourth candidate, $x^2 + x + 1$. Recall that *h2* is our original polynomial, reduced modulo 2. We want to know if this is divisible by the irreducible quadratic. We use the `PolynomialDivision` function, which returns the quotient and remainder when given two polynomials.

```
q = quads[[4]]
```

```
PolynomialDivision[h2, q]
```

We see that *h2* is the square of *q*.

```
q q == h2
```

These two steps show that there are two quadratic factors, so this polynomial is reducible over $\mathbb{Z}_2$. Hence, we need to try a new prime since we *cannot* make any conclusion about reducibility over $\mathbb{Q}$ when we know that we have reducibility over $\mathbb{Z}_p$. Next we try $p = 3$.

```
h3 = Poly[Z[3], h]
Map[PolynomialEvaluation[h3, #] &, {0, 1, 2}]
```

This also yields a reducible polynomial (since we have a zero). Try again with $p = 5$.

```
h5 = Poly[Z[5], h]
Map[PolynomialEvaluation[h5, #] &, {0, 1, 2, 3, 4}]
```

We can see that at least there are no linear factors, but now we have to consider quadratic ones. Below, with $p = 5$, we use the "brute force" method used with $p = 2$. Here, we create all possible *monic* quadratic polynomials over $\mathbb{Z}_5$. (To be monic, the leading coefficient is the unity of the ring, 1. You may wonder, and rightly so, why we need to consider only monic polynomials. Look in your text for an answer or ask your professor if you can't find out why.)

Q9. Before computing the list of *possible* monic quadratics, count how many there will be. How many? How did you arrive at your result?

```
quads = Table[x² + a x + b, {a, 0, 4}, {b, 0, 4}] // Flatten //
    Map[Poly[Z[5], #] &, #] &
```

With each of these quadratics, let's divide *h5* by the candidate and look at the remainder on division.

```
Map[PolynomialRemainder[h5, #] &, quads]
```

Observing that there are no remainders consisting of the 0 polynomial, we see that there are no quadratic factors. Therefore, since *h5* has no linear factors and no quadratic factors, it is irreducible. Then, by the Mod 5 Irreducibility Test, $h(x)$ is also irreducible over $\mathbb{Q}$.

■ 7.4.5 Theorem 2—$2x^4 - 2x^3 - 17x^2 + 25x - 7$ over $\mathbb{Z}$

In this case, we wish to know whether a polynomial is factorable over $\mathbb{Z}$, the integers. Consider the following theorem, which may be of use.

Theorem 2. Let $f(x)$ be in $\mathbb{Z}[x]$. If $f(x)$ is reducible over $\mathbb{Q}$, then it is reducible over $\mathbb{Z}$.

Let $f(x)$ be defined as follows.

```
Clear[f, x]
f = x⁴ + 2 x³ + 3 x² + 6 x + 9
```

Q10. Give as complete a factorization for $f(x)$ as possible. Is it reducible over $\mathbb{Z}$? Why or why not? Hint: Use the Rational Root Theorem to look for zeros. If there are quadratic factors, they must look like $(x^2 + ax + c)(x^2 + bx + d)$. Since the middle coefficients of f are relatively small and all positive, it is reasonable to *guess* that $c = d = 3$ so that if it factors into two quadratics, it looks like $(x^2 + ax + 3)(x^2 + bx + 3)$. Try expanding this product and comparing coefficients.

▪ 7.4.6 Eisenstein's Criterion—$3x^8 - 4x^6 + 8x^5 - 10x + 6$ over $\mathbb{Q}$

First we define our polynomial.

```
f = 3 x⁸ - 4 x⁶ + 8 x⁵ - 10 x + 6
```

Perhaps we should think about this one from another perspective at first. Suppose we graph this over some domain.

```
Plot[f, {x, -3, 3}];
```

It looks as if the interesting part of the function occurs in a smaller domain.

```
Plot[f, {x, -1.65, 1.1}];
```

Q11. What does the graph say about the number of *real* zeros? What does it say about the number of rational zeros? Considering the graph and the degree of the polynomial, what does it say about the number of pure complex (i.e., nonreal) zeros?

Consider the following.

```
Clear[h]
h = Poly[Z[5], f]
Map[PolynomialEvaluation[h, #] &, Elements[Z[5]]]
```

This indicates that there are no linear factors. There could, however, be quadratic, cubic or quartic factors. If there was a cubic factor, there could also be a quintic or another cubic and a quadratic. It could get quite complicated. Fortunately, for some very special polynomials whose coefficients behave in a particular fashion, there is a theorem that can be used to determine irreducibility.

Eisenstein's Criterion. Let $f(x) = a_n x^n + a_{n-1} x^{n-1} + \cdots + a_2 x^2 + a_1 x + a_0$ be a polynomial in $\mathbb{Z}[x]$ (with a_n not zero). If there is a prime p such that p *does not* divide the leading coefficient a_n but *does* divide every other coefficient, and we also have p^2 *not* dividing a_0, then $f(x)$ is irreducible over $\mathbb{Q}$.

Let's consider the polynomial *f* again.

```
f
CoefficientList[f, x]
```

Note that the prime $p = 2$ does indeed divide every coefficient except the leading 3 and that $p^2 = 4$ does not divide 6. Therefore, by Eisenstein's Criterion, *f* is irreducible. Although Eisenstein's Criterion is fairly easy to implement by hand, the following will walk you through the steps.

```
EisensteinsCriterionQ[f, Mode → Textual]
```

■ 7.4.7 Over a finite set—$x^4 + 4$ over $\mathbb{Z}_7$

We now consider a polynomial over $\mathbb{Z}_7$.

```
Clear[p]
p = Poly[Z[7], x⁴ + 4]
Map[PolynomialEvaluation[p, #] &, Elements[Z[7]]]
```

This code defines the polynomial and then shows that there is no zero for the polynomial over $\mathbb{Z}_7$. Since we are interested in determining irreducibility over $\mathbb{Z}_7$, we cannot use the Mod *p* Irreducibility Test. (Why?) Having no zero indicates that there is no linear factor (and hence no corresponding irreducible cubic). There is still the possibility of two irreducible quadratic factors. Let's check this out. We assume that we have two quadratic factors.

```
Clear[a, b, c, d, e, f]
PossibleFactors = {a x² + b x + c, d x² + e x + f}
```

Here is the product of these factors.

```
prod = Apply[Times, PossibleFactors]
Expand[prod]
```

We know that *a* and *d* are nonzero and hence invertible. (If either was zero, they would not be quadratics.) We can factor out both of them and assume that they are equal to 1.

```
a = d = 1;
t = Expand[prod]
```

Let's look at our original polynomial and compare it to *t* with identical factors of *x* collected together.

```
p
Collect[t, x]
```

Note that we need $cf = 4$, $ce + bf = 0$, $c + be + f = 0$ and $b + e = 0$. Additionally, we want the modulus to be 7.

```
equations =
  {c f == 4, c e + b f == 0, c + b e + f == 0, b + e == 0, Modulus == 7}
```

How can we get the constant term of 4? What are the values for c and f? We know that $2 * 2 = 4$. Is there any other possibility? Yes. Since $\mathbb{Z}_7$ is a field, there are 5 other possibilities (such as $3 * 6 = 4$). One way to see all of them is to look at the multiplication table for $\mathbb{Z}_7$.

```
MultiplicationTable[Z[7], Mode → Visual];
```

The unordered pairs that have a product of 4 (mod 7) are {4, 1}, {2, 2}, {3, 6}, and {5, 5}. We now solve the system of equations arising from each possibility of values for c and f having a product of 4. Notice that `Rest[equations]` represents a list of all but the first equation in `equations`. First we test when we have $c = 4$ and $f = 1$.

```
s1 = Solve[Rest[equations] /. {c → 4, f → 1}, {e, b}]
```

There is no solution for this case. Next we try $c = 2$ and $f = 2$.

```
s2 = Solve[Rest[equations] /. {c → 2, f → 2}, {e, b}]
```

Here we get two solutions, although we will see that they are really the same.

Q12. Given that we now already know our polynomial is reducible, why might we want to consider the cases where $c = 5$ and $f = 5$ or $c = 3$ and $f = 6$?

Let's consider the remaining two cases. First we consider $c = 3$ and $f = 6$.

```
s3 = Solve[Rest[equations] /. {c → 3, f → 6}, {e, b}]
```

Again, there is no solution in this case. Finally, we try $c = 5$ and $f = 5$.

```
s4 = Solve[Rest[equations] /. {c → 5, f → 5}, {e, b}]
```

Again, no solution. Let's return to the second case. The first solution is

```
s2[[1]]
```

Here are the results of the possible factors for these values of b, c, e, and f.

```
PossibleFactors /. {b → 5, c → 2, e → 2, f → 2}
```

The second solution:

```
s2[[2]]
```

We repeat the process used with the second solution.

```
PossibleFactors /. {b → 2, c → 2, e → 5, f → 2}
```

Therefore, there really is only one solution. The difference in our two results is the ordering of the two factors. In conclusion, we see that f is reducible over the field $\mathbb{Z}_7$ in exactly one way: into two quadratics.

■ 7.4.8 Over the reals—$x^4 - 2x^3 + x^2 + 1$ over $\mathbb{R}$

This situation is a little different because we are considering factoring a polynomial over the reals.

```
Clear[p]
p = x^4 - 2 x^3 + x^2 + 1
```

First we consider rational factorizations. The Rational Root Test indicates that the only possibilities for zeros are ± 1.

```
p /. x → {1, -1}
```

This indicates that there are no linear factors over $\mathbb{Q}$. How about the mod p test?

```
h = Poly[Z[2], p]
Map[PolynomialEvaluation[h, #] &, Elements[Z[2]]]
```

This also shows that there are no linear factors (no surprise), but what about quadratic factors? We have already seen in section 7.4.4 that the only irreducible quadratic factor for h is $x^2 + x + 1$.

```
Poly[Z[2], Expand[(x^2 + x + 1) (x^2 + x + 1)]]
```

This shows that h is reducible mod 2, but unfortunately, it does not address the question about reducibility in $\mathbb{Q}$. We could continue with another prime, but this will at best indicate irreducibility over $\mathbb{Q}$ and will not address the irreducibility over $\mathbb{R}$ at all. Let's try some other approaches.

■ First attempt

We use some techniques from calculus: determining the derivative and setting it equal to zero to find possible relative extrema.

```
dp = D[p, x]
dzeros = Solve[dp == 0]
```

Now we evaluate p at the zeros.

```
p /. dzeros
```

This indicates that we have two relative minima at the value 1 and a relative maximum at the value $\frac{17}{16}$.

> **Q13**. Why can we make this conclusion?

> **Q14**. Why does this tell us that this polynomial has no linear factors?

■ **Second attempt**

Perhaps we should plot it!

```
Plot[p, {x, -2, 2}, PlotRange → {0, 3}];
```

The graph seems to indicate that there are no real zeros, since the function does not cross the *x*-axis.

```
NSolve[p == 0]
```

This confirms that the solutions are complex and thus nonreal. Hence, there are no linear factors.

> **Q15**. Determine whether we can write *p* as a product of two irreducible quadratic factors over the reals. Caution: This takes a little bit of work.

7.5 More polynomials for practice

Consider the following collection of polynomials over various fields. In each case, determine whether the polynomial is reducible or irreducible over the stated field. When reducible, factor it as completely as possible. In each case, state the theorem(s) you use and show your work. Some are fairly easy, while others may take some work. For your convenience, all the theorems used in this lab are collected together in a list in section 7.6.

> **Q16**. $x^3 + x^2 + x + 1$ over $\mathbb{Q}$

> **Q17**. $x^4 + x^2 - 6$ over $\mathbb{Q}$

> **Q18**. $4x^3 + 3x^2 + x + 1$ over $\mathbb{Z}_5$

Q19. $25\,x^5 - 9\,x^4 + 3\,x^2 - 12$ over $\mathbb{Q}$

Q20. $x^4 - 3\,x^3 + 2\,x^2 + 4\,x - 1$ over $\mathbb{Z}_5$

7.6 Toolbox of theorems

Theorem 1: Let F be a field. A polynomial of degree 2 or 3 over F is reducible over F if and only if it has a zero in F.

Rational Root Theorem: Suppose $f(x) = a_n\,x^n + a_{n-1}\,x^{n-1} + \cdots + a_2\,x^2 + a_1\,x + a_0$ is a polynomial in $\mathbb{Z}[x]$ (with a_n not zero). If r and s are relatively prime and $f(\frac{r}{s}) = 0$, then $r\,|\,a_0$ and $s\,|\,a_n$.

Mod p Irreducibility Test: Let p be a prime and suppose that $f(x)$ is a polynomial over $\mathbb{Z}$ (the integers) with degree greater than or equal to 1. Let $h(x)$ be the polynomial in $\mathbb{Z}_p[x]$ obtained from $f(x)$ in $\mathbb{Z}[x]$ by reducing all the coefficients of $f(x)$ modulo p. If the degree of $h(x)$ equals the degree of $f(x)$ and if $h(x)$ is irreducible over $\mathbb{Z}_p$, then $f(x)$ is irreducible over $\mathbb{Q}$ (the rationals).

Theorem 2. Let $f(x)$ be in $\mathbb{Z}[x]$. If $f(x)$ is reducible over $\mathbb{Q}$, then it is reducible over $\mathbb{Z}$.

Eisenstein's Criterion. Let $f(x) = a_n\,x^n + a_{n-1}\,x^{n-1} + \cdots + a_1\,x + a_0$ be a polynomial in $\mathbb{Z}[x]$ (with a_n not zero). If there is a prime p such that p *does not* divide the leading coefficient a_n, but *does* divide every other coefficient, and we also have p^2 *not* dividing a_0, then $f(x)$ is irreducible over $\mathbb{Q}$.

7.7 Final perspective

We have been purposely avoiding a very powerful built-in function of *Mathematica* that does much of the work shown in this lab in one swoop.

```
Factor[x⁴ - 2 x³ - 7 x² - 11 x/3 - 4/3]

Factor[2 x³ + 3 x² - 1]

Factor[x⁴ - 3 x² + 2 x + 1]

Factor[x⁴ + 4, Modulus → 7]
```

Isn't this disgusting? All the sweat and blood, not to mention the joy, is removed by this powerful function `Factor`.

The function `ModpIrreducibilityQ` may be illustrative as well.

```
ModpIrreducibilityQ[x⁴ - 3 x² + 2 x + 1, Mode → Textual]
```

```
ModpIrreducibilityQ[12 + 5 x + 8 x² + 11 x³ + 3 x⁴, Mode → Textual]
```

```
Factor[12 + 5 x + 8 x² + 11 x³ + 3 x⁴, Modulus → 11]
```

```
RationalRootTheorem[6 x³ - 5 x² - 7 x + 4]
Factor[6 x³ - 5 x² - 7 x + 4]
```

Q21. Evaluate the following input cell. What is a reasonable conjecture based on your observations? Can you prove it?

```
TableForm[Table[
    {Prime[n], Factor[x⁴ + 1, Modulus → Prime[n]]}, {n, 1, 25}]]
```

Ring Lab 8

Roots of Unity

🖿 8.1 Prerequisites

No other lab needs to be completed before attempting this lab. However, experience with cyclic groups (see Group Lab 6) may prove beneficial.

🖿 8.2 Goals

The main goal of this lab is to become familiar with the *roots of unity*, the roots of polynomials of the form $x^n - 1$.

🖿 8.3 Introduction

This whole lab focuses on the polynomial $x^n - 1$ (for a positive integer n) or factors thereof. This is a fairly simple polynomial. You first saw $x^2 - 1$ and $x^3 - 1$ in first-year algebra.

> **Q1.** Quick! Do you remember how to factor $x^3 - 1$ (over $\mathbb{Q}$)? How is it done?

You can use *Mathematica* to check your answer to question 1.

```
Clear[x]
Factor[x^3 - 1]
```

Was your answer correct? What about other values of n? Let's make a table.

```
TableForm[Table[Factor[x^n - 1], {n, 1, 17}],
    TableSpacing → {0.5, 0}]
```

Q2. What observations can you make from this table?

Q3. Modify the `Table` command to get a listing of values from 18 to some higher value. Do your observations still hold? Do you have any new ones or modifications?

Let's look at $x^{63} - 1$ and consider some questions regarding it.

```
Factor[x^63 - 1]
```

What things might be of interest? Consider the following list, for starters.

1. How many factors are there?
2. What coefficients are used?
3. What is the highest degree of any factor?
4. What is the list of all the degrees that occur among the factors?

The `Factor` command easily factors this polynomial, but what is returned is just the product of all the factors with no way of accessing the pieces of the factorization. The function `FactorList` gives us a *list* of the factors, so we can access them.

```
FactorList[x^63 - 1]
```

Note that this function returns pairs of the form $\{g(x), m\}$, where $g(x)$ is a factor of the polynomial and m is an integer. This integer represents the *multiplicity* of the factor. This is just a fancy term for expressing how many times a factor occurs in the factorization. For an example, the following should illustrate the concept of multiplicity.

```
FactorList[Expand[(x - 3)^4 (x + 1) (x + 2)^3 x^8]] //
    TraditionalForm
```

To get at just the polynomial factors, we need to transpose the list so that the factors are in the first row rather than the first column. (Version 3 of *Mathematica* introduces {1, 1} into the factor list; since we are not interested in it, we drop this term.) We can then ask for the first row. Here is our result for the polynomial $x^{63} - 1$:

```
factors = Drop[First[Transpose[FactorList[x^63 - 1]]], 1]
```

Now we can pursue some of the questions in the list. The first question was to count the number of factors. For $x^{63} - 1$, of course, we can simply step through and count, but we want to look for a way of doing this in *Mathematica*. What we really want to know is how many elements there are in the list `factors`, which can be measured by the `Length` function.

```
Length[factors]
```

The second question concerned the coefficients. The function `Coefficient-List` does exactly what we want.

```
CoefficientList[factors, x]
```

It may be of interest to remove any duplicates and not distinguish from which factor the values were obtained. The `Flatten` function puts all these factors into one list.

```
Flatten[CoefficientList[factors, x]]
```

Now we use the `Union` function, which sorts the factors and removes any duplicates.

```
Union[Flatten[CoefficientList[factors, x]]]
```

Questions 3 and 4 deal with the degree of the factors, namely the highest degree of all the factors and the degree of each factor. The latter can be easily arrived at with the `Exponent` function.

```
Exponent[factors, x]
```

Once we have this list, it is easy to find the maximum by applying the `Max` function.

```
Exponent[factors, x] // Max
```

Let's put all these steps into one function.

```
PolyInfo[x^(n_Integer)?Positive - 1] :=
 PolyInfo[x^n - 1] =
Module[{factors, exps},
    factors =
    Drop[First[Transpose[FactorList[x^n - 1]]], 1];
  exps = Exponent[factors, x];
  {n, Length[factors],
    Union[Flatten[CoefficientList[factors, x]]],
    Max[exps], Union[exps]}]
```

Here is a test of this function.

```
FactorList[x^12 - 1]
```

```
PolyInfo[x^12 - 1]
```

Note that the information is in the form {*n*, # factors, coefficients used, highest degree, all degrees}. Here is a table showing some results.

```
TableForm[Table[PolyInfo[xⁿ - 1], {n, 2, 37}],
    TableDepth → 2, TableSpacing → {0.5, 1}, TableHeadings →
    {None, {"n", "#fac", "coefs", "max deg", "degrees\n"}}]
```

> **Q4**. In question 2 of the list, you were asked what observations you could make from what you had seen. Do you have any additions, corrections, or comments? Make some conjectures.

> **Q5**. Change the range of values for *n* in the Table command to test your conjectures with more data. In particular, make sure you test your conjectures with values of *n* that exceed 100 (examples to consider might be $n = 210$ or $n = 165$). (Hint: If testing $n = 210$, do not test from $n = 2$ to $n = 210$, but use a smaller range, centered at 210.)

> **Q6**. What is the relationship between *n* and the number of factors of $x^n - 1$?

> **Q7**. What is the relationship between the degrees of the factors and the highest degree among the factors?

> **Q8**. What conclusion(s) can you draw about the highest degree of a factor of $x^n - 1$ and *n*?

▣ 8.4 A closer look—graphically

For the moment, we focus on $x^6 - 1$.

```
factors = Rest[First[Transpose[FactorList[x⁶ - 1]]]]
```

As we have seen, zeros are intimately related to factors. The first two factors have the zeros 1 and -1 associated with them. What about the last two factors? What are the zeros? Let's check out the second quadratic factor (the fourth factor in the list) by using the Solve command.

```
solns = Solve[factors[[4]] == 0]
```

These are complex zeros (though they may not look like it). What are the real and imaginary parts of these zeros?

```
Map[{Re[#], Im[#]} &, x /. solns]
```

Q9. These numbers, in pairs, should look familiar. To what are they related or where have you seen them before? Find the zeros for the third factor and find the real and imaginary parts as above.

Q10. Look again at the factorization of $x^6 - 1$. The first two factors (together) yielded two zeros and the last two factors had two zeros each, yielding a total of six zeros. Graph (not necessarily with *Mathematica*) these six zeros in the complex plane, using the vertical axis as the imaginary axis and the horizontal for the real axis. In other words, the complex number $a + b\,i$ is to be mapped to the point (a, b) in the plane. Can you give a precise description of the resulting graph?

Perhaps it may be worth looking at the zeros of other polynomials. Here we use the `Solve` command on several others.

```
Solve[x⁸ - 1 == 0]
Solve[x¹⁰ - 1 == 0]
Solve[x¹⁶ - 1 == 0]
```

Since the exact answers may not be so informative, let's use `NSolve` instead, resulting in approximate decimal answers.

```
NSolve[x⁸ - 1 == 0]
NSolve[x¹⁰ - 1 == 0]
NSolve[x¹⁶ - 1 == 0]
```

Let's plot the zeros of this last equation. First we need our list of zeros.

```
zeros = x /. NSolve[x¹⁶ - 1 == 0]
```

Next we need to convert each zero to its real and imaginary part and make them *Mathematica* graphics-type points. Finally, we graph them. Do not concern yourself with the *Mathematica* details.

```
Show[Graphics[{RGBColor[0, 0, 1], PointSize[0.025],
    (Point[{Re[#1], Im[#1]}] &) /@ zeros}],
  Axes → True, AspectRatio → Automatic];
```

This sure looks beautiful! Note, however, that this graph has *not* considered the individual factors in the factorization of the polynomial $x^{16} - 1$:

```
Rest[FactorList[x¹⁶ - 1]]
```

How do the plotted zeros relate to the individual factors just given? What we would like to do is find the zeros for *each* factor and then plot them so that we can discriminate them from each other. We need to slightly modify the code and write a function that can take any positive integer n as input and output a graph of the zeros of the *n*th roots of unity, colored by factor. (Again, don't concern yourself with the *Mathematica* details.)

```
RootsOfUnityZeros[n_Integer?Positive] :=
   Module[{factors =
   First[Transpose[Drop[FactorList[x^n - 1], 1]]]], zeros,
     len, pts, sizedpts, width = 1.2, p = Cyclotomic[n, x]},
 factors = Join[Complement[factors, {p}], {p}];
 zeros = x /. Map[NSolve[# == 0] &, factors];
 len = Length[zeros];
 pts = Map[Map[Point[{Re[#], Im[#]}] &, #] &, zeros, 2];
 sizedpts = Transpose[{Table[{Hue[i / len], PointSize[0.015 +
     i * 0.008]}, {i, len}], pts}];
 Show[Graphics[Map[Flatten, sizedpts]], Axes → True,
     AspectRatio → Automatic, PlotRange → {{-width, width},
     {-width, width}}, PlotLabel → "n = " <> ToString[n]]]
 (* no output - just a definition *)
```

Let's try this on a few examples. Depending on the memory of your *Mathematica* Front End, you may want to expand or narrow the range of values in the following loop by adjusting the values of lowk and highk. (Note: You may wish to delete all previous graphics cells before continuing, if you are getting low on memory.)

```
lowk = 4;
highk = 11;
Do[RootsOfUnityZeros[k], {k, lowk, highk}]
```

In each graph, dots of the same color come from the same factor.

Q11. What observations can you make from these graphs?

Q12. Do you see any groups (or a ring) lurking behind the scenes?

When we find the roots of the polynomial $x^n - 1$, we are finding the solutions (zeros) to the equation $x^n - 1 = 0$ or, more simply, $x^n = 1$. In other words, we are finding all the *n*th roots of 1, the unity, hence, the *n*th roots of unity. Since 1 is always a solution, starting with 1 let's number the roots and work counterclockwise, labeling the first with 0. (You might ask yourself why we start counting/labeling with 0.) Following is the modified code to reflect the numbering/counting scheme.

```
RootsOfUnityZeros[n_Integer?Positive, powers] :=
   Module[{factors =
   First[Transpose[Drop[FactorList[x^n - 1], 1]]]], zeros,
     len, pts, sizedpts, width = 1.3, p = Cyclotomic[n, x]},
 factors = Join[Complement[factors, {p}], {p}];
 zeros = x /. Map[NSolve[# == 0] &, factors];
 len = Length[zeros];
```

```
pts = Map[Map[Point[{Re[#], Im[#]}] &, #] &, zeros, 2];
sizedpts = Transpose[{Table[{Hue[i / len], PointSize[0.015 +
    i * 0.008]}, {i, len}], pts}];
Show[Graphics[{Map[Flatten, sizedpts], {RGBColor[0, 0, 1],
    Table[Text[i, 1.2 {Cos[i / n 2 Pi], Sin[i / n 2 Pi]}],
    {i, 0, n - 1}]}}],
  Axes -> True, AspectRatio -> Automatic,
  PlotRange -> {{-width, width}, {-width, width}},
  PlotLabel -> "n = " <> ToString[n]]]
```

Now let's try it out. (Note: You again may wish to delete all previous graphics cells before continuing, if you are getting low on memory.)

```
(* adjust lowk and/or highk
  depending on memory considerations *)
lowk = 4;
highk = 11;
Do[RootsOfUnityZeros[k, powers], {k, lowk, highk}]
```

> **Q13.** You should be seeing two distinct groups. Can you name them? Do you now know why we started counting at 0?

Instead of simply labeling the points 0 through $n - 1$, let's use these numbers but divide them by n, which is the number of roots (and the degree of the polynomial). Here we slightly modify the code again to reflect this change.

```
RootsOfUnityZeros[n_Integer?Positive, fractions] :=
  Module[{factors =
    First[Transpose[Drop[FactorList[x^n - 1], 1]]], zeros,
    len, pts, sizedpts, width = 1.3, p = Cyclotomic[n, x]},
  factors = Join[Complement[factors, {p}], {p}];
  zeros = x /. Map[NSolve[# == 0] &, factors];
  len = Length[zeros];
  pts = Map[Map[Point[{Re[#], Im[#]}] &, #] &, zeros, 2];
  sizedpts = Transpose[{Table[{Hue[i / len],
      PointSize[0.015 + i * 0.008]}, {i, len}], pts}];
  Show[Graphics[{Map[Flatten, sizedpts],
      {RGBColor[0, 0, 1], Table[Text[InputForm[i / n], 1.2
        {Cos[i / n 2 Pi], Sin[i / n 2 Pi]}], {i, 0, n - 1}]}}],
    Axes -> True, AspectRatio -> Automatic,
    PlotRange -> {{-width, width}, {-width, width}},
    PlotLabel -> "n = " <> ToString[n]]]
```

Let's try it one more time. (Note: You again may wish to delete all previous graphics cells before continuing, if you are getting low on memory.)

```
(* adjust lowk and/or highk
   depending on memory considerations *)
lowk = 4;
highk = 11;
Do[RootsOfUnityZeros[k, fractions], {k, lowk, highk}]
```

> **Q14.** Can you find any connections between the fractions used and the colors of the dots to which they correspond? (You may wish to try another range of values.)

🔲 8.5 Another look—algebraically

In section 8.3 we used the function `FactorList` but often had to manipulate it to get just the list of factors. Since we wish to do this often, let's write a function to do it for us.

```
FactorsOfUnity[n_Integer?Positive] :=
 First[Transpose[Rest[FactorList[x^n - 1]]]]
```

Here is how it works.

```
FactorsOfUnity[4]
```

Let's try a few others.

```
Map[FactorsOfUnity, {2, 4, 8, 16, 32}] // ColumnForm
(* here n = 2, 4, 8, 16, 32 *)

Map[FactorsOfUnity, {3, 6, 18, 36}] // ColumnForm

Map[FactorsOfUnity, {5, 20}] // ColumnForm

Map[FactorsOfUnity, {14, 28}] // ColumnForm

Map[FactorsOfUnity, {3, 6, 12, 24}] // ColumnForm
```

> **Q15.** Look at the relationship(s) between the numbers `FactorsOfUnity` is acting on and the results of the factorizations given. What conjectures or conclusions can you make? Can you give any explanations for your conclusions?

Note: Some of the ideas for this lab came from chapter 17 of *Exploring Mathematics with Mathematica* by Theodore W. Gray and Jerry Glynn (Addison-Wesley, 1991).

Ring Lab 9

Cyclotomic Polynomials

▦ 9.1 Prerequisites

Before working on this lab, you should have completed Ring Lab 8, on roots of unity.

▦ 9.2 Goals

The goal for this lab is to formulate recursive and nonrecursive definitions of the cyclotomic polynomials and discover some of their properties.

▦ 9.3 Introduction

Recall that in Ring Lab 8 we focused on the polynomial $x^n - 1$ (for a positive integer n). Here, we continue to focus on this polynomial—or, more accurately, factors of it.

First let's remind ourselves what the factorizations of these polynomials look like.

```
Clear[x]
TableForm[Table[{n, Factor[x^n - 1]}, {n, 1, 11}]]
```

We also redefine two functions we created in Ring Lab 8.

```
RootsOfUnityZeros[n_Integer?Positive,
    fractions, opts___?OptionQ] := Module[
   {factors = First[Transpose[Drop[FactorList[x^n - 1], 1]]],
    zeros, len, pts, sizedpts,
    width = 1.3, p = Cyclotomic[n, x]},
  factors = Join[Complement[factors, {p}], {p}];
  zeros = x /. (NSolve[#1 == 0] &) /@ factors;
  len = Length[zeros];
  pts = Map[(Point[{Re[#1], Im[#1]}] &) /@ #1 &, zeros, 2];
  sizedpts = Transpose[
```
$$\{\text{Table}[\{\text{Hue}[\frac{i}{\text{len}}], \text{PointSize}[0.015 + i\ 0.008]\}, \{i, \text{len}\}],$$
```
      pts}]; Show[Graphics[{Flatten/@sizedpts,
```
$$\{\text{RGBColor}[0, 0, 1], \text{Table}[\text{Text}[\text{InputForm}[\frac{i}{n}],$$
$$1.2\ \{\text{Cos}[\frac{i\ 2\ \pi}{n}], \text{Sin}[\frac{i\ 2\ \pi}{n}]\}]\}, \{i, 0, n - 1\}]\}\}],$$
```
  Axes -> True, AspectRatio -> Automatic,
  PlotRange -> {{-width, width}, {-width, width}},
  PlotLabel -> "n = " <> ToString[n], opts]]
UnityFactors[n_Integer?Positive] :=
  First[Transpose[Drop[FactorList[x^n - 1], 1]]]
```

Also, we remind ourselves how the zeros of these polynomials are distributed.

```
Do[RootsOfUnityZeros[k, fractions], {k, 2, 11}]
```

Our goal is to consider what is called a *cyclotomic polynomial*. For each positive integer n, there is a polynomial $\Phi_n(x)$ called the nth cyclotomic polynomial. Since cyclotomic polynomials are quite important, there is a built-in function for them in *Mathematica*. Here are the fifth and sixth cyclotomic polynomials.

```
{Cyclotomic[5, x], Cyclotomic[6, x]}
(* x simply indicates with what
   indeterminate to express the polynomial *)
```

The cyclotomic polynomials are related to the roots of unity. We were reminded earlier that these numbers are all on the unit circle. (Note that *cyclo* refers to a circle and *tomic* indicates a cutting.) The nth roots of unity take the form $u(n, k) = \cos(2\pi k / n) + i \sin(2\pi k / n) = e^{2\pi i k/n}$, where $k = 0, 1, 2, \ldots, n - 1$.

If you have ever browsed through *The Mathematica Book*, you may have come across the description of the cyclotomic polynomials. The description is that the nth cyclotomic polynomial in x is the product of all linear factors of the form $x - u(n, k)$ for which k is relatively prime to n (i.e., GCD[n, k] = 1). These are the factors corresponding to the so-called "primitive *n*th roots of unity" that are labeled red in the graphs.

$$\mathtt{u[n\_,\ k\_]\ :=\ Cos}\Big[\frac{\mathbf{2\ \pi\ k}}{\mathbf{n}}\Big]\mathtt{+\ I\ Sin}\Big[\frac{\mathbf{2\ \pi\ k}}{\mathbf{n}}\Big]$$

For example, there are two primitive sixth roots of unity, $u(6, 1)$ and $u(6, 5)$.

```
{Expand[(x - u[6, 1]) (x - u[6, 5])], Cyclotomic[6, x]}
```

Q1. a. Determine $\Phi_8(x)$, the eighth cyclotomic polynomial in x, by expanding the proper linear factors.

  b. Without generating $\Phi_{128}(x)$ itself, determine its degree.

Our goal at this point is to develop an alternate way of building the cyclotomic polynomials without using the complex numbers. We will be asking you to look at many patterns to see if you can come up with this alternate definition. First let's compare $x^n - 1$ and the *n*th cyclotomic polynomial in tandem. Note that the output consists of *n*, then the factorization of $x^n - 1$, followed by $\Phi_n(x)$.

```
TableForm[Table[{{n, Factor[xⁿ - 1], Cyclotomic[n, x],
        "------------------------"}}, {n, 1, 11}]]
```

Q2. At least one observation regarding the relationship between $\Phi_n(x)$ and the factorization of $x^n - 1$ should be obvious. What is it?

Q3. Suppose *n* is a prime number. How can you formulate $\Phi_n(x)$ in terms of $x^n - 1$?

In question 3, you should have found a method of formulating $\Phi_n(x)$ in terms of $x^n - 1$, when *n* is prime. What if *n* is composite? (Note: We simply *define* $\Phi_1(x)$ to be $x - 1$, just as we *define* 0! to be 1.) Let's extend our table and make some more comparisons.

```
TableForm[Table[{{n, Factor[xⁿ - 1], Cyclotomic[n, x],
        "------------------------"}}, {n, 12, 30}]]
```

It appears that $\Phi_n(x)$ is a factor of $x^n - 1$. Recall from the last lab that `UnityFactors` gave just the list of factors of the polynomial $x^n - 1$. Let's test some random indices and see if $\Phi_n(x)$ really is a factor of $x^n - 1$. (We use `MemberQ` to see if `Cyclotomic[n, x]` belongs to the list of factors of $x^n - 1$.)

```
n = Random[Integer, {31, 120}]
MemberQ[UnityFactors[n], Cyclotomic[n, x]]
```

We should perhaps try this a number of times.

```
Table[{n = Random[Integer, {31, 120}],
    MemberQ[UnityFactors[n], Cyclotomic[n, x]]}, {20}]
```

One more time with other values of *n*:

```
Table[{n = Random[Integer, {121, 180}],
    MemberQ[UnityFactors[n], Cyclotomic[n, x]]}, {15}]
```

This is not a proof, but the evidence seems to indicate that $\Phi_n(x)$ divides $x^n - 1$. Let's assume that this is the case and that there exists some function (depending on *n*), $g_n(x)$, such that $\Phi_n(x) g_n(x) = x^n - 1$. We already know that $g_n(x) = x - 1$ when *n* is prime. What about when *n* is composite?

> **Q4**. Although premature for any definitive solution, do you have any thoughts on the nature of $g_n(x)$ for composite *n*?

9.4 Search for $g_n(x)$

Since we are assuming that we have $\Phi_n(x) g_n(x) = x^n - 1$, perhaps we should solve for $g_n(x)$ and explore the quotient of the two pieces we can calculate. (Note, of course, that *Mathematica* knows $\Phi_n(x)$, but we do not, yet.) Let's consider this with *n* = 6.

$$\text{Simplify}\left[\frac{x^6 - 1}{\text{Cyclotomic}[6, x]}\right]$$

We want to factor this so that we can know what the pieces look like.

```
Factor[%]
```

We should define and then try out a function to do all these steps at once.

$$\text{gFunction}[n\_] := \text{Factor}\left[\text{Simplify}\left[\frac{x^n - 1}{\text{Cyclotomic}[n, x]}\right]\right]$$

```
TableForm[Table[{n, gFunction[n]}, {n, 1, 14}]]
```

This may not be so revealing. Perhaps we should also add the *n*th cyclotomic polynomial to the table. Study the following results and look for patterns.

```
TableForm[Table[{{n, gFunction[n], Cyclotomic[n, x],
    "-----------------------------"}}, {n, 1, 14}]]
```

Extend the table, if you wish, by looking at six more cases (in conjunction with the previous ones).

```
TableForm[Table[{{n, gFunction[n],
        Cyclotomic[n, x], "------------------"}}, {n, 15, 20}]]
```

Q5. What is the nature of $g_n(x)$ for composite n? For a hint, given an index n, consider the (proper) divisors d of n and the corresponding collection of $\Phi_d(x)$. Can they be used to construct $g_n(x)$? Look carefully at the first 15 examples and apply the hint and see what you can come up with.

Q6. Now we want to give a recursive definition for $\Phi_n(x)$. In other words, you should be able to write $\Phi_n(x) = \text{RHS}$, where the RHS depends on $\Phi_k(x)$ for one or more values of k that are less than n.

▥ 9.5 Some properties of $\Phi_n(x)$

We now consider some properties of the cyclotomic polynomials. First we remind ourselves of some of these polynomials and precede each with its degree.

```
TableForm[Table[{{n, Exponent[p = Cyclotomic[n, x], x],
        p, "----------------------"}}, {n, 2, 15}]]
```

■ 9.5.1 Each cyclotomic polynomial is monic

Q7. Based on what you know about the function $\Phi_n(x)$, why is the nth cyclotomic polynomial monic? (Recall that *monic* means that the leading coefficient, the one associated with the term of highest degree, has the value of 1.)

■ 9.5.2 Degrees of cyclotomic polynomials

Q8. Compare the index n of the function $\Phi_n(x)$ with its degree. You should be able to see an interesting relationship and be able to express the degree of $\Phi_n(x)$ as a "nice" function of n. What is it?

■ 9.5.3 Irreducibility of the cyclotomic polynomial

In Ring Lab 7 we explored the notion of irreducibility. Which, if any, of the cyclotomic polynomials are irreducible over the rationals? Try the following to factor a random cyclotomic polynomial.

```
TableForm[Table[{n = Random[Integer, {2, 50}],
    Factor[Cyclotomic[n, x]]}, {10}]]
```

> **Q9**. Repeat and/or modify the range of the random index until you are ready to make a conjecture regarding when (i.e., for what *n*) $\Phi_n(x)$ is irreducible over the rationals.

Let's choose a prime, say 29, and consider $\Phi_n(x)$ for $n = 29$. We call the polynomial *h* for this example.

```
Clear[h]
h[x_] = Cyclotomic[29, x]
```

Now consider $h(x + 1)$.

```
h[x + 1]
```

Of course *we* knew that we wanted this polynomial expanded, but we need to tell *Mathematica* explicitly to do so.

```
Expand[%]
```

Now examine the coefficients more closely.

```
coeffs = CoefficientList[%, x]
```

Since $h(x)$ is monic, so is $h(x + 1)$. Therefore the leading coefficient is 1. Is there any significance to the rest of the coefficients? Suppose we ignore the last (i.e., leading) coefficient and ask for the greatest common divisor of the balance.

```
Apply[GCD, Drop[coeffs, -1]]
```

This says that 29 divides each of the coefficients except the leading one. Note further that 29^2 does not divide the constant term, 29. We can now use Eisenstein's Criterion to conclude that the polynomial $h(x + 1)$ is irreducible over the rationals. Consequently, the polynomial $h(x)$ is also irreducible over the rationals. This illustrates that a cyclotomic polynomial with prime index is irreducible. It is also true for composite indices, but the proof requires some understanding of extension fields. (See Ring Lab 11 for an introduction to quadratic extension fields.)

▪ 9.5.4 Graphs of roots of the cyclotomic polynomial

As we saw in section 9.5.3, the cyclotomic polynomial is irreducible over the rationals. It is, of course, reducible over the complex numbers. In fact, we saw this in Ring Lab 8. As was pointed out earlier, the zeros of the nth cyclotomic polynomial are a subset of the whole set of nth roots of unity.

Let's remind ourselves what these roots look like. The roots of the cyclotomic polynomial are the red ones plotted in the complex plane. (Note: You may wish to delete all previous graphics cells before continuing, due to memory demand.)

```
Do[RootsOfUnityZeros[k, fractions], {k, 5, 20, 3}]
```

> **Q10.** The zeros of the cyclotomic polynomials (the red dots) are labeled with fractions of the form k/n for certain values of k. What are the k values? How do they relate to n? You may wish to modify the preceding cell to see more examples.

> **Q11.** In each of the graphs, the full collection of zeros forms a group under multiplication. To what group is this group naturally isomorphic? What is the relationship between the zeros of the cyclotomic polynomials and this group?

As a noted earlier, the roots of the nth cyclotomic polynomial are often called the *primitive nth roots of unity.* Can you imagine why?

■ 9.5.5 $\Phi_n(x)$ and $\Phi_{2n}(x)$ for odd n

Here we wish to look for a relationship, if any, between $\Phi_n(x)$ and $\Phi_{2n}(x)$. First we look at a table listing some pairs.

```
TableForm[Table[{{k, Cyclotomic[k, x], 2 k, Cyclotomic[2 k, x],
    "---------------------------", " "}}, {k, 3, 13, 2}]]
```

You may wish to expand this table to encompass other values. A plot of an example might also be revealing. You may wish to modify your value of n to see other plots. Φ_n is plotted in red and Φ_{2n} in blue.

```
n = 9;
Plot[Evaluate[{Cyclotomic[n, x], Cyclotomic[2 n, x]}], {x,
    -2, 2}, PlotStyle → {RGBColor[1, 0, 0], RGBColor[0, 0, 1]}];
```

Note that this plot reveals that there are no real zeros (at least in the domain shown). We can still, however, compare the complex zeros. You may need to enlarge the output below to view it.

```
n = 11;
gr1 = RootsOfUnityZeros[
    11, fractions, DisplayFunction → Identity];
gr2 = RootsOfUnityZeros[22, fractions,
    DisplayFunction → Identity];
Show[GraphicsArray[{gr1, gr2}]];
```

Q12. What conclusion(s), if any, can you make about how $\Phi_n(x)$ and $\Phi_{2n}(x)$ are related?

▪ 9.5.6 $\Phi_p(x)$ and $\Phi_{p^k}(x)$ for prime p

Suppose p is a prime and k is an integer with $k \geq 1$. We wish to compare $\Phi_p(x)$ and $\Phi_{p^k}(x)$ for various primes p. We then make a table consisting of a 5-tuple $\{p, k, p^k, \Phi_p(x), \Phi_{p^k}(x)\}$, where p runs through the first six primes and k is a random integer between 2 and 4. Study the results.

```
TableForm[Table[k = Random[Integer, {2, 4}];
    {{p = Prime[n], k, p^k, Cyclotomic[p, x], Cyclotomic[p^k, x],
    "---------------------"}}, {n, 1, 6}]]
```

Q13. Since k is generated randomly, you may wish to evaluate the cell again if it is not clear what relationship is being expressed. Determine how you can express $\Phi_{p^k}(x)$ in terms of the pth cyclotomic polynomial, if possible.

▪ 9.5.7 $\Phi_n(x)$ and $\Phi_m(x)$, where m and n have similar prime decompositions

Suppose m is an integer whose prime-power decomposition is given as $p_1 p_2 \ldots p_k$ and n is an integer whose prime-power decomposition is given as $p_1^{e_1} p_2^{e_2} \ldots p_k^{e_k}$ where $e_1, e_2, \ldots e_k$ are the positive integral exponents belonging to primes p_1 through p_k. We wish to compare $\Phi_m(x)$ and $\Phi_n(x)$.

For investigative purposes, suppose $m = 2 * 3 * 5$ and $n = 2^a \, 3^b \, 5^c$ for some powers a, b, and c. We make a table consisting of a 4-tuple $\{\{a, b, c\}, n, \Phi_m(x), \Phi_n(x)\}$, where a, b, c are random integers between 2 and 4. Study the results.

```
m = 2 3 5;
TableForm[
  Table[{a, b, c} = Table[Random[Integer, {2, 4}], {3}];
    n = 2^a 3^b 5^c; {{{{"a = " <> ToString[a],
        "b = " <> ToString[b], "c = " <> ToString[c]}},
      "2^" <> ToString[a] <> " 3^" <> ToString[b] <> " 5^" <>
      ToString[c] <> " = " <> ToString[n], Cyclotomic[m, x],
      Cyclotomic[n, x], "--------------------"}}, {6}]]
```

> **Q14.** Since a, b, and c are generated randomly, you may wish to evaluate the cell again if it is not clear what relationship is being expressed. What relationship is there between $\Phi_n(x)$ and $\Phi_m(x)$ for the n as obtained above and $m = 30$? Generalize for arbitrary m and n satisfying the hypothesis given at the beginning of this section.

■ 9.5.8 $\Phi_n(1)$

We have seen many cyclotomic polynomials $\Phi_n(x)$ and even seen a few plotted, but we haven't investigated viewing them as polynomial functions and evaluating them at specific points. Here we consider $\Phi_n(1)$ for various values of n. Consider the table of values of $\Phi_n(1)$ for n generated randomly between 1 and 150. Study the results.

```
TableForm[Table[n = Random[Integer, {1, 150}];
  {n, Cyclotomic[n, x] /. x → 1}, {20}],
 TableHeadings → {None, {"n", "\!\(\Φ\_n\) (1)\n"}},
 TableSpacing → {0.5, 3}]
```

> **Q15.** What conclusions can you make? (Evaluate again, if necessary.) State them as conjectures.

Evaluate the following cell.

```
TableForm[Table[{n, Cyclotomic[n, x] /. x → 1}, {n, 1, 25}],
 TableHeadings → {None, {"n", "Φ_n (1)\n"}},
 TableSpacing → {0.5, 3}]
```

> **Q16.** Does the result confirm or deny your previous conjecture? Reformulate, if necessary.

■ 9.5.9 How $\Phi_n(x)$ is related to the Moebius function, yielding a nonrecursive definition

The Moebius function (often denoted with the lower-case Greek letter μ) is a number-theoretic function defined as follows.

$$\mu(n) = 1, \text{ if } n = 1;$$
$$\mu(n) = (-1)^k, \text{ if } n \text{ is a product of } k \text{ distinct primes};$$
$$\mu(n) = 0, \text{ if } p^2 \text{ divides } n \text{ for some prime } p.$$

> **Q17.** For $n = 1, 2, 3, 4, 6$, and 12, calculate $\mu(n)$. (Note that these are all divisors of 12.)

Since the Moebius function is important for various parts of mathematics, it is a built-in function in *Mathematica*. The following table can be used to confirm your answers to question 17.

```
Map[MoebiusMu, {1, 2, 3, 4, 6, 12}]
```

Applying the `MoebiusMu` function to the divisors of an integer proves itself useful here. In *Mathematica*, we can obtain the list of divisors of an integer using the `Divisors` function.

```
Divisors[12]
```

With $n = 12$, what we now want to do is raise the polynomial $x^{12/d} - 1$ to the power $\mu(d)$ for each divisor d. Here is one way of doing it.

```
n = 12;
polys = ((x ^ (n / #1) - 1) ^ MoebiusMu[#1] & ) /@ Divisors[n]
```

Now we want to multiply these polynomials.

```
Apply[Times, polys]
```

```
Simplify[%]
```

```
Cyclotomic[n, x]
```

> **Q18.** In the previous few input cells, replace 12 with 20 and evaluate each cell. What is your result? Any conjectures? Try it again with another value or two.

> **Q19.** In question 6 you were asked to determine a recursive definition for $\Phi_n(x)$. Now give a nonrecursive definition.

■ 9.5.10 Disclaimer/warning

You are not finished! Nearly all the conclusions you have reached in this lab were based on looking at patterns and formulating conjectures. Your next step should be to prove at least some of your conjectures. Your instructor will no doubt assign at least some of these proofs to you.

Ring Lab 10

Quotient Rings of Polynomials

10.1 Prerequisites

To complete this lab, you should be familiar with the ring of polynomials over a field, the division property for polynomials over a field, and the definitions of homomorphism, kernel, and ideal. Finally, you should be familiar with the First Isomorphism Theorem for ring homomorphisms (Ring Lab 5).

10.2 Goals

Extensions of finite fields are generally motivated by the need to solve polynomial equations. These extensions are actually quite concrete in the sense that they arise from quotient rings, where the ideal from which cosets are formed is the kernel of a familiar homomorphism. In this case, the homomorphism is the remainder function for division by a fixed polynomial (the modulus). In this lab we introduce quotient rings, and in Ring Lab 11 we explore how they are used to construct roots of polynomials.

10.3 Polynomials over a field

First we read in the *Mathematica* code we need for this lab.

```
Needs["AbstractAlgebra`Master`"];
SwitchStructureTo[Ring];
```

Given any field, *F*, we ultimately want to be able to solve any polynomial equation over it. For example, *F* might be the integers modulo 3.

```
F = Z[3]
```

It is clear, by simple hand-evaluation of $m(x) = x^2 + x + 2$, that none of the elements of *F* are roots of $m(x)$. This can also be done using just the built-in *Mathematica* functions. First define the polynomial function.

```
Clear[m1, x]
m1[x_] := x^2 + x + 2
(* we call it m1 here because we use m below *)
```

Then evaluate the function at each element in *F*.

```
{m1[0], m1[1], m1[2]}
```

For larger fields, it may be easier to do this another way.

```
Map[m1, Elements[F]]
```

Reducing mod 3 (recall what ring we are in), we get

```
Mod[%, 3]
```

Q1. Why does this last output show that $m(x)$ does not have any roots in *F*?

We can also use the functions built into the packages that were read in. First define the polynomial over *F* using the `Poly` function.

```
m = Poly[F, x^2 + x + 2]
```

Then use the `PolynomialEvaluation` function on the domain elements.

```
Map[PolynomialEvaluation[m, #] &, Elements[F]]
```

Q2. Does the polynomial $m(x) - 1$ (which is really $m(x) + 2$) have any roots in *F*?

We conclude that the field *F* does *not* contain the roots of $m(x)$. In this lab we concentrate on the construction of a field that contains the roots of $m(x)$. In Ring Lab 11 we discuss the factorization of $m(x)$ in this new field, along with other issues involving finite field extensions.

We use *V* to denote the smallest extension of *F* that contains all the roots of $m(x)$. The first step in constructing *V* is to consider the whole ring of polynomials over *F*.

```
P = PolynomialsOver[F]
```

> **Q3.** In the packages these labs are based on, a `Ringoid` has three arguments: the list of elements, the addition function, and the multiplication function. Why isn't *P* a `Ringoid` in this sense? Is *P* a ring?

▣ 10.4 A homomorphism based on `PolynomialRemainder`

The homomorphism θ that we consider here depends on *m*. If the definition of *m* is altered, a different homomorphism and, at least superficially, a different extension is constructed. Generally, the smallest degree we consider for the polynomial function *m* is two, as in this example.

The function θ takes any polynomial in *P* and returns the polynomial that is the remainder on division by *m* (our polynomial $x^2 + x + 2$).

```
θ = PolynomialRemainder[P, #, m] &
```

You may wish to remind yourself how the function `PolynomialRemainder` works.

```
? PolynomialRemainder
```

For a quick test of θ, let's apply it to *m* and to $m + 1 = x^2 + x$. (Think about the answer before evaluating.)

```
θ[m]
θ[Poly[F, x² + x]]
```

> **Q4.** a. What happens when we apply θ to $m * x + 2 = x^3 + x^2 + 2x + 2$? Think about it and then calculate it.
>
> b. What happens when we apply θ to $m * (x + 2) + x$? Think about it and then calculate it.

The division property of polynomials over a field assures us that the remainder (the value of our function θ) will have its degree less than the divisor's degree. So in our examples, the remainders are all linear polynomials. To illustrate with a computation, we take *some* of the elements of *P*, the ones of degree three or less, and map the `PolynomialRemainder` function θ over them. (Note: If your computer is relatively slow, you may want to change the following expression to `SomePolynomials = PolynomialsUpToDegreeN[F, 2]`, generating only polynomials of degree two or less.)

```
SomePolynomials = PolynomialsUpToDegreeN[F, 3]
```

Next we map θ over this list. Duplicate images are removed and we can verify by examination that all nine linear polynomials over $\mathbb{Z}_3$ are in this list.

```
θRange = Map[θ, SomePolynomials] // Union
```

> **Q5.** Why are there nine possible remainders? If $m(x)$ were a cubic (instead of a quadratic) polynomial, how many different elements would be in the range of θ? (In other words, how many different remainders would there be if $m(x)$ were a cubic?)

Let's review what we have done. Since the set of all polynomials over F (which we call P) is infinite in size, we cannot apply our function θ to the whole of P. So we took a finite subset of P (SomePolynomials—those of degree less than four) and applied θ to this subset to get a sense of how θ worked. The result over this set is the set we called θRange. We now rename this set linPolys, which is the set of linear polynomials over F.

```
linPolys = θRange
```

By considering θ as a function from P into the linear polynomials (linPolys), we want to establish θ as a homomorphism. For θ to be a homomorphism, more fundamental than showing that the operations are preserved, is the need to make sure that both the domain and codomain are rings. The domain is clearly a ring—it is the ring of polynomials over $\mathbb{Z}_3$. It is not so clear that the set of remainders, linPolys, is also a ring. How should addition and multiplication be defined?

> **Q6.** Consider the set of remainders from division by $m(x)$, linPolys. Why can't the usual polynomial multiplication be used on linPolys to form a ring?

The First Homomorphism Theorem gives us a clue how to operate in linPolys. This is because the range is isomorphic to P/K, where K is the kernel of θ. Recall that given an isomorphism $f : R \rightarrow S$, the inverse map (which exists, since f is a bijection), $f^{-1} : S \rightarrow R$, is also an isomorphism. We want to consider this inverse isomorphism, the function that maps an element g in linPolys to the coset $g + K$ in P/K.

Here are the elements of K that come from the list SomePolynomials. Note that they are obtained by selecting those whose image under θ is the zero of P.

```
partialK = Select[SomePolynomials, θ[#] == Zero[P] &]
```

Q7. What do these polynomials all have in common? For a hint, take one of these polynomials over *F* (remember, they were all formed with `Poly`) and try the `PolynomialDivision` function on it with *m*(*x*).

```
? PolynomialDivision

p = partialK[[1]]
PolynomialDivision[p, m]
```

Let's take a pair of elements from *P* and form two cosets from them. We select polynomials that are also in `linPolys` so that the sum and product of their cosets give us a hint at how addition and multiplication are defined in `linPolys`.

```
p = Poly[F, x + 1]
q = Poly[F, 2 x + 1]
```

Since *K* is infinite (you should ask yourself why this is true), we cannot form the full coset *p* + *K*, but we can form a portion of *p* + *K* by using `partialK`.

```
pCoset = Map[Addition[P][p, #] &, partialK]
```

We do the same for *q* + *K*.

```
qCoset = Map[Addition[P][q, #] &, partialK]
```

We add the two cosets by adding arbitrary representatives of `pcoset` and `qcoset`. There are many possibilities here, but the easiest pair to add is *p* and *q*.

```
sum = Addition[P][p, q]
θ[sum]
```

Now we pick random representatives and go through the same process.

```
Print["Rep. from pCoset: ", pRep = RandomElement[pCoset]]
Print["Rep. from qCoset: ", qRep = RandomElement[qCoset]]
Print["Sum in P: ", sum = Addition[P][pRep, qRep]]
Print["Image of this sum under θ: ", θ[sum]]
```

Q8. Consider three more pairs of representatives from `pCoset` and from `qCoset` by evaluating the cell three more times. What do you observe?

Q9. By changing `Addition` to `Multiplication`, multiply several random pairs of representatives as in the previous exercise. (See below.) What do you observe? Do the same using *p* and *q* as representatives.

```
Print["Rep. from pCoset: ", pRep = RandomElement[pCoset]]
Print["Rep. from qCoset: ", qRep = RandomElement[qCoset]]
Print["Product in P: ", prod = Multiplication[P][pRep, qRep]]
Print["Image of this product under θ: ", θ[prod]]
```

It can be proven that the observations that you (hopefully) made are not coincidence. It is reasonable to consider that the sum of *p* and *q* is 2 and their product is an "obvious value." The general definitions of the operations on `linPolys` are based on these observations.

10.5 Defining a quotient ring of polynomials

For any polynomial $m = \text{Poly}[R, a_n x^n + a_{n-1} x^{n-1} + \cdots + a_2 x^2 + a_1 x + a_0]$, with each a_i in some ring *R* with unity, the expression `QuotientRing[R, m]` generates a quotient ring of *R* with *modulus m*. The elements of the ring are the polynomials of degree less than the degree of *m*. With *m* and *F* defined above, we define *V* to be the quotient ring of *F* mod *m*.

```
V = QuotientRing[F, m]
```

10.5.1 Addition

Addition in *V* is simply addition of polynomials.

```
{p, q} = {Poly[F, 2 x + 1], Poly[F, 2 x]}
```

```
Addition[V][p, q]
```

Let's try this addition with a few random pairs.

```
{p, q} = RandomElements[V, 2]
Addition[V][p, q]
```

Q10. Why is *V* closed under the usual addition of polynomials?

10.5.2 Multiplication

Multiplication in *V* is much like multiplication on the integers mod *n*. To multiply polynomials *p* and *q*, you perform the usual multiplication (as in *P*) and then divide by the modulus, retaining the remainder for the value of the product. In other words, you apply θ to the product. For example,

```
p = Poly[F, x + 1]
q = Poly[F, x + 2]
r = Multiplication[P][p, q]
```

```
θ[r]
```

This result, of course, can also be achieved as follows.

```
Multiplication[V][p, q]
```

Remember, $\theta(r)$ is computed by dividing the product, r, by m and then returning the remainder. The complete division result, quotient and remainder, is given by `PolynomialDivision`. (However, only the remainder is used for the product in V.)

```
PolynomialDivision[P, r, m, Mode → Textual]
```

Q11. Compute the following in V:
   a. $(2x + 1) + (x + 2)$
   b. $x\,x = x^2$
   c. $x\,x\,x\,x = x^4$
   d. $x\,x\,x\,x\,x\,x\,x\,x = x^8$
   e. multiplicative inverse of $2x + 1$

▥ 10.6 The `PolynomialRemainder` function θ is indeed a homomorphism

Now that we have established the operations on V, we can verify that θ is a homomorphism. Here we verify only that the homomorphism properties are true for a pair of random third degree elements in P.

```
{p, q} = Table[RandomElement[P, 3], {2}]
```

First the addition property,

```
θ[Addition[P][p, q]] == Addition[V][θ[p], θ[q]]
```

then the multiplication property,

```
θ[Multiplication[P][p, q]] == Multiplication[V][θ[p], θ[q]]
```

Q12. Repeat these steps several times. Are you convinced that θ is a homomorphism? Why?

Q13. Consider the situation where the multiplication on *V* is defined to be `Multiplication[V][p, q] = 0` for all *p* and *q* (with no change to the addition). Is θ still a homomorphism?

▦ 10.7 Is *V* a field?

You probably have seen a theorem that states that the range or image of a (commutative) ring under a homomorphism is a (commutative) ring. Therefore, *V* is a commutative ring. To show that it is also a field, we need to verify that it has a unity and that every nonzero element has an inverse.

Q14. Does *V* have a unity? If so, what is it and why is it the unity?

Q15. While we're at it, does *V* have a zero? If so, what is it and why is it the zero?

Therefore, we know that *V* has most of the field properties. The multiplication table for *V* can answer the question regarding multiplicative inverses.

```
MultiplicationTable[V, Mode → Visual];
```

Q16. Is *V* a field? On what do you base your conclusion?

▦ 10.8 Is *V* what we claimed?

Recall that at the outset we claimed that we would find a field (and call it *V*) that would be the smallest field containing all the roots of $m = x^2 + x + 2$. We have defined *V*, but we have not yet shown that it contains any roots of *m*.

It is also true that we can do addition and multiplication in *V*. Therefore, we should be able to take any element in *V*, square it, add this to itself, and then add the constant polynomial 2 to this result. In other words, we should be able to do arithmetic in this field.

Next we define a function *f* that does what we indicated in the previous paragraph.

```
Clear[f, y]
f[y_] :=
    Multiplication[V][y, y] + y + Poly[F, 2, Indeterminate → x]
```

Q17. What connection is there between the polynomial function $m = x^2 + x + 2$ defined at the outset and f? Justify your answer.

Now we want to evaluate f at each of the elements in V. In other words, we want to Map f over V. We form pairs of the form $\{x, f(x)\}$ so that we can see the domain elements that create the range elements.

```
Map[{#, f[#]}&, Elements[V]]
```

Note that two elements, Poly[F, x] and Poly[F, 2x + 2], yield zero.

```
root1 = Poly[F, x]
root2 = Poly[F, 2 x + 2]

f[root1]
f[root2]
```

Q18. Does the V we created have all the roots of m? Could m have more than two roots over V? (Remember that V is an integral domain, since it is a field.) Is it possible that a smaller field contains these roots? Justify your answers.

Ring Lab 11

Quadratic Field Extensions

▣ 11.1 Prerequisites

To complete this lab, you should be familiar with the construction of quotient rings of the ring of polynomials over a field F. You should also be familiar with irreducible polynomials over a field. This lab does not presume any other prior knowledge of field extensions. Doing Ring Lab 10 first would be helpful, but it is not necessary.

▣ 11.2 Goals

The goal of this lab is to provide some experience in working with quadratic field extensions to make the general study of finite field extensions easier to understand.

▣ 11.3 The general problem

Not every polynomial with real coefficients has real roots; the simplest example is $p(x) = x^2 + 1$. To find the roots of $p(x)$, we extend the real numbers to the complex numbers. That is, we construct a field that contains the real numbers but also includes some new elements that are roots of $p(x)$. This process of extending a field can be used to find roots of any nonconstant polynomial over any field. When the polynomial is quadratic, the smallest extension containing a root of this polynomial takes on a simple form.

The general problem we will consider is to find the roots of a quadratic polynomial $p(x) = a\,x^2 + b\,x + c$, where a, b, and c are elements of a field F, with $a \neq 0$. Recall that t is a root of $p(x)$ if $p(t) = 0$.

> **Q1.** Prove that if $p(x)$ has one root in F, then it also has a second root (possibly identical to the first) in F.

The case when $p(x)$ has its two roots in F is not very interesting. Therefore, consider the case where $p(x)$ has no roots in F; that is, $p(x)$ is irreducible over F. Suppose we have a "larger" field that contains F and also contains a root, z, of $p(x) = a\,x^2 + b\,x + c$.

> **Q2.** Prove that all positive powers of z can be written in the form $s\,z + t$, where s and t are elements of F. Hint: Start with z^2 and then proceed by induction.

The result expressed in question 2 tells us exactly what the elements of the smallest extension field look like. A few loose ends may need to be tied together, but you should be able to see that this new field, a quadratic extension of F, is

$$F[z] = \{s\,z + t \mid s, t \in F\},$$

where

$$(s\,z + t) + (s'\,z + t') = (s + s')\,z + (t + t')$$

and

$$(s\,z + t)\,(s'\,z + t') = r\,z + u$$

for some r and u. How do we get the result $r\,z + u$? This calculation is described in Ring Lab 10, so here we just review the process. First think of $(s\,z + t)\,(s'\,z + t')$ simply as a product of two polynomials. We expand this product into a quadratic and then divide by $p(x)$. The remainder from this division is our product, $r\,z + u$.

As an example, suppose that F is the field of rational numbers, $\mathbb{Q}$, and $p(x) = x^2 - x - 1$. If W is a root of $p(x)$, the product of $2\,W + 5$ and $3\,W - 7$ in $\mathbb{Q}[W]$ is $7\,W - 29$.

```
Clear[W]
PolynomialRemainder[(2 W + 5) (3 W - 7), W² - W - 1, W]
```

> **Q3.** In the quadratic extension of the rational numbers that contains W, determine the values of W^n, for $n = 2, 3, 4, 5, 6, 7$. Can you identify a pattern? An example of how to calculate the value of W^3 follows.

```
PolynomialRemainder[W³, W² - W - 1, W]
```

Alternatively, we can ask *Mathematica* to find the real roots of $p(x)$.

```
Clear[x]
Solve[x² - x - 1 == 0, x]
```

We select the first root provided and call it *w*. Are Q[*w*] and Q[*W*] the same? Not really, but they are isomorphic.

```
w = x /. First[Solve[x² - x - 1 == 0, x]]
```

Now we can verify that the product $(2w + 5)(3w - 7)$ is consistent with the product computed involving *W*'s. To compare $(2w + 5)(3w - 7)$ and $-29 + 7w$, we need to expand both sides.

```
Expand[(2 w + 5) (3 w - 7)] == Expand[-29 + 7 w]
```

Q4. The isomorphism hinted at would map $a + bW$ to $a + bw$, where *w* is the first root given to us by `Solve`. Could we have used the second root? If *W* maps into the second root, what would map into *w*?

▓ 11.4 An extension of $\mathbb{Z}_3$ using *Mathematica*

To continue we read in the *Mathematica* code needed for the rest of the lab.

```
Needs["AbstractAlgebra`Master`"];
SwitchStructureTo[Ring];
```

Consider the polynomial $p(x) = x^2 + x + 2$ over the integers modulo 3, $\mathbb{Z}_3$.

Q5. Verify that none of the elements in $\mathbb{Z}_3$, {0, 1, 2}, are roots of $p(x)$.

The extension of $\mathbb{Z}_3$ that contains a root *z* of $p(x)$ can be generated with `Quo-tientRing` (see Ring Lab 10). We shorten the name of $\mathbb{Z}_3[z]$ to *V*.

```
p = Poly[Z[3], x² + x + 2]
V = QuotientRing[Z[3], p]
```

When you work with *V*, be aware of the form of its elements. They all appear as ordinary polynomial expressions, but in fact their internal form is more complicated. We have to use the `Poly` function to create them. For example, consider the third element in the list, appearing as $2x$.

```
third = Elements[V][[3]]
```

Is this simply $2x$? In other words, can we just use $2x$ in place of it?

```
2 x === third
```

No. They have different internal forms. For instance, $2x$ to *Mathematica* is

```
FullForm[2 x]
```

which is just the product of 2 and *x*, as we think of it in ordinary usage. The *Mathematica* internal form for the third element in *V*, 2*x*, is quite different.

```
FullForm[third]
```

This form is significantly more complicated. The reason is that it represents a formal polynomial whose underlying ring is embedded in the structure, as well as several other pieces of data. (Purely optional: This data structure has the head `AbstractAlgebra`RingExtensions`Private`poly` with two arguments. The first is a list with four arguments and the second is the list of coefficients, starting with the constant term, then linear, and so on. The arguments in the first list consist of the underlying ring (in internal form), whether we should view the polynomial from right to left or left to right, the indeterminate that should be used, and finally whether the coefficients are numeric or not. Thankfully, the user does not need to worry about these details.)

Therefore, to use the third element, we either pick it off the list of elements, as we did earlier or we create it anew with the `Poly` function. In this case, this polynomial can be obtained by

```
newthird = Poly[Z[3], 2 x]
```

It is important that all polynomials are created in this fashion when they are over a ring besides the integers, rationals, reals, or complex numbers. Note that this is now identical to `third`.

```
newthird === third
```

What is more important is the *mathematical* concept of what this 2*x* represents. You can just think of it as an abstract element in a new ring, or you can think of it as the coset representative of the coset $2x + \langle x^2 + x + 2 \rangle$, since this element really comes from the quotient ring $\mathbb{Z}_3[x]/\langle x^2 + x + 2 \rangle$.

What have we done so far? Let's recap for a moment. We have a polynomial $p = x^2 + x + 2$ that does not have any roots in $\mathbb{Z}_3$. We formed the quotient ring $\mathbb{Z}_3[x]/\langle x^2 + x + 2 \rangle$ and called it *V*. We also know (from Ring Lab 10) that this *V* contains a root (call it *z*) of *p*, as well as containing $\mathbb{Z}_3$ itself. (How does it contain $\mathbb{Z}_3$?) We therefore also call *V* by the name $\mathbb{Z}_3[z]$. To verify that *z* (whatever it turns out to be) is a root of *p*, we need to represent *p*'s coefficients as elements of *V*. In other words, we can think of a new *p*, call it *pV*, whose coefficients 1, 1, and 2 are constant polynomials in *V*. We also want the multiplication $x*x$ that yields x^2 to be taking place in *V*. Hence, we have a new polynomial function as follows.

```
pV[y_] :=
    Multiplication[V][y, y] + y + Poly[Z[3], 2, Indeterminate → x]
```

The constant term 2 in p is reflected by using `Poly[Z[3], 2, Indetermi-nate → x]`. The use of `Indeterminate → x` is just for consistency. Since constants do not use an indeterminate, adding this option indicates that we are viewing this polynomial in the indeterminate x.

We have stated that z is a root of p (or of this new version of p, pV), but what is z? We know we have a root if, when we evaluate a polynomial, we get 0. Since z is supposed to be found in V (and we already know that one third of the elements in V can be dismissed as candidates—why?), we should perhaps map our new polynomial pV over our elements in V and look for 0 (i.e., `Poly[Z[3], 0]`).

```
valuesOfpV = Map[pV, Elements[V]]
```

We obtain 0 twice. What values of V yielded the zeros? Let's look at V again.

```
Elements[V]
```

Combining and transposing,

```
Transpose[{Elements[V], valuesOfpV}] // TableForm[#,
    TableHeadings → {None, {"v ∈ V", "pV[v]\n"}}] &
```

makes it clear that our candidates for z are the elements (polynomials) x and $2x + 2$.

The formation and evaluation of pV is not particularly pleasant. There is a function called `EvaluationInExtension` that can do these tasks. Here is how it works.

```
? EvaluationInExtension

EvaluationInExtension[V, ModulusPolynomial[V], Poly[Z[3], x]]
```

We can also give it a list of elements to evaluate.

```
EvaluationInExtension[V, ModulusPolynomial[V], Elements[V]]
```

Q6. Let $q = x^2 + 2x + 2$ over $\mathbb{Z}_3$. Show that q is irreducible over $\mathbb{Z}_3$. Calculate `V2 = QuotientRing[Z[3], q]`. Find a zero for q in $V2$.

Recall that to compute a product in V, such as $x^2 = (x)(x)$, we can use `Multiplication[V]`.

```
Multiplication[V][Poly[Z[3], x], Poly[Z[3], x]]
```

For higher powers, it may be easier to use the `ElementToPower` function.

```
? ElementToPower

ElementToPower[V, Poly[Z[3], x], 2]
```

> **Q7.** Compute more powers of x (x^2, x^3, ..., x^9 ...) until you can identify a pattern. (Hint: Using the `Table` function with `ElementToPower` is the cleanest way of doing this.)

If we consider another irreducible quadratic polynomial over $\mathbb{Z}_3$, will we need a further extension, or will our original extension contain the second irreducible's roots?

> **Q8.** Verify that $q(x) = x^2 + 1$ has no roots in the $\mathbb{Z}_3$ but has both its roots in V.

11.5 Theorems motivated from this lab

The calculations we have performed here suggest several theorems. To prove the theorems, there are many details that need to be shown. However, the examples in this lab certainly show that the theorems are plausible.

> **Q9.** We state these theorems with a few key words left for you to fill in.
>
> **Theorem 1.** If $p(x)$ is irreducible over F, $E = F[z]$ is a quadratic extension of F containing a root z of $p(x)$, and y is the second root of $p(x)$, then $F[z]$ is _____ to $F[y]$.
>
> **Theorem 2.** If w is a real root of $p(x) = x^2 + x + 1$, then for each positive integer n, $w^n = a_n + b_n w$, where a_n and b_n are consecutive _____.
>
> **Theorem 3.** If F is a finite field and $F[z]$ is a quadratic extension of F, then the set of nonzero elements of $F[z]$ with multiplication is a _____ group of order _____ and z is a _____ of the group.
>
> **Theorem 4.** If F is a finite field, a quadratic polynomial $p(x)$ is irreducible over F, and z is a root of $p(x)$ in an extension E of F, then every _____ polynomial over F has its roots in $F[z]$.

Note: These theorems can all be generalized in several directions, so the theorem in your text may not match them exactly.

Ring Lab 12

Factoring in $\mathbb{Z}[\sqrt{d}]$

📖 12.1 Prerequisites

You should have an elementary understanding of divisors and factoring with integers. It may also be helpful to be familiar with the ring $\mathbb{Z}[\sqrt{d}]$.

📖 12.2 Goals

The goal of this lab is to explore the notion of factoring numbers in $\mathbb{Z}[\sqrt{d}]$ for various integers d. In particular, we want to see when this factorization is unique (in some sense) and when it is not.

📖 12.3 Introduction to divisibility

To work on this lab, we load the packages that define the needed functionality.

```
Needs["AbstractAlgebra`Master`"];
SwitchStructureTo[Ring];
```

This lab focuses on a particular class of rings that are extensions of the integers. Let d be a fixed integer and consider the set $\{r + s\sqrt{d} \mid r, s \in \mathbb{Z}\}$. We denote this set by $\mathbb{Z}[\sqrt{d}]$. If $d = 2$, the following illustrates a subset of $\mathbb{Z}[\sqrt{2}]$, when we restrict r and s to $\{-2, -1, ..., 3\}$.

```
Adjoin[Range[-2, 3], √2]
```

> **Q1.** What is the result of letting $d = 4$? In other words, what does $\mathbb{Z}[\sqrt{4}]$ look like? What about $\mathbb{Z}[\sqrt{9}]$? What about $\mathbb{Z}[\sqrt{16}]$? Can you think of any restrictions on d that we may wish to impose?

Note that when we let $d = -1$, we obtain $\mathbb{Z}[\sqrt{-1}]$, the Gaussian integers; it is frequently denoted $\mathbb{Z}[i]$.

Throughout the rest of the lab, we assume that D is an integral domain. If r and s are two elements in D, with r nonzero, we say that r *divides* s (or r is a *factor* of s) if $s = rt$ for some t in D. When r divides s, it is denoted $r \mid s$ and we say that r is a *divisor* of s. Note that the units of D are those elements that are divisors of the unity of D.

> **Q2.** For each ring given, determine whether $r \mid s$. If yes, indicate the value of t such that $s = rt$.
>   a. In $\mathbb{Z}$, is it the case that $5 \mid 15$?
>   b. In $\mathbb{Z}$, is it the case that $15 \mid 5$?
>   c. In $\mathbb{Z}[\sqrt{3}]$, is it the case that $(2 + \sqrt{3}) \mid (-7 - 6\sqrt{3})$?
>   d. In $\mathbb{Z}[\sqrt{3}]$, is it the case that $(4 - 5\sqrt{3}) \mid (-7 - 6\sqrt{3})$?
>   e. In $\mathbb{Z}[\sqrt{-5}]$, is it the case that $3 \mid 9$?
>   f. In $\mathbb{Z}[\sqrt{-5}]$, is it the case that $(2 + \sqrt{-5}) \mid 9$?
>   g. In $\mathbb{Z}[\sqrt{-5}]$, is it the case that $(2 - \sqrt{-5}) \mid 9$?
>   h. What is the product of $2 + \sqrt{-5}$ and $2 - \sqrt{-5}$?
>   i. In $\mathbb{Z}[\sqrt{-5}]$, is it the case that $3 \mid (2 + \sqrt{-5})$?

> **Q3.** If r is a unit in D and x is any other element, what can you say about $r \mid x$? Justify your answer.

When we are working with (ordinary) integers, it is fairly easy to know when r divides s. Essentially, we want to know if the quotient $\frac{s}{r}$ is an integer. We can define a function `DividesQ` by using this approach.

```
DividesQ[6, 18]
DividesQ[6, 16]

? DividesQ
```

When we work over the ring $\mathbb{Z}[\sqrt{d}]$, we extend this function by adding the option `Radical → d`. The following shows that $1 + \sqrt{2}$ is a unit in $\mathbb{Z}[\sqrt{2}]$.

```
DividesQ[1 + √2 , 1, Radical → 2]
```

This result indicates that there is an element t in $\mathbb{Z}\left[\sqrt{2}\,\right]$ such that $\left(1 + \sqrt{2}\,\right)t = 1$. What is t? Let's perform the division.

$$\frac{1}{1 + \sqrt{2}}$$

```
Simplify[%]
```

Note that `Simplify` does not return an element in the form $a + b\sqrt{2}$, so we need to turn to another function.

```
ZdDivide[1, 1 + √2 ]
```

We see that t is indeed an element in $\mathbb{Z}\left[\sqrt{2}\,\right]$.

▦ 12.4 Associates, irreducibility, and norms

We need to introduce (or review) some definitions. Given elements r and s in D, we say that they are *associates* if there is a unit u in D such that $r = s\,u$. A nonzero element r in D is called *irreducible* if r is not a unit, and whenever $r = b\,c$ (for elements b and c in D), then b or c is a unit. (In other words, the only divisors of r are units and associates of r.)

Q4. a. What are the associates of 5 over the integers?
b. What are the associates of 5 over $\mathbb{Z}[i]$?
c. Are $3 + 4\sqrt{2}$ and $5 - \sqrt{2}$ associates over $\mathbb{Z}\left[\sqrt{2}\,\right]$?

Q5. a. Is 5 irreducible over the integers? Why or why not?
b. Is 6 irreducible over the integers? Why or why not?
c. Is 5 irreducible over $\mathbb{Z}[i]$? Why or why not?

The function `ZdAssociatesQ` can confirm whether a pair of numbers are associates over $\mathbb{Z}\left[\sqrt{d}\,\right]$.

```
? ZdAssociatesQ
```

```
ZdAssociatesQ[2, 3 + 4 √2 , 5 - √2 ]
```

For negative d, the function `ZdIrreducibleQ` specifies whether an element is irreducible. (The problem is more difficult to answer for positive d.)

```
ZdIrreducibleQ[-1, 5]
```

Since 5 is not irreducible, we should be able to factor this over $\mathbb{Z}[i]$, the `Gauss-ianIntegers`.

```
FactorInteger[5, GaussianIntegers → True]
```

The last idea introduced in this section is the *norm* function. We define the function N: $\mathbb{Z}[\sqrt{d}] \to$ {nonnegative integers} by $N(a + b\sqrt{d}) = |a^2 - db^2|$. (Note: Some authors prefer a slightly different definition.) In *Mathematica*, we use the function `ZdNorm` to compute the norm. Let's consider some examples over $\mathbb{Z}[\sqrt{5}]$.

```
examples = Adjoin[{-1, 0, 1, 2}, √5]

Map[{#, ZdNorm[#]} &, examples] //
  TableForm[#, TableHeadings →
    {None, {"x", "N(x)\n"}}, TableSpacing → {0.5, 3}] &
```

There are four important properties about the norm function that should be observed; check your text for details.

1. $N(x) = 0$ if and only if $x = 0$,
2. $N(xy) = N(x) N(y)$ for all x and y,
3. $N(x) = 1$ if and only if x is a unit, and
4. If $N(x)$ is prime (in $\mathbb{Z}$), then x is irreducible in $\mathbb{Z}[\sqrt{d}]$.

Q6. In the list of examples of elements from $\mathbb{Z}[\sqrt{5}]$, which are units and which can be readily seen as being irreducible?

Consider $x = 2 + 3\sqrt{-6}$ and $y = 2 - 3\sqrt{-6}$ as elements in $\mathbb{Z}[\sqrt{-6}]$. Let's verify property 2.

```
x = 2 + 3 √-6
y = 2 - 3 √-6
ZdNorm[x y] === ZdNorm[x] ZdNorm[y]
```

Q7. Let $x = 1 + 3\sqrt{-5}$ and $y = 1 - 3\sqrt{-5}$. Verify that property 2 is true with these values of x and y. Repeat using $x = 2$ and $y = 23$.

Be careful how you read property 4; the converse is not true, in general. For example, with $x = 1 + 3\sqrt{-5}$, used in question 7, the norm of x is 46, which is composite.

```
ZdNorm[1 + 3 √-5 ]
```

This does *not* imply that x is *not* irreducible! In fact, in section 12.6, we show that x is indeed irreducible.

12.5 Units in $\mathbb{Z}\left[\sqrt{d}\,\right]$

Recall from the third property of the norm function that x is a unit if and only if $N(x) = 1$. Can we use this tact to determine the units of $\mathbb{Z}\left[\sqrt{d}\,\right]$? Let $x = a + b\sqrt{d}$ be an element in $\mathbb{Z}\left[\sqrt{d}\,\right]$. Then $N(x) = |a^2 - db^2|$. First, let's consider the case where $d < -1$ and let $k = -d$. Then we have $N(x) = a^2 + kb^2$, with $k > 1$. Since $k > 1$ and we are assuming that $N(x) = 1$ (since we want x to be a unit), then we must have $b = 0$, for otherwise $a^2 + kb^2 \geq k > 1$. Therefore, $a^2 = 1$, so the only units are 1 and -1.

Let's consider a geometric argument for the same question. To be specific, suppose we consider $k = 2$ (i.e., $d = -2$). What we are really considering is whether there are integral solutions to the equation $a^2 + 2b^2 = 1$. But this is just the equation of an ellipse, so let's look at its graph. We need to first read in a package that allows us to plot equations implicitly.

```
Needs["Graphics`ImplicitPlot`"]
```

Here we plot the ellipse determined by this equation.

```
Clear[a, b]
gr1 = ImplicitPlot[a² + 2 b² == 1, {a, -2, 2},
    AspectRatio → Automatic, PlotRange → {{-2, 2}, {-1, 1}}];
```

Next we graph a backdrop for our ellipse, consisting of points with integer coordinates in the neighborhood of the ellipse.

```
gr2 = IntegerLatticeGrid[{-2, 2}, {-1, 1},
    PlotStyle → {PointSize[0.025], RGBColor[0, 0, 1]},
    AspectRatio → Automatic];
```

Putting them together results in the following.

```
Show[{gr2, gr1}, AspectRatio → Automatic];
```

It should be clear that the only points with integer coordinates that lie on the ellipse occur where $a = 1$ and $b = 0$ and where $a = -1$ and $b = 0$.

This figure illustrates that when $d = -2$ the only units are 1 and -1. What about smaller values of d? Consider the following series of graphics. After evaluating the cell, you may wish to double-click on one of the graphics and adjust the animation speed by typing 1 through 9 (1 slow, 9 fast).

```
Do[ImplicitPlot[a² - d b² == 1, {a, -2, 2},
    AspectRatio → Automatic, PlotRange → {{-1, 1}, {-1, 1}},
    PlotLabel → "d = "<>ToString[d]], {d, -3, -15, -2}];
```

> **Q8.** What is the conclusion? If $d < -2$, how many units are there? What are they? Justify your answer.

What if $d = -1$? We then have $N(x) = a^2 - (-1) b^2 = 1$, or $N(x) = a^2 + b^2 = 1$.

```
gr1 = ImplicitPlot[a² + b² == 1, {a, -2, 2},
    AspectRatio → Automatic, PlotRange → {{-2, 2}, {-2, 2}},
    DisplayFunction → Identity];
Show[{gr2, gr1}, AspectRatio → Automatic,
    DisplayFunction → $DisplayFunction];
```

> **Q9.** When $d = -1$, how many units are there? What are they? Justify your answer.

Next we consider $d > 1$. When $d = 2$ we have $N(x) = |a^2 - 2 b^2| = 1$. Therefore, we need $a^2 - 2 b^2$ to have the value 1 or -1. This simply gives rise to two hyperbolas, $a^2 - 2 b^2 = 1$ and $a^2 - 2 b^2 = -1$. They are now graphed to gain some geometric insight.

```
c = 4;
gr1 = ImplicitPlot[{a² - 2 b² == 1, a² - 2 b² == -1},
    {a, -c, c}, AspectRatio → Automatic, PlotStyle →
        {{Thickness[0.02], Green}, {Thickness[0.02], Magenta}},
    PlotRange → {{-c, c}, {-c, c}}];
```

Here is the graph with an integer lattice backdrop (actually, a "foredrop" this time), showing the two together.

```
gr2 = IntegerLatticeGrid[{-c, c},
    {-c, c}, PlotStyle → {PointSize[0.025], Blue},
    AspectRatio → Automatic, DisplayFunction → Identity];
Show[{gr1, gr2}, DisplayFunction → $DisplayFunction];
```

> **Q10.** Are you able to "see" any units? (Remember, we are looking for units in $\mathbb{Z}[\sqrt{2}] = \{a + b\sqrt{2} \mid a, b \in \mathbb{Z}\}$ that have integer coordinates and must satisfy one of the equations giving rise to a hyperbola.) List the ones you see. Use the function ZdUnitQ if you wish to verify that it is a unit; see the following example.

The function `ZdUnitQ` can be used to confirm or deny whether an element is a unit or not.

```
? ZdUnitQ
```

```
ZdUnitQ[2, 2 + √2 ]
```

Consider the element $x = 3 + 2\sqrt{2}$; this is a unit in $\mathbb{Z}\big[\sqrt{2}\,\big]$.

```
ZdUnitQ[2, x = 3 + 2 √2 ]
```

Since x is a unit, then any power of x is also a unit. (Appropriate use of properties 2 and 3 is one method of verifying this statement.) Let's examine the powers of this element. Here are the first five powers.

```
pts1 = Table[Expand[x^k], {k, 1, 5}]
```

We can verify that these are indeed units.

```
Map[ZdUnitQ[2, #] &, pts1]
```

These elements appear to be growing without bound. Here is geometric view of what is happening, showing only the parts of the hyperbolas in the first quadrant and the powers of x colored according to `Hue` (the "rainbow").

```
convertPts[lst_] :=
  lst /. {a_ + b_ √2 :> {a, b}, a_ + √2 :> {a, 1}}
pts1 = convertPts[pts1];
n = Length[pts1];
c = 3400;
gr1 = ImplicitPlot[{a^2 - 2 b^2 == 1, a^2 - 2 b^2 == -1},
    {a, 0, c}, AspectRatio → Automatic,
    PlotStyle → {Green, Magenta}, PlotRange → {{0, c}, {0, c}},
    Epilog → Table[{Hue[k/n], PointSize[0.03],
      Map[Point, pts1[[k]], {0}]}, {k, 1, n}]];
```

Here are the five colors (in order) of the dots plotted.

```
Show[Graphics[
    Table[{Hue[i/5], Rectangle[{i, 0}, {i + 1, 1}]}, {i, 5}]]];
```

(You might ask where the yellow is on the plot.) Let's try another element, $x = -1 + \sqrt{2}$, which is also a unit, and its first ten powers.

```
ZdUnitQ[2, x = -1 + √2 ]
pts2 = Table[Expand[x^k], {k, 1, 10}]
(ZdUnitQ[2, #1] &) /@pts2
```

Here is the graph of the hyperbolas with the ten powers of this x plotted.

```
pts2 = convertPts[pts2];
n = Length[pts2];
c = 3400;
gr1 = ImplicitPlot[{a² - 2 b² == 1, a² - 2 b² == -1}, {a, -c, c},
    AspectRatio → Automatic, PlotStyle → {Green, Magenta},
    PlotRange → {{-c, c}, {-c, c}}, Epilog → Table[{Hue[ k/n ],
        PointSize[0.03], Map[Point, pts2[[k]], {0}]}, {k, 1, n}]];
```

As n increases in the power x^n, the points are colored according to the following scheme, where the power n is located in the rectangle.

```
Show[Graphics[Table[{Hue[ i/10 ], Rectangle[{i, 0}, {i + 1, 1}],
    Black, Text[i, {0.5 + i, 0.5}]}, {i, 10}]]];
```

Q11. What is fundamentally different between the powers of these two elements ($3 + 2\sqrt{2}$ and $-1 + \sqrt{2}$), as shown by the graphs?

As can be seen, $\mathbb{Z}[\sqrt{2}]$ has an infinite number of units. This is true whenever $d > 1$.

12.6 Factoring 46 in $\mathbb{Z}[\sqrt{-5}]$

In question 7, we investigated the norm of the following four numbers.

```
x = 1 + 3 √-5
y = ZdConjugate[x]
a = 2
b = 23
ZdNorm /@ {x, y, a, b}
```

Note further that $x\,y = a\,b = 46$.

```
Expand[x y] == 46
```

This expansion shows that x, y, a, and b are all divisors of 46. Are there any more?

```
ZdDivisors[-5, 46]
```

We are really only interested in the nontrivial ones.

> **ZdDivisors[-5, 46, NonTrivialOnly → True]**

That 46 has only four nontrivial divisors (over $\mathbb{Z}\left[\sqrt{-5}\,\right]$) is not significant in itself. If we consider the integer 80, we know that it has many divisors as well.

> **IntegerDivisors[80, NonTrivialOnly → True]**

We know that 80 can be written as $2*40$, $4*20$, $5*16$, or $8*10$, but in each case one or both factors can be further factored (into primes) until one obtains the result $2*2*2*2*5$, or 2^4*5.

> **FactorInteger[80]**

Except for the order in which these factors are written, or perhaps exchanging a factor with its associate, this product is unique. This is a result of the Fundamental Theorem of Arithmetic. The question to pursue here is whether an analogous situation holds in $\mathbb{Z}\left[\sqrt{-5}\,\right]$ and other rings. In other words, even though $\left(1+3\sqrt{-5}\right)\left(1-3\sqrt{-5}\right)=46=2*23$ in $\mathbb{Z}\left[\sqrt{-5}\,\right]$ shows that we have two factorizations that appear different, we want to make sure that we have factored 46 into irreducibles. Also, we want to see if $1+3\sqrt{-5}$ is possibly an associate of 2 or 23 and if $1-3\sqrt{-5}$ is an associate of the other. Thus, we have two steps to take.

> 1. Determine whether 2, 23, $1+3\sqrt{-5}$, and $1-3\sqrt{-5}$ satisfy our definition of being irreducible.

> 2. Determine if any of $\{2, 23\}$ are associates with any of $\left\{1+3\sqrt{-5}, 1-3\sqrt{-5}\right\}$.

Recall that for r to be irreducible over an integral domain D, it must be nonzero, not a unit, and if we ever have $st=r$ (for s, $t \in D$), we must have either s or t a unit. Clearly none of the four are zero, and when we calculated their norms we observed that none had norm 1, so they are not units. First let's consider 2 and show that it is irreducible. Its norm is four.

> **ZdNorm[2]**

Therefore, if $2=st$ for some s and t in $D=\mathbb{Z}\left[\sqrt{-5}\,\right]$, then $N(s)\,N(t)=4$, by the second property of norms. To show that 2 is irreducible, we need to show that either s or t is a unit and consequently has norm 1 (by the third property). Therefore, since the divisors of 4 are $\{1, 2, 4\}$, we want to show that we cannot have $N(s)=N(t)=2$, and then we will know that 2 is irreducible.

> **Q12**. Why do we need to show only that we cannot have $N(s) = 2$ to show that
> 2 is irreducible?

If we let $s = a + b\sqrt{-5}$, then we see that $N(s) = a^2 + 5b^2$. For $N(s) = 2$, we must have integral coordinates for a point somewhere on the ellipse $a^2 + 5b^2 = 2$. Inspecting this algebraically, clearly if b is not zero, then the left-hand side is already at least 5, which is not possible (since we still have to add a^2 to $5b^2$, while the right-hand side is 2). Geometrically, we can see this as follows.

```
ZdPossibleNorms[-5, 2, Mode → Visual]
```

This function shows the ellipse $a^2 + 5b^2 = 2$, as well as showing all points whose norm is less than or equal to 2 (and the output is a list of all norms that are possible that are less than or equal to 2). We can also just use the function `ZdPossibleNormQ`.

```
ZdPossibleNormQ[-5, 2]
```

Now we know that 2 is indeed an irreducible element. Next we attack 23 and investigate whether it is irreducible. Let's assume that it is not irreducible and see if we can find a contradiction. Therefore $23 = st$, where s and t are in D. Consequently, $N(23) = N(s)N(t)$, and we need to find elements s and t whose norm is a divisor of the norm of 23. Not that we need *Mathematica* to help us in this case (the arithmetic is easy), but to illustrate the general idea, we want to look for the nontrivial divisors of 23^2.

```
IntegerDivisors[ZdNorm[23], NonTrivialOnly → True]
```

The only possible divisor is 23, so both $N(s)$ and $N(t)$ must be 23. Is it possible to have this norm in the ring $D = \mathbb{Z}[\sqrt{-5}]$?

```
ZdPossibleNormQ[-5, 23]
```

Let's "see" why not.

```
ZdPossibleNorms[-5, 23, Mode → Visual]
```

The value 23 is not obtainable as a norm, and therefore there are no s and t such that $N(s) = N(t) = 23$, so either s or t is a unit and 23 is irreducible.

There are two elements left to check for irreducibility, $1 + 3\sqrt{-5}$ and $1 - 3\sqrt{-5}$. We proceed in the same manner.

```
IntegerDivisors[ZdNorm[1 + 3 √-5], NonTrivialOnly → True]
IntegerDivisors[ZdNorm[1 - 3 √-5], NonTrivialOnly → True]
```

Since the only possible nontrivial divisors of the norms are 2 and 23, and we have seen that neither of these norms is possible, we know that both the elements $1 + 3\sqrt{-5}$ and $1 - 3\sqrt{-5}$ are irreducible.

Q13. Here is an optional question. Define the function v by $v(x)$ is the number of norm values possible in $\mathbb{Z}[\sqrt{-5}]$ that are less than or equal to x. Determine $v(100)$. How much larger is $v(200)$ than $v(100)$? (In other words, compare the number of possible norm values from 1 to 100 in contrast to the number of possible norm values from 101 to 200.) How does $v(300)$ compare to $v(200)$? Is there a monotonic pattern? You may wish to use the following to help you, with the Length function to measure the output. (Eliminate the Mode → Visual option if the graphics take too long—they are not needed for the count.)

```
ZdPossibleNorms[-5, 100, Mode → Visual]
```

We have shown that the four divisors of 46 are all irreducible. Next we want to know if any of $\{2, 23\}$ are associates with $\{1 + 3\sqrt{-5}, 1 - 3\sqrt{-5}\}$, step 2 in our outline. Recall that for 2 to be an associate of $1 + 3\sqrt{-5}$, it must have a unit u such that $2 = u * (1 + 3\sqrt{-5})$. If so, then $N(2) = N(u) N(1 + 3\sqrt{-5})$, or $4 = 1 * 46$, which is clearly not possible. The following shows that no possible arrangement (matching across the two sets) for pairing up associates will be fruitful.

```
Map[ZdNorm, {2, 23}]
Map[ZdNorm, {1 + 3 √-5 , 1 - 3 √-5 }]
```

(Note that checking for associates could have also been done as follows.)

```
ZdAssociatesQ[-5, 2, 1 + 3 √-5 ]
ZdAssociatesQ[-5, 23, 1 + 3 √-5 ]
```

What is the conclusion? In $\mathbb{Z}[\sqrt{-5}]$ we have $46 = 2 * 23$ and $46 = (1 + 3\sqrt{-5}) * (1 - 3\sqrt{-5})$. These four divisors are all irreducible and the two factorizations do not involve associates (as would $46 = 2 * 23$ and $46 = (-2) * (-23)$). Consequently, we have two distinct, unrelated factorizations of the number 46. This verifies that $\mathbb{Z}[\sqrt{-5}]$ is *not* a Unique Factorization Domain (UFD).

Historical note: There was a short time in the nineteenth century when Fermat's Last Theorem was considered to have been proven, but there was a flaw in the proof. This flaw, discovered by Kummar, traced back to an assumption that factorization was unique in rings such as $\mathbb{Z}[\sqrt{-5}]$.

🔖 12.7 Is $\mathbb{Z}\left[\sqrt{-6}\right]$ a UFD?

Since we have shown that $\mathbb{Z}\left[\sqrt{-5}\right]$ is not a UFD (Unique Factorization Domain), we may wonder about other related rings. Let's consider a neighbor (in some sense), $D = \mathbb{Z}\left[\sqrt{-6}\right]$. To show that D is not a UFD, we need to find an element r in D that has two distinct factorizations. Since we may want to "automate" the process, we label each step along the way with variables to hold the values, so we can readily try other rings.

In this example, we are using $d = -6$.

```
d = -6
```

Let's pick two random—but not too large—coefficients.

```
{a, b} = Table[Random[Integer, {1, 5}], {2}]
```

We use these values to form x and let y be its conjugate.

```
x = a + b √d
y = ZdConjugate[x]
```

Let z be their product.

```
z = Expand[x y]
```

We use `nrm` for the norm of this product.

```
nrm = ZdNorm[z]
```

Let's take a look at all the divisors of z.

```
ZdDivisors[d, z, DivisorsComplete → True]
```

We can view these divisors by pairing up associates.

```
ZdDivisors[d, z,
  Combine → Associates, DivisorsComplete → True]
```

We can also pair up the divisors so that the product of each pair is z.

```
ZdDivisors[d, z, Combine → Products, DivisorsComplete → True]
```

From here, we should look for two pairs whose product is z, all divisors are irreducible, and none is an associate of the others.

> **Q14.** Find two pairs that satisfy these conditions. If there are not two pairs, go back and generate a new x and y and see if two pairs arise that satisfy these conditions.

Ring Lab 13

Finite Fields

▣ 13.1 Prerequisites

To complete this lab, you should be familiar with the construction of a quotient ring of a ring of polynomials over a field F, as described in Ring Lab 10. You should also be familiar with `Morphoids` of rings.

▣ 13.2 Goals

The goal of this lab is to introduce some of the properties of finite fields. Until now, we have mostly explored the general situation of polynomials over a ring mod an arbitrary polynomial. Now we assume that the base ring is a field and that the polynomial is irreducible over that field. These assumptions are made easier to enforce by using the `FiniteFields` package in `AbstractAlgebra`.

▣ 13.3 Creation of finite fields

First we make sure that the `AbstractAlgebra` packages are available and `Rings` is the `DefaultStructure`.

```
Needs["AbstractAlgebra`Master`"];
SwitchStructureTo[Rings];
```

The package `AbstractAlgebra`FiniteFields.m` generates finite fields in a fashion similar to the way `QuotientRing` does but with restrictions on the allowable inputs so that only fields are generated. For example, the ring of polynomials over $\mathbb{Z}_3$ mod the polynomial $x^2 + 2$ can be created with `QuotientRing`. It has $3^2 = 9$ elements.

```
Clear[x]
R = QuotientRing[Z[3], Poly[Z[3], x² + 2]]
```

As we can see by the result of `Inverses`, R is not a field.

```
Inverses[R, Operation → Multiplication]
```

This is because the modulus polynomial, $x^2 + 2$, can be factored as $(x + 1)(x + 2)$ over $\mathbb{Z}_3$. The built-in function `Factor` can show this.

```
Factor[x² + 2, Modulus → 3]
```

We can also verify this reducibility as follows.

```
IrreduciblePolyOverZpQ[Poly[Z[3], x² + 2], 3]
```

The ring R has zero divisors. For example, the product of $x + 1$ and $x + 2$ is 0.

```
Multiplication[R][Poly[Z[3], x + 1], Poly[Z[3], x + 2]]
```

Q1. Evaluate the following cell, which creates the multiplication table for R. Are there any other zero divisors?

```
MultiplicationTable[R, Mode → Visual];
```

In the rest of this lab we can avoid this type of ring by using the `GF` function. `GF[n]` returns the Galois field (hence, GF) of order n (if $n = p^d$ for some prime p), while `GF[n, poly]` creates the finite field using the specified irreducible polynomial. Here is a field with nine elements. They are the same polynomials of degree one or less that are contained in R, but the multiplication operation is different

```
Clear[F]
F₁ = GF[9]
```

In F_1, the product of $x + 1$ and $x + 2$ is *not* zero.

```
Multiplication[F₁][x + 1, x + 2]
```

Another advantage to using `GF` is that the elements of the rings that are created are simply ordinary polynomial expressions, as opposed to the more complex structures created with the `Poly` function.

F_1 is a field because its modulus polynomial is irreducible over $\mathbb{Z}_3$.

```
FieldIrreducible[F₁]
Factor[%, Modulus → 3]
```

An alternate irreducible polynomial is also allowed; one simply adds it as a second argument to GF. GF does not allow a reducible polynomial, however.

```
GF[9, x² + 2]
```

Here is a valid alternate irreducible polynomial over $\mathbb{Z}_3$.

```
F₂ = GF[9, x² + 2 x + 2]
```

You can check a polynomial to see if it is irreducible with IrreduciblePoly-OverZpQ.

```
Map[IrreduciblePolyOverZpQ[#1, 3] &, {x² + 2, x² + 2 x + 2}]
```

▣ 13.4 Finite field theorems and illustrations

We list some of the basic theorems in the theory of finite fields. Your text probably contains their proofs (or the proofs may be exercises). We illustrate these theorems with examples to give you a better feel for what they say and what their implications are. The first theorem severely restricts the possible size (or order) of a finite field.

Theorem 1. Every finite field is of order p^k for some prime p and positive integer k.

Because of Theorem 1, GF does not accept an order such as 6.

```
GF[6]
```

For a given order, there is really no variety among fields of that order.

Theorem 2. All finite fields of order p^k, p a prime, are isomorphic.

The same field can appear in many different forms and be created in different ways. We illustrate this theorem by generating GF[8] with two different irreducible polynomials over $\mathbb{Z}_2$, $p_1 = x^3 + x + 1$ and $p_2 = x^3 + x^2 + 1$. We should be able to create an isomorphism between these fields.

```
F₃ = GF[8, p1 = x³ + x + 1]
```

```
F₄ = GF[8, p2 = x³ + x² + 1]
```

One reasonable guess might be that the identity function is an isomorphism, since the domains are identical.

```
id = FormMorphoid[Identity, F₃, F₄]
```

This morphism is certainly one-to-one and onto, but it does not satisfy the morphism property for some pairs. For example, x^2 and x is such a pair.

```
PreservesQ[id, {x², x}, Mode → Visual]
```

Another reasonable guess might be that the squaring function that takes each element in F_3 to its square (in F_4).

```
θ = FormMorphoid[ElementToPower[F₃, #1, 2] &, F₃, F₄]
```

This map is one-to-one.

```
OneToOneQ[θ]
```

Q2. Is there any need to check whether θ is onto? Why or why not?

In the end, this attempt fails also.

```
MorphismQ[θ]
```

Q3. Give one concrete example showing why θ is not a morphism.

A final attempt begins by looking for the roots of p_1 in F_3. This is a bit tricky since we need to use the operations of F_3 to evaluate a polynomial expression. Here is a function that does this calculation.

```
evalp1[p_] :=
 Addition[F₃][1, Addition[F₃][p, ElementToPower[F₃, p, 3]]]
 (* note that this defines f (p) = 1 + p + p^3 *)
```

Now we evaluate p_1 at each element of F_3. The first column consists of elements in F_3 and the second column is p_1 evaluated at the elements.

```
TableForm[Map[{#1, evalp1[#1]} &, Elements[F₃]]]
```

We can see that the roots of $p_1 = x^3 + x + 1$ are x, x^2, and $x^2 + x$. (Why is this the case?) Now consider $p_2 = x^3 + x^2 + 1$, the irreducible polynomial for F_4. Let's see if the roots for p_2 are in F_3. Again, the function to do the evaluation is a bit messy.

```
evalp2[p_] := Addition[F₃][1, Addition[F₃][
   ElementToPower[F₃, p, 2], ElementToPower[F₃, p, 3]]]
```

Now we evaluate the modulus polynomial p_2 at each element of F_3.

```
TableForm[Map[{#1, evalp2[#1]} &, Elements[F₃]]]
```

All three roots of $p_2 = x^3 + x^2 + 1$ are also in F_3. Notice that they are the roots of p_1, with 1 added to each of them. With this observation, we make an attempt to find a morphism by generating a function using the rule $x \mapsto x + 1$. After applying this rule, we simplify the expression. Therefore, the image of x^2 will be $(x + 1)^2$, which expands to $x^2 + 1$, since $x + x = (1 + 1)x = 0$.

Q4. By hand, determine each of the remaining images according to the function just described and illustrated.

Now that we have a function in mind to work with, we use `FormMorphoid-Setup` to determine a position list for forming a `Morphoid`.

```
FormMorphoidSetup[F₃, F₄];
```

With this graphic, you can describe a function using a list such as {a1, a2, ..., a8}. A list of this sort implies that the image of the first entry on the left is the one in position a_1 on the right, the image of the second entry on the left is the one in position a_2 on the right, and so on. For example, the identity function is defined by the list {1, 2, 3, 4, 5, 6, 7, 8}. Based on your answer to question 4, fill in the list corresponding to this attempt at an isomorphism. Evaluate the cell after you have filled in the list.

```
τ = FormMorphoid[
      {replace this with a list consisting of an ordering of 1..8},
      F₃, F₄]
```

Your function will be one-to-one and onto if the position list is a permutation of {1, 2, 3, 4, 5, 6, 7, 8}. Now let's see if we have an isomorphism. It should work if you answered question 4 correctly. You should get `True` for your next output!

```
MorphismQ[τ]
```

Q5. It follows from Theorem 2 that the two fields of order 9, F_1 and F_2 that were generated earlier, are isomorphic. Is the identity function an isomorphism between them? If so, are there more isomorphisms? If not, find at least one isomorphism.

The nonzero elements of any field form a multiplicative group. In the finite case, the group is particularly simple.

Theorem 3. The multiplicative group of nonzero elements of a finite field is cyclic.

```
G = FormGroupoid[Units[F₃], Multiplication[F₃]]
```

```
CyclicQ[G]
```

The linear monomial term, x in this case, happens to be a generator for GF[n].
To view the powers of x, you can use PowerList.

```
TableForm[PowerList[F₃]]
```

Note that this result indicates that x^6 is $x^2 + 1$. Also, notice that PowerList acts
on the whole finite field and always starts with {0, 0}, which is a special case
since 0 is not in the group. One could evaluate Rest[PowerList[F_3]] to get
only the powers of x.

> **Q6.** To what more common group is G isomorphic? What cyclic groups of
> order 20 or less are isomorphic to the group of units of some finite field?

The concept of a subfield is similar to that of a subring. A subfield of a field F is a
subset of F that is a field, using the operations of F.

> **Q7.** Find a subfield of GF[16] containing four elements. Use Theorem 3 to
> explain why GF[16] has no subfield of order 8.

The previous question generalizes to the following theorem about subfields of a
finite field.

Theorem 4. If F is a finite field of order p^k, then F has a subfield of order p^j if
and only if j is a divisor of k.

For example, the field GF[2^6] has proper subfields of order 2, $2^2 = 4$, and $2^3 = 8$,
but not 2^4 or 2^5. The only proper subfield of GF[2^5] has 2 elements.

> **Q8.** If p is a prime, what are the divisors of $p^n - 1$? What connection does this
> have with Theorem 4?

Theorem 5. The function $f : \text{GF}[p^n] \to \text{GF}[p^n]$ defined by $f(a) = a^p$ is an automor
phism of GF[p^n], called the *Frobenius automorphism*.

Earlier, we examined a similar function between two versions of GF[2^3] that
were based on different irreducible polynomials, and the function was not a
morphism. However, the Frobenius automorphism has the same field for both
domain and codomain. We can easily create the Frobenius automorphism. The
following Morphoid is identical to θ, defined earlier, except that the codomain is
changed from F_4 to F_3. We convert it to the rules format to see the images of each
element more easily.

```
f = ToRules[FormMorphoid[ElementToPower[F₃, #1, 2] &, F₃, F₃]]
```

The following shows that f is an isomorphism.

```
OneToOneQ[f] && MorphismQ[f]
```

Theorem 6. The group of automorphisms of GF[p^n] is the set $\{f, f^2, ..., f^n = i\}$, where f is the Frobenius automorphism.

We can create this group of automorphisms easily enough and then look at its table.

```
Auts = GenerateGroupoid[{f},
    MorphoidComposition[#1, #2] &, WideElements → True]

CayleyTable[Auts, Mode → Visual, ShowKey → False];
```

Q9. We see that the automorphism group of GF[8] is isomorphic to $\mathbb{Z}_3$. What about the automorphism group of GF[16]? To what is it isomorphic?

Part III

User's Guide

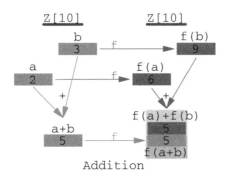

Addition

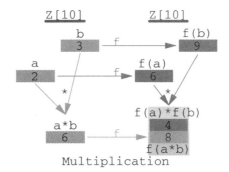

Multiplication

Chapter 1

Introduction to

`AbstractAlgebra`

This guide is written with the assumption that the reader has at least minimal familiarity with groups, rings, and homomorphisms; consult an abstract algebra text for details of any unfamiliar algebraic concept. A bibliography in the Appendix contains some suggested references. The purpose of this guide is to provide details for (and illustrations of) many of the structures and functions used in the packages in `AbstractAlgebra`. Many of these structures and functions are also used in the laboratory notebooks in *Exploring Abstract Algebra with Mathematica*. For updates to this guide, updates to the packages, as well as other related resources, the web page http://www.central.edu/eaam.html (which is mirrored at http://www.uml.edu/Dept/Math/eaam/eaam.html) can be consulted.

1.1 Packages in `AbstractAlgebra`

Typically, to use a function from `AbstractAlgebra`, the `Master` package needs to be evaluated. This loads all the *names* of the functions that are found in the packages and is done as follows.

```
In[1]:= Needs["AbstractAlgebra`Master`"]
```

Reading in the `Master` package causes all the names in all the packages to be recognized. When a particular function is used, the appropriate package is then called. For example, if we want to know the orders of the elements in the dihedral group D_3, we provide as input `Orders[Dihedral[3]]`.

```
In[2]:= Orders[Dihedral[3]]
```

Out[2]= {{1, 1}, {Rot, 3}, {Rot^2, 3},
 {Ref, 2}, {Rot ** Ref, 2}, {Rot^2 ** Ref, 2}}

The function to form the dihedral group D_3 is found in the Groupoids package, while the function to calculate the orders is found in the GroupProperties package. They both call the Core package (as well as several standard *Mathematica* packages). This is all intended to be transparent to the user. The packages are organized to minimize the amount of code that needs to be read in and retained in memory. Consequently, this guide follows the *logical* structure, not the *physical* structure, of the packages. For those interested, however, the following describes the *Mathematica* packages included in AbstractAlgebra, including dependencies; by reading in any one of these packages, the packages on which it depends are called automatically.

Core	basic collection of functions that are used throughout the other packages
FiniteFields	functions to work with finite fields, constructed as quotient rings of the form $\mathbb{Z}_n[x]/\langle p(x)\rangle$ —calls RingExtensions
Groupoids	built-in groupoids—calls Core
GroupProperties	functions to work specifically with groups—calls Core
Joint	functions useful for both groups and rings—calls Core
LabCode	specialized functions that are used only in the group and ring labs in *Exploring Abstract Algebra with Mathematica*—calls GroupProperties, Groupoids, and Joint
Matrices	functions to work with matrices over both standard rings ($\mathbb{Z}, \mathbb{Q}, \mathbb{R}$, and $\mathbb{C}$) and arbitrary rings—calls RingExtensions
Morphisms	functions to work with homomorphisms between groups or rings—calls Joint and GroupProperties
Permutations	functions to work with permutations, cycles, and permutation groups—calls Core
RingExtensions	functions to work with ring extensions, including polynomials and functions over a ring—calls RingProperties
Ringoids	built-in ringoids—calls Core
RingProperties	functions to work specifically with rings—calls Joint
Zd	functions to work with divisibility-related issues in $\mathbb{Z}[\sqrt{d}]$— calls Core

Packages in the AbstractAlgebra directory.

Although loading the Master package is recommended, and this guide assumes that it is loaded, individual packages can be loaded and their specific functions used.

- Suppose we want to load the Zd package.

In[3]:= **Needs["AbstractAlgebra`Zd`"]**

- To obtain the list of names in this package, evaluate the following.

In[4]:= **Names["AbstractAlgebra`Zd`*"]**

Out[4]= {Associates, Combine, DivisesQ, DivisorsComplete,
        IntegerDivisors, Negations, NonTrivialOnly, Products,
        Radical, ValuesHavingGivenNorm, ZdAssociatesQ,
        ZdCombineAssociates, ZdConjugate, ZdDivide,
        ZdDividesQ, ZdDivisors, ZdIrreducibleQ, ZdNorm,
        ZdPossibleNormQ, ZdPossibleNorms, ZdQ, ZdUnitQ}

- To learn about the ZdDivisors function, use the ? approach (or, of course, read the pertinent portion of this documentation).

In[5]:= **? ZdDivisors**

> ZdDivisors[d, x, (opts)], when d is negative, returns
> all the divisors of the number x in Z[Sqrt[d]].
> When d is positive, ZdDivisors[d, x, max] returns
> all divisors of x in Z[Sqrt[d]] whose norm is less
> than or equal to the norm of the integer max.
> Available options are Combine, NonTrivialOnly,
> and DivisorsComplete. See them for more details.

1.2 Basic structures used in **AbstractAlgebra**

■ 1.2.1 Overview

The goal of the packages in AbstractAlgebra is to provide an environment for working with groups, rings, fields, and homomorphisms. The fundamental structures are called Groupoids, Ringoids, and Morphoids. These structures need not necessarily represent bona fide groups, rings, or morphisms. A Groupoid consists of a set (List) of elements followed by an operation, whereas a Ringoid consists of a set of elements followed by two operations. A Morphoid consists of a function (or list of rules) followed by the domain and codomain, which are either both groupoids or both ringoids. In this guide, we use the term *groupoid* when referring to the mathematical structure consisting of a set and an operation and use Groupoid when we refer to the *Mathematica* structure that embodies the mathematical groupoid. We make a similar distinction between *ringoid* and Ringoid; the mathematical underpinnings for a Morphoid is simply a function between two groupoids or two ringoids. Note that the operations in a Groupoid or Ringoid need not be binary operations in the usual mathematical sense (where given a pair of inputs from a set, the image also is required to belong to the set).

■ For an example of a groupoid, consider the set {2, 3, 4} and the built-in *Mathematica* operator `Plus`, which performs addition.

```
In[6]:= G = Groupoid[{2, 3, 4}, Plus]
```

```
Out[6]= Groupoid[{2, 3, 4}, Plus]
```

This structure, defined as *G*, constitutes a groupoid, although it clearly does not form a group. Also, `Plus` is not a binary operation (over this set) in the normal sense of the word, since 2 + 3 = 5 and 5 is not a member of the underlying set {2, 3, 4}.

■ Similarly, we can form a ringoid as follows. Again, the operations in a `Ringoid` do not need to be binary operations in the normal sense, as illustrated here.

```
In[7]:= R = Ringoid[{2, 3, 4}, Plus, Times]
```

```
Out[7]= Ringoid[{2, 3, 4}, Plus, Times]
```

■ Consider the function from $\mathbb{Z}_{12}$ to $\mathbb{Z}_6$ by sending x to $x \bmod 6$. Here is an example of a `Morphoid` based on this function.

```
In[8]:= f = Morphoid[Mod[#, 6]&, Z[12], Z[6]]
```

```
Out[8]= Morphoid[Mod[#1, 6] &, Groupoid[
          {0, 1, 2, 3, 4, 5, 6, 7, 8, 9, 10, 11}, Mod[#1 + #2, 12] &],
          Groupoid[{0, 1, 2, 3, 4, 5}, Mod[#1 + #2, 6] &]]
```

Note that this returns a structure of type `Morphoid` and three arguments consisting of the function and the two groups. (We could have made this a ring morphism by indicating that we intended $\mathbb{Z}_{12}$ and $\mathbb{Z}_6$ to be ringoids.)

A better (and strongly recommended) mechanism to form either a `Groupoid`, a `Ringoid` or a `Morphoid` is given in section 1.2.2.

■ 1.2.2 How to form `Groupoids`, `Ringoids` and `Morphoids`

In section 1.2.1 we formed the `Groupoid` *G* by

```
In[9]:= G = Groupoid[{2, 3, 4}, Plus]
```

```
Out[9]= Groupoid[{2, 3, 4}, Plus]
```

The recommended method is to use the `FormGroupoid` function as follows.

```
In[10]:= G = FormGroupoid[{2, 3, 4}, Plus]
```

```
Out[10]= Groupoid[{2, 3, 4}, -Operation-]
```

Although the output is often the same (though not quite, in this case), there are some internal steps that are taken when the FormGroupoid function is called that may later prove useful. Additionally, one can add options to the FormGroupoid function that provide information about the groupoid being formed. The syntax for the Form-Groupoid function requires the first parameter to be a list of elements and the second parameter to be the (binary) operation used. An optional third parameter indicates the symbol used to represent the operation; the default value is *, if none is given. After these parameters, a variety of options can be added, as indicated in what follows.

In[11]:= **? FormGroupoid**

> FormGroupoid[els, op, opsym, opts] is the basic command
> for forming a Groupoid consisting of the list els
> governed by the operation op. The symbol opsym
> defaults to '*' if not specified. The available options
> for opts are WideElements, IsAGroup, Generators,
> GroupoidDescription, GroupoidName, FormatOperator,
> FormatElements, MaxElementsToList, KeyForm, and
> CayleyForm. See each one for more information.
>
> This function, rather than just wrapping Groupoid around
> a list and an operation, is strongly recommended.

Each of the options for FormGroupoid has default values in these packages, but any of them can be overridden with other values, or other default values can be defined. Although we discuss in detail the options and illustrate how they are used in chapter 2, we now briefly illustrate the use of a few of these options.

In[12]:= **Options[FormGroupoid]**

Out[12]= {CayleyForm → OutputForm, FormatElements → False,
FormatOperator → True, Generators → {},
GroupoidDescription → , GroupoidName → TheGroup,
IsAGroup → False, KeyForm → InputForm,
MaxElementsToList → 50, WideElements → False}

- Suppose we want to define the group of integers mod n, the group denoted $\mathbb{Z}_n$. It is already built into the packages, defined under the name Z; here we call our new version newZ. This is how we might define it.

In[13]:= **newZ[n_Integer ? Positive] :=**
             FormGroupoid[Range[0, n - 1],
        Mod[#1 + #2, n] &, "+", IsAGroup → True,
        FormatOperator → False, Generators → {1},
           GroupoidDescription → "Integers under addition mod n",
        GroupoidName → StringJoin["Z[", ToString[n], "]"]]

- To form $\mathbb{Z}_5$, we could use the following.

In[14]:= **G = newZ[5]**

Out[14]= Groupoid[{0, 1, 2, 3, 4}, Mod[#1 + #2, 5] &]

- Since FormatOperator is set to False, the binary operator Mod[#1 + #2, 5]& is given, instead of hiding behind -Operation-, as seen earlier. The IsAGroup option indicates that any tests to see if it is a group can be skipped and True will be immediately returned by GroupQ.

In[15]:= **GroupQ[G]**

Out[15]= True

- The value of the option GroupoidName shows up when we view a Cayley table.

In[16]:= **CayleyTable[G, Mode → Visual]**

Out[16]= {{0, 1, 2, 3, 4}, {1, 2, 3, 4, 0},
{2, 3, 4, 0, 1}, {3, 4, 0, 1, 2}, {4, 0, 1, 2, 3}}

- This name also shows up when asking for information learned about *G*, as does the description and the results of some tests performed on *G*. We use the GroupInfo command to retrieve this information.

In[17]:= **GroupInfo[G]**

Out[17]= {Z[5], Integers under addition mod n, this is a group}

There are several other means for forming groupoids, including GenerateGroupoid and FormGroupoidByTable. They are discussed in detail in chapter 2, but we give a brief illustration here. GenerateGroupoid is used by giving a set of generators and a binary operator, while FormGroupoidByTable requires a list of elements followed by the Cayley table.

- Here is an illustration of GenerateGroupoid.

In[18]:= **G2 =**
  GenerateGroupoid[{{{1, 0}, {0, 7}}, {{4, 0}, {0, 1}}},

```
      Mod[#1 . #2, 9] &, WideElements → True,
      FormatElements → True, GroupoidName → "What am I?",
      GroupoidDescription → "generated by two 2-by-2
         matrices under matrix multiplication mod 9"]
```

Out[18]= Groupoid[{-Elements-}, -Operation-]

- Here is an example of `FormGroupoidByTable`.

In[19]:= **Clear[a, b, c]**
            G3 = FormGroupoidByTable[
            {a, b, c}, {{a, c, b}, {b, a, c}, {c, b, a}}]

Out[20]= Groupoid[{a, b, c}, -Operation-]

- We can view the Cayley tables of these two groupoids. Are either of these groups?

In[21]:= **CayleyTable[{G2, G3}, Mode → Visual];**

```
      KEY for What am I?: label used → element: {g1 → {{1,
         0}, {0, 1}}, g2 → {{1, 0}, {0, 4}}, g3 → {{1, 0},
         {0, 7}}, g4 → {{4, 0}, {0, 1}}, g5 → {{4, 0}, {0,
         4}}, g6 → {{4, 0}, {0, 7}}, g7 → {{7, 0}, {0,
         1}}, g8 → {{7, 0}, {0, 4}}, g9 → {{7, 0}, {0, 7}}}
```

Note that a key is given (for G2) since the option `WideElements → True` indicates that the elements should not be placed in the table. In this case, a key is set up, associating shorter generic labels with the elements in the groupoid.

- Another way of seeing the elements of the first `Groupoid` is to simply ask for them with the `Elements` function.

In[22]:= **Map[MatrixForm, Elements[G2]]**

$$Out[22]= \ \{\begin{pmatrix} 1 & 0 \\ 0 & 1 \end{pmatrix}, \begin{pmatrix} 1 & 0 \\ 0 & 4 \end{pmatrix}, \begin{pmatrix} 1 & 0 \\ 0 & 7 \end{pmatrix}, \begin{pmatrix} 4 & 0 \\ 0 & 1 \end{pmatrix},$$
$$\begin{pmatrix} 4 & 0 \\ 0 & 4 \end{pmatrix}, \begin{pmatrix} 4 & 0 \\ 0 & 7 \end{pmatrix}, \begin{pmatrix} 7 & 0 \\ 0 & 1 \end{pmatrix}, \begin{pmatrix} 7 & 0 \\ 0 & 4 \end{pmatrix}, \begin{pmatrix} 7 & 0 \\ 0 & 7 \end{pmatrix}\}$$

Forming ringoids is very similar to forming groupoids. Analogous to `FormGroupoid` is `FormRingoid`. More details can be found in the chapter 3.

- The default values for the options are similar to those for `FormGroupoid`, as is the method of using them. Here is an illustration.

```
In[23]:= R1 = FormRingoid[{-1, 0, 1}, Plus, Times,
            FormatOperator → False, RingoidName → "Ringoid example"]
```

```
Out[23]= Ringoid[{-1, 0, 1}, Plus, Times]
```

- By looking at the Cayley tables, we can try to determine if this is a ring.

```
In[24]:= CayleyTables[R1, Mode → Visual]
```

```
Out[24]= {{{-2, -1, 0}, {-1, 0, 1}, {0, 1, 2}},
          {{1, 0, -1}, {0, 0, 0}, {-1, 0, 1}}}
```

- The `RingQ` function can confirm our suspicion.

```
In[25]:= RingQ[R1]
```

```
Out[25]= False
```

- Similar to `GroupInfo`, `RingInfo` returns what has been learned about a ringoid.

```
In[26]:= RingInfo[R1]
```

Out[26]= {Ringoid example,
     the set is not closed under this addition,
     the set is closed under multiplication,
     this is NOT a ring}

While individual groups or rings can be studied by examining the set of elements, often additional results can be learned by considering homomorphisms from one group (or ring) to another. In these packages, homomorphisms are simply called morphisms and the `Morphoid` is the underlying *Mathematica* structure for working with these. There are several ways of using the `FormMorphoid` function, which is the standard means of constructing morphisms.

- The default values of the options for `FormMorphoid` are as follows.

In[27]:= **Options[FormMorphoid]**

Out[27]= {Mode → Computational, FormatFunction → False}

In chapter 4 we look at the details of these options as well as the various means of forming `Morphoids`. Here let's construct several `Morphoids` to illustrate the possibilities. If a specific function can be easily specified, then using this function is often the easiest method of forming a `Morphoid`.

- For example, consider the groups $\mathbb{Z}_6$ and $\mathbb{Z}_3$, and the function that takes $x \in \mathbb{Z}_6$ and maps it to $x \bmod 3$ in $\mathbb{Z}_3$.

In[28]:= **f1 = FormMorphoid[Mod[#, 3] &,**
         Z[6, Structure → Group], Z[3, Structure → Group]]

Out[28]= Morphoid[Mod[#1, 3] &, -Z[6]-, -Z[3]-]

- We can specify the same morphism by using a list of rules of the form $x \to y$, where x is in the domain and y is in the codomain.

In[29]:= **f2 =**
         FormMorphoid[{0 → 0, 1 → 1, 2 → 2, 3 → 0, 4 → 1, 5 → 2},
         Z[6, Structure → Group], Z[3, Structure → Group]]

Out[29]= Morphoid[
     {0 → 0, 1 → 1, 2 → 2, 3 → 0, 4 → 1, 5 → 2}, -Z[6]-, -Z[3]-]

- While the method of defining these `Morphoids` is different, as `Morphoids` they are equal.

In[30]:= **EqualMorphoidQ[f1, f2]**

Out[30]= True

- Note that we can convert a `Morphoid` created by a function to one created by rules.

In[31]:= **f3 = ToRules[f1]**

Out[31]= Morphoid[
    {0 → 0, 1 → 1, 2 → 2, 3 → 0, 4 → 1, 5 → 2}, -Z[6]-, -Z[3]-]

- We can also start from scratch in defining a Morphoid by first defining the underlying function.

In[32]:= **Clear[g, x]**
       g[x_] := Mod[x, 3]

- Now we use *g* to define the morphism.

In[34]:= **f4 = FormMorphoid[g,**
       Z[6, Structure → Group], Z[3, Structure → Group]]

Out[34]= Morphoid[g[#1] &, -Z[6]-, -Z[3]-]

- Since we know that D_3 is isomorphic to S_3, we may want to set up a specific isomorphism to reflect this. The function FormMorphoidSetup helps us do so. (We will see later how to exploit the output of this function to set up a morphism.)

In[35]:= **FormMorphoidSetup[Dihedral[3], Symmetric[3]];**

Domain			Codomain
1	1	1	{1, 2, 3}
Rot	2	2	{1, 3, 2}
Rot^2	3	3	{2, 1, 3}
Ref	4	4	{2, 3, 1}
Rot**Ref	5	5	{3, 1, 2}
Rot^2**Ref	6	6	{3, 2, 1}

▓ 1.3 How to use a Mode

One common thread that weaves through many of the functions in AbstractAlgebra is the use of the Mode option. By default, functions are set to use the Computational mode (though this can be changed for some functions). This means that the function simply produces the desired computation, with no further elaboration. As has been illustrated, adding Mode → Visual to some functions adds a visual or graphical image, as well as the computation. In addition to the Visual mode, there is often a Textual mode and sometimes a Visual2 mode.

Computational	default mode for all functions (unless modified by the user with SetOptions); in this mode, the output is simply the result of the computation
Textual	when available, provides additional printed information about the function, often providing a general definition of the function as well as information about the specific application involving the given parameters
Visual	when available, provides additional visual or graphical information about the algebraic concept represented by the function.
Visual2	if there is a second visualization for a function, this mode is used

Values for the Mode option.

- Consider the following example that illustrates the use of the various modes. Here we use the group U_{18}, which consists of the positive integers less than 18 that are relatively prime to 18, under multiplication mod 18. The group U_n can also be considered as the group of units in the ring $\mathbb{Z}_n$.

In[36]:= **U[18]**

Out[36]= Groupoid[{1, 5, 7, 11, 13, 17}, Mod[#1 #2, 18] &]

- Here we use the Computational mode (the default) for both forming U_{18} and testing if it is cyclic.

In[37]:= **CyclicQ[U[18]]**

Out[37]= True

- By specifying the mode to be Textual, we obtain information in addition to the result of the Computational output.

In[38]:= **CyclicQ[U[18], Mode → Textual]**

A Groupoid G is said to be cyclic if there
    exists an element g in G such that for all h in
    G there exists an integer n such that h = g^n.
In this case, U[18] is
    indeed cyclic, being generated by 5.

Out[38]= True

- The Visual mode works similarly; both here and with the Textual mode, the output is True, as given in the Computational mode.

In[39]:= **CyclicQ[U[18], Mode → Visual]**

```
U[18]              x * y
  x\y  1   5   7  11  13  17
   1   1   5   7  11  13  17
   5   5   7  17   1   ▓  13
   7   7  17  13   5   1  11
  11  11   1   5  13  17   7
  13  13  11   1  17   7   5
  17  17  13  11   7   5   1

   n  1  2  3  4  5  6
 g^n  5  7 17 13     1
```

Out[39]= True

- If a function does not support a particular mode, the user is notified, either with a message or, as in standard *Mathematica* usage, by returning the request.

In[40]:= **MaxTaker[5, Mode → Visual]**

> *Mode::notavail :*
> *The function MaxTaker does not support the Visual mode.*

Out[40]= Groupoid[{1, 2, 3, 4, 5}, Max]

▧ 1.4 Using `Visual` mode with "large" elements

The elements in some groupoids or ringoids are typographically larger than in other structures. For example, when working with permutations in S_3, the elements are too large to fit easily into the graphical Cayley table. In such situations, a key is set up to convert from generic elements in the table to the actual elements in the groupoid or ringoid.

ElementToKey[*G*, *el*]	return the key value used for the element *el* in the structure *G*, when a key is set up for functions such as `CayleyTable`
KeyToElement[*G*, *key*]	return the element in the structure *G* that corresponds to *key*

Interacting with the key.

- Note the list of rules that define the key, associating generic elements g_i with the various permutations.

In[41]:= **CayleyTable[G = Symmetric[3], Mode → Visual];**

> KEY for S[3]: label used → element: {g1 →
> {1, 2, 3}, g2 → {1, 3, 2}, g3 → {2, 1, 3},
> g4 → {2, 3, 1}, g5 → {3, 1, 2}, g6 → {3, 2, 1}}

S[3]   x * y

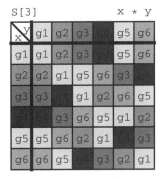

- The key corresponding to the permutation {3, 2, 1} is g6.

In[42]:= **ElementToKey[G, {3, 2, 1}]**

Out[42]= g6

- To reverse the direction, we can specify the key and obtain the corresponding element. The key can be given as a symbol or a string. Since it is possible that the symbol g6 already has a value (from a previous computation), using a string is generally better. Note that if the symbol has already been defined, its value is returned.

In[43]:= **{KeyToElement[G, g6], KeyToElement[G, "g6"],**
      g6 = 8, KeyToElement[G, g6]}

Out[43]= {8, {3, 2, 1}, 8, 8}

- Generally, when the key is displayed, the elements are written using Input-Form.

In[44]:= **CayleyTables[GF[8], Mode → Visual];**

      KEY for Add(GF[8]): label used → element:
       {g1 → 0, g2 → x^2, g3 → x, g4 → x + x^2, g5 →
       1, g6 → 1 + x^2, g7 → 1 + x, g8 → 1 + x + x^2}

Add.   x + y   Mult.   x * y

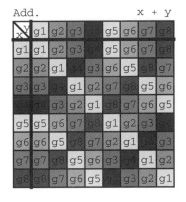   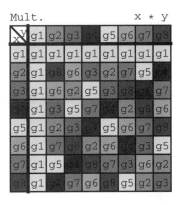

- We can override this default by using the `KeyForm` option in `CayleyTable`.

In[45]:= **CayleyTables[GF[8],**
    Mode → Visual, KeyForm → StandardForm];

KEY for Add(GF[8]): label used → element:
{g1 → 0, g2 → x^2, g3 → x, g4 → $x + x^2$, g5 →
1, g6 → $1 + x^2$, g7 → $1 + x$, g8 → $1 + x + x^2$}

1.5 How to change the `Structure`

There are several functions whose usage overlaps both groups and rings. For example, depending on the context, we can view $\mathbb{Z}_n$ as either a group or a ring. Accordingly, when working with groups, the option `Structure` defaults to the value `Group` when defining Z, but defaults to the value `Ring` when working with rings. At any point, it has a value of either `Group` or `Ring`.

- The global variable `DefaultStructure` holds the current value.

In[46]:= **DefaultStructure**

Out[46]= Group

SwitchStructureTo[Group]	DefaultStructure becomes Group to work with groupoids
SwitchStructureTo[Ring]	DefaultStructure becomes Ring to work with ringoids

Switching structures.

- We can change the value of `DefaultStructure` (as well as some other values) by using the `SwitchStructureTo` function.

In[47]:= **SwitchStructureTo[Ring]**

Out[47]= Ring

- If the default value is the desired value, then we can use Z without mention of the intended structure.

In[48]:= **Z[5]**

Out[48]= Ringoid[{0, 1, 2, 3, 4}, Mod[#1 + #2, 5] &, Mod[#1 #2, 5] &]

- If, however, we are in the Rings context and wish to use Z to create a group, we simply add the option Structure → Group.

In[49]:= **Z[5, Structure → Group]**

Out[49]= Groupoid[{0, 1, 2, 3, 4}, Mod[#1 + #2, 5] &]

- This method is fine for one evaluation, but there are other approaches for multiple evaluations. If only Z (or some other function) is affected, use the SetOptions command. For example, if the default is Ring and we wish to consider the group $\mathbb{Z}_n$, we use the following.

In[50]:= **SetOptions[Z, Structure → Group]**

Out[50]= {Mode → Computational, Structure → Group}

- Now we do not need to add the Structure option to obtain a group.

In[51]:= **Z[5]**

Out[51]= Groupoid[{0, 1, 2, 3, 4}, Mod[#1 + #2, 5] &]

- We are still in the Rings environment, however.

In[52]:= **DefaultStructure**

Out[52]= Ring

- If the focus is to switch from considering rings to considering groups, use the SwitchStructureTo function.

In[53]:= **SwitchStructureTo[Group]**

Out[53]= Group

- The Structure option works similarly for variations of the Z function.

In[54]:= **Z[10, 2, Structure → Ring]**

Out[54]= Ringoid[{0, 2, 4, 6, 8}, Mod[#1 + #2, 10] &, Mod[#1 #2, 10] &]

Although the only acceptable values for the Structure option of Z and other functions are Group and Ring, the SwitchStructureTo function allows the variants Groups, Groupoid, Rings, and Ringoid.

Chapter 2

Groupoids

▩ 2.1 Forming `Groupoids`

As we saw in chapter 1, there are several means by which `Groupoids` can be formed.
Here we consider these methods in detail and consider all the available options.

`FormGroupoid[{`*els*`},` *op*`]`	use when the list of elements and the operation are known
`GenerateGroupoid[` `{`*gens*`},` *op*`]`	use when a list of generators and the operation are known
`FormGroupoidByTable[` `{`*els*`}, {`*row1*`,` *row2*`, ...}]`	use when the list of elements and Cayley table are known
`FormGroupoidFromCycles[` `{`*cycles*`}]`	use when a list of permutations written as cycles is known
`RandomGroupoid[`*n*`,` *k*`]`	use when a random groupoid of order *n* is desired; *k* takes on a value from 1 to 3 specifying the level of randomness

Methods of forming `Groupoids`.

The syntax for the `FormGroupoid` function requires the first parameter to be a list of
elements and the second parameter to be the (binary) operation. An optional third parameter is used to indicate the symbol used to represent the operation; the default value is * if
none is given. After these parameters, we can add a variety of options.

- The third parameter is optional and defaults to *.

```
In[1]:= {OperatorSymbol[ FormGroupoid[{2, 3}, Plus]],
         OperatorSymbol[ FormGroupoid[{2, 3}, Plus, "+"]]}

Out[1]= {*, +}
```

■ 2.1.1 `FormGroupoid`

`FormGroupoid[` `{els}, op]`	form a `Groupoid` with elements {*els*} and operation *op* using ∗ as the operator symbol
`FormGroupoid[` `{els}, op, sym]`	form a `Groupoid` with elements {*els*} and operation *op* using *sym* as the operator symbol
`FormGroupoid[` `{els}, op, opts]`	form a `Groupoid` with elements {*els*} and operation *op* using the options given by *opts*
`FormGroupoid[` `{els}, op, sym, opts]`	form a `Groupoid` with elements {*els*} and operation *op* using *sym* as the operator symbol and the options *opts*

Variations in using `FormGroupoid`.

option name	default value	
`CayleyForm`	`OutputForm`	specifies how to format elements appearing in a Cayley table
`FormatElements`	`False`	displays elements or formats them to -Elements-
`FormatOperator`	`True`	displays the operator or formats it to -Operation-
`Generators`	`{}`	list of generators of the groupoid, if known
`GroupoidDescr` `iption`	`" "`	brief description of the groupoid
`GroupoidName`	`"TheGroup"`	name of the groupoid, if available
`IsAGroup`	`False`	specifies whether we can assume the groupoid is a group
`KeyForm`	`InputForm`	specifies how to format elements appearing in a key when a Cayley table is formed with a groupoid that has `WideElements` set to `True`
`MaxElementsTo` `List`	`50`	specifies the number of elements to list before the elements are automatically formatted
`WideElements`	`False`	specifies whether the elements are considered too wide to fit into a Cayley table, initiating a key to be set up

Options for `FormGroupoid`.

■ Typically, we want the elements of a groupoid formatted if there are too many elements to list or the elements themselves are large and take a lot of space to display. `MaxElementsToList` handles the first problem and is illustrated later. When a groupoid is defined, we need to determine case by case whether the listing of the elements should be suppressed. In this example, we unnecessarily suppress the elements.

```
In[2]:= G1 = FormGroupoid[{2, 4, 6},
        Plus, "+", FormatElements → True,
        FormatOperator → False, GroupoidName → "ex. 1",
        GroupoidDescription → "illustrates suppressed elements"]

Out[2]= Groupoid[{-Elements-}, Plus]
```

- When InnerAutomorphismGroup is used, the elements are usually formatted.

```
In[3]:= G2 = InnerAutomorphismGroup[Dihedral[4]]

Out[3]= Groupoid[{-Elements-}, -Operation-]
```

- We can always view the elements by using the Elements function.

```
In[4]:= Elements[G2]

Out[4]= {Morphoid[Conjugation by 1, -D[4]-, -D[4]-],
         Morphoid[Conjugation by Rot, -D[4]-, -D[4]-],
         Morphoid[Conjugation by Ref, -D[4]-, -D[4]-],
         Morphoid[Conjugation by Rot**Ref, -D[4]-, -D[4]-]}
```

- Although FormatElements defaults to False, if the number of elements exceeds MaxElementsToList, the list of elements is formatted. For some groups with wider elements (permutations, for example), it may be desirable to set the default value lower than 50.

```
In[5]:= {Z[50], Z[51]}

Out[5]= {Groupoid[{0, 1, 2, 3, 4, 5, 6, 7, 8, 9, 10, 11, 12, 13, 14,
            15, 16, 17, 18, 19, 20, 21, 22, 23, 24, 25, 26, 27, 28,
            29, 30, 31, 32, 33, 34, 35, 36, 37, 38, 39, 40, 41,
            42, 43, 44, 45, 46, 47, 48, 49}, Mod[#1 + #2, 50] &],
         Groupoid[{-Elements-}, Mod[#1 + #2, 51] &]}
```

- Every Groupoid needs to be formed with at least a set of elements and an operation. Sometimes the *Mathematica* representation of the binary operation conveys insight into how the operation works. For example, when we consider $\mathbb{Z}_5$, the operation is shown as Mod[#1 + #2, 5]&. Even if we do not know that & indicates that a pure function is being used and that #1 and #2 indicate locations for formal parameters, it should be fairly clear that we are combining two inputs under mod 5 addition.

```
In[6]:= Z[5]

Out[6]= Groupoid[{0, 1, 2, 3, 4}, Mod[#1 + #2, 5] &]
```

- On the other hand, there are occasions where the actual *Mathematica* definition is only a distraction to understanding the operation. For example, consider the following direct product of the groups $\mathbb{Z}_5$ and U_4. Here the operation is suppressed.

In[7]:= **G3 = DirectProduct[Z[5], U[4]]**

Out[7]= Groupoid[{{0, 1}, {0, 3}, {1, 1}, {1, 3}, {2, 1}, {2, 3}, {3, 1}, {3, 3}, {4, 1}, {4, 3}}, -Operation-]

- W can view the operation by using the Operation function; it should be clear why the operation is suppressed in this case.

In[8]:= **Operation[G3]**

Out[8]= MapThread[#1[#2, #3] &,
            {Operation /@ {Groupoid[{0, 1, 2, 3, 4}, Mod[#1 + #2, 5] &],
             Groupoid[{1, 3}, Mod[#1 #2, 4] &]}, #1, #2}] &

- By default, when a groupoid does not have WideElements set to True, the actual elements appear in the body of the Cayley table (and are colored according to the elements listed on the top and left sides).

In[9]:= **CayleyTable[G1, Mode → Visual]**

For each element, a different color is used.
The entries in the table corresponding to the
elements are then colored and labeled accordingly.

ex. 1   x + y

Out[9]= {{4, 6, 8}, {6, 8, 10}, {8, 10, 12}}

- When direct products are formed, the option WideElements → True is automatically added, which causes a key to be used when CayleyTable is called.

In[8]:= **CayleyTable[G3, Mode → Visual];**

```
KEY for Z[5] x U[4]: label used → element:
   {g1 → {0, 1}, g2 → {0, 3}, g3 → {1, 1}, g4 →
   {1, 3}, g5 → {2, 1}, g6 → {2, 3}, g7 → {3,
   1}, g8 → {3, 3}, g9 → {4, 1}, g10 → {4, 3}}
```

Z[5] x U[4]   x * y

y / x	g1	g2	g3	g4	g5	g6	g7	g8	g9	g10
g1	g1	g2	g3	g4	g5	g6	g7	g8	g9	g10
g2	g2	g1	g4	g3	g6	g5	g8	g7	g10	g9
g3	g3	g4	g5	g6	g7	g8	g9	g10	g1	g2
g4	g4	g3	g6	g5	g8	g7	g10	g9	g2	g1
g5	g5	g6	g7	g8	g9	g10	g1	g2	g3	g4
g6	g6	g5	g8	g7	g10	g9	g2	g1	g4	g3
g7	g7	g8	g9	g10	g1	g2	g3	g4	g5	g6
g8	g8	g7	g10	g9	g2	g1	g4	g3	g6	g5
g9	g9	g10	g1	g2	g3	g4	g5	g6	g7	g8
g10	g10	g9	g2	g1	g4	g3	g6	g5	g8	g7

- We add the option WideElements → True whenever a groupoid is being formed whose elements may be too wide to fit in the table. Here we form a new groupoid, G4, based on the elements of the symmetric group S_3 but with a twisted operation: given two triples (permutations in S_3 are given as ordered triples), the result of this operation is the triple {maximum of the elements in the first coordinate, minimum of the second coordinates, add 1 to the absolute value of the difference of the third coordinates}.

```
In[9]:= CayleyTable[(G4 = FormGroupoid[Elements[Symmetric[3]],
      {Max[First[#1], First[#2]], Min[#1[[2]], #2[[2]]],
      Abs[Last[#1] - Last[#2]]+1}&, WideElements → True]), Mode
      → Visual]

      KEY for TheGroup: label used → element: {g1 →
         {1, 2, 3}, g2 → {1, 3, 2}, g3 → {2, 1, 3},
         g4 → {2, 3, 1}, g5 → {3, 1, 2}, g6 → {3, 2, 1}}

      MIA indicates that an element is not in the domain, so
         it cannot be keyed; see output for actual values.
```

TheGroup   x * y

y / x	g1	g2	g3		g5	g6
g1	MIA	MIA	MIA	MIA	g5	MIA
g2	MIA	MIA	MIA	MIA	MIA	MIA
g3	MIA	MIA	MIA	g3	g5	MIA
	MIA	MIA	g3		g5	g6
g5	g5	MIA	g5	g5	MIA	g5
g6	MIA	MIA	MIA	g6	g5	g6

```
Out[9]=  {{{1, 2, 1}, {1, 2, 2}, {2, 1, 1}, {2, 2, 3},
          {3, 1, 2}, {3, 2, 3}}, {{1, 2, 2}, {1, 3, 1},
          {2, 1, 2}, {2, 3, 2}, {3, 1, 1}, {3, 2, 2}},
         {{2, 1, 1}, {2, 1, 2}, {2, 1, 1}, {2, 1, 3},
          {3, 1, 2}, {3, 1, 3}}, {{2, 2, 3}, {2, 3, 2},
          {2, 1, 3}, {2, 3, 1}, {3, 1, 2}, {3, 2, 1}},
         {{3, 1, 2}, {3, 1, 1}, {3, 1, 2}, {3, 1, 2},
          {3, 1, 1}, {3, 1, 2}}, {{3, 2, 3}, {3, 2, 2},
          {3, 1, 3}, {3, 2, 1}, {3, 1, 2}, {3, 2, 1}}}}
```

- While the Cayley table makes it clear that this is not a group, we can also ask whether it is.

```
In[12]:= GroupQ[G4]

Out[12]= False
```

- When forming a groupoid, if it is known that the structure is a group, we can specify it by using the IsAGroup option. This may save time for future operations that require testing whether the structure is indeed a group.

```
In[13]:= {Timing[GroupQ[FormGroupoid[
            {1, -1, I, -I}, Times, IsAGroup → True]]],
          Timing[GroupQ[FormGroupoid[{1, -1, I, -I}, Times]]]}

Out[13]= {{0.0333333 Second, True}, {0.1 Second, True}}
```

- When a groupoid is formed and WideElements is set to True, we may also wish to consider the form the key should take. Usually the default of Input-Form works well, though occasionally OutputForm is better.

```
In[14]:= G5 = FormGroupoid[{{{1, 0}, {0, 1}}, {{1, 0}, {0, 2}},
            {{1, 0}, {0, 3}}, {{1, 0}, {0, 4}}, {{4, 0}, {0, 1}},
            {{4, 0}, {0, 2}}, {{4, 0}, {0, 3}}, {{4, 0}, {0, 4}}},
            Mod[#1.#2, 5] &, WideElements → True, KeyForm → TeXForm]
```

Out[14]= Groupoid[{{{1, 0}, {0, 1}}, {{1, 0}, {0, 2}}, {{1, 0}, {0, 3}},
        {{1, 0}, {0, 4}}, {{4, 0}, {0, 1}}, {{4, 0}, {0, 2}},
        {{4, 0}, {0, 3}}, {{4, 0}, {0, 4}}}, -Operation-]

- Rarely would we use KeyForm → MatrixForm, even if matrices are involved; TeXForm is even worse. Is G5 a group? If so, to what group is it isomorphic?

In[15]:= **CayleyTable[G5, Mode → Visual];**

   KEY for TheGroup: label used → element: {g1 → \{ \{ 1,
      0\} ,\{ 0,1\} \} , g2 → \{ \{ 1,0\} ,\{ 0,2\} \} ,
      g3 → \{ \{ 1,0\} ,\{ 0,3\} \} , g4 → \{ \{ 1,
      0\} ,\{ 0,4\} \} , g5 → \{ \{ 4,0\} ,\{ 0,1\} \} ,
      g6 → \{ \{ 4,0\} ,\{ 0,2\} \} , g7 → \{ \{ 4,
      0\} ,\{ 0,3\} \} , g8 → \{ \{ 4,0\} ,\{ 0,4\} \} }

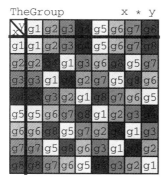

■ 2.1.2 GenerateGroupoid

In addition to the GenerateGroupoid function, there are two other general means of forming groupoids. These are GenerateGroupoid and FormGroupoidByTable.

In contrast to FormGroupoid, for which we provide a list of elements and an operation, GenerateGroupoid requires a list of generators and an operation. The cyclic and dihedral groups are formed using this function. The syntax for GenerateGroupoid is similar to that for FormGroupoid.

GenerateGroupoid[{*gens*}, *op*]	form a Groupoid with generators {*gens*} and operation *op* using ∗ as the operator symbol
GenerateGroupoid[{*gens*}, *op*, *sym*, *opts*]	form a Groupoid with generators {*gens*} and operation *op* using *sym* as the operator symbol and the options *opts*

Variations in using GenerateGroupoid.

The options for `GenerateGroupoid` are also similar to those for `FormGroupoid`, with one additional option.

option	default value	
SizeLimit	25	specifies the maximum number of elements that will be generated

Additional option for `GenerateGroupoid`.

If, while generating a groupoid, the size of the set exceeds the bound `SizeLimit`, the process is aborted. The motive for this option is to avoid inadvertently creating arbitrarily large sets.

- We could define $\mathbb{Z}_n$ using `GenerateGroupoid` (but `FormGroupoid` is actually used).

In[16]:= **newZ[n_] := GenerateGroupoid[{1}, Mod[#1 + #2, n] &];**
          newZ[5]

Out[16]= Groupoid[{0, 1, 2, 3, 4}, -Operation-]

- For another example, first observe that U_{20} is not cyclic and has eight elements.

In[17]:= **{U[20], CyclicQ[U[20]]}**

Out[17]= {Groupoid[{1, 3, 7, 9, 11, 13, 17, 19}, Mod[#1 #2, 20] &],
          False}

- U_{20}, however, can be generated by the elements 3 and 11, but *not* by the elements 3 and 7.

In[18]:= **{Elements[GenerateGroupoid[{3, 11}, Mod[#1 #2, 20] &]] ===**
              Elements[U[20]],
          GenerateGroupoid[{3, 7}, Mod[#1 #2, 20] &,
          FormatOperator → False]}

Out[18]= {True, Groupoid[{1, 3, 7, 9}, Mod[#1 #2, 20] &]}

- To what group is the following isomorphic?

In[19]:= **GenerateGroupoid[**
          {{{1, 0}, {0, 7}}, {{4, 0}, {0, 1}}}, Mod[#1.#2, 9] &]

Out[19]= Groupoid[{{{1, 0}, {0, 1}},
              {{1, 0}, {0, 4}}, {{1, 0}, {0, 7}}, {{4, 0}, {0, 1}},
              {{4, 0}, {0, 4}}, {{4, 0}, {0, 7}}, {{7, 0}, {0, 1}},
              {{7, 0}, {0, 4}}, {{7, 0}, {0, 7}}}, -Operation-]

■ 2.1.3 `FormGroupoidByTable`

The third general means of forming a groupoid is with the `FormGroupoidByTable` function. As the name implies, this function takes as inputs a set of elements and a Cayley table (as opposed to an operation). This is useful when the operation of the groupoid can be more easily represented by a table then by a specific function of two variables.

`FormGroupoidByTable[` `{els}, table]`	form a groupoid with elements *{els}* and operation implicit in *table*, using ∗ as the operator symbol
`FormGroupoid ByTable[` `{els}, table, sym, opts]`	form a groupoid with elements *{els}* and operation implicit in *table*, using *sym* as the operator symbol and the options *opts*

Variations in using `FormGroupoidByTable`.

- For example, if we wish to consider the Klein 4 group using the elements *{e, a, b, c}*, then it is easier to give the table to reflect the operation than to devise a function that establishes it. Indeed, the built-in group `Klein4` is done in this manner.

In[20]:= **G = Klein4**

> *Klein4::warning :*
> *The elements e, a, b, c are considered strings and thus*
> *need to have double quotes around them when being used.*

Out[20]= Groupoid[{e, a, b, c}, –Operation–]

- The foregoing warning simply instructs the user to treat the elements as strings when using them in functions. Thus, if we wish to determine the inverse of *b* in this group, we would need to enter it as `"b"`.

In[21]:= **GroupInverse[G, "b"]**

Out[21]= b

- The options for `FormGroupoidByTable` are similar to those of `Form-Groupoid`; here is an illustration using some options.

In[22]:= **CayleyTable[FormGroupoidByTable[**
 {2, 3, 6}, {{2, 3, 3}, {3, 2, 6}, {6, 6, 2}},
 "+", GroupoidName → "example",
 IsAGroup → False], Mode → Visual]

```
example          x + y
```

y\x	2	3	6
2	2	3	3
3	3	2	6
6	6	6	2

Out[22]= {{2, 3, 3}, {3, 2, 6}, {6, 6, 2}}

■ 2.1.4 `FormGroupoidFromCycles` and `RandomGroupoid`

There are two other means of forming groupoids, both having specialized uses. The first is `FormGroupoidFromCycles` and the second is `RandomGroupoid`.

FormGroupoidFrom⁖ Cycles [*cycles, opts*]	form a groupoid of permutations based on the list of cycles (or products of cycles) in *cycles* using the options given by *opts*

Using `FormGroupoidFromCycles`.

The options for `FormGroupoidFromCycles` are similar to `FormGroupoid`, with the addition of the `ProductOrder` option, since we are working with permutations.

option	*default value*	
ProductOrder	RightToLeft	specifies the direction in which the permutations should be multiplied

Additional option for `FormGroupoidFromCycles`.

- Here we form a groupoid by listing the elements as either cycles or products of cycles (using @ as an infix operator for multiplying cycles).

In[23]:= **G = FormGroupoidFromCycles[{Cycle[1], Cycle[1,3,2] @ Cycle[4,6,5] @ Cycle[7,8], Cycle[1,3,2] @ Cycle[4,6,5] @ Cycle[8], Cycle[1,2,3] @ Cycle[4,5,6] @ Cycle[8], Cycle[1,2,3] @ Cycle[4,5,6] @ Cycle[7,8], Cycle[7,8]}]**

Out[23]= Groupoid[{{1, 2, 3, 4, 5, 6, 7, 8},
        {3, 1, 2, 6, 4, 5, 8, 7}, {3, 1, 2, 6, 4, 5, 7, 8},
        {2, 3, 1, 5, 6, 4, 7, 8}, {2, 3, 1, 5, 6, 4, 8, 7},
        {1, 2, 3, 4, 5, 6, 8, 7}}, -Operation-]

■ If we use a random check on associativity (using `RandomAssociativeQ`), the function `ProbableGroupQ` checks to see if *G* is (probably) a group.

In[24]:= **[ProbableGroupQ[G], GroupInfo[G]]**

Out[24]= {True, {TheGroup, this is a group of permutations,
       the set is closed under the operation,
       the left identity is {1, 2, 3, 4, 5, 6, 7, 8},
       the right identity is {1, 2, 3, 4, 5, 6, 7, 8},
       the identity is {1, 2, 3, 4, 5, 6, 7, 8},
       every element has an inverse,
       the operation is probably associative with
         these elements, this is (probably) a group}}

`RandomGroupoid` can be used to generate groupoids of various random orders and characteristics.

`RandomGroupoid [n, 1]`	form a groupoid of order *n* based on a random table of n^2 elements from the list of generic elements $\{g_1, g_2, ..., g_n\}$
`RandomGroupoid [n, 2]`	form a random groupoid of order *n* where each row is a permutation of *n* elements
`RandomGroupoid [n, 3]`	form a random groupoid of order *n* where each column is a permutation of *n* elements

Variations in using `RandomGroupoid`.

■ Here we illustrate the differences among the three types of random groupoids of order three available with this function.

In[25]:= **CayleyTable[{RandomGroupoid[3, 1], RandomGroupoid[3, 2],**
        RandomGroupoid[3, 3]}, Mode → Visual];

▥ 2.2 Structure of Groupoids

The *Mathematica* equivalent of a groupoid is an object whose head is Groupoid and has two arguments. The first argument is the list of elements, and the second is the operation for the groupoid. (Actually, there is a third argument that is appended on creation that is suppressed and is used for internal purposes. This is further impetus for defining groupoids with one of the functions designed for this purpose. While giving true confessions, the real head is the symbol groupoid, which is private to the Core package, but is formatted to look like Groupoid.)

Elements[G]	return the elements in the groupoid G
Elements[$\{G_1, G_2, \ldots G_n\}$]	return a list of the elements in the groupoids $\{G_1, G_2, \ldots G_n\}$
Domain[G]	identical to Elements[G]
Operation[G]	return the operation in the groupoid G
Operation[$\{G_1, G_2, \ldots G_n\}$]	return a list of the operations in the groupoids $\{G_1, G_2, \ldots G_n\}$

Functions for working with the structure of a Groupoid.

- The elements in any groupoid are easily retrieved.

In[26]:= **Elements[AlternatingGroup[4]]**

Out[26]= {{1, 2, 3, 4}, {1, 3, 4, 2}, {1, 4, 2, 3}, {2, 1, 4, 3},
{2, 3, 1, 4}, {2, 4, 3, 1}, {3, 1, 2, 4}, {3, 2, 4, 1},
{3, 4, 1, 2}, {4, 1, 3, 2}, {4, 2, 1, 3}, {4, 3, 2, 1}}

- For two or more groupoids, simply provide a list of groupoids to either Elements or Operation. (Many functions in AbstractAlgebra can take a list of arguments like this.)

In[27]:= **{Elements[{Dihedral[3], TwistedZ[13]}],**
      Operation[{Z[9], U[12]}]}

Out[27]= {{{1, Rot, Rot^2, Ref, Rot ** Ref, Rot^2 ** Ref},
{0, 1, 2, 3, 4, 5, 6, 7, 8, 9, 10, 11}},
{Mod[#1 + #2, 9] &, Mod[#1 #2, 12] &}}

Although the data structure Groupoid consists of just elements and an operation, there are several other functions related to the structure of a groupoid.

- Every Groupoid has a name, even if it is the generic name TheGroup.

In[28]:= **Map[GroupoidName, someGroupoids =**
{Z[10], DihedralGroup[5], RootsOfUnity[8],
MinTaker[5], FormGroupoid[{3, 4}, Plus]}]

Out[28]= {Z[10], D[5], RootsOfUnity[8], MinTaker[5], TheGroup}

- Additionally, every Groupoid has a symbol for its operator, defaulting to * if not specified.

In[29]:= **Map[OperatorSymbol, someGroupoids]**

Out[29]= {+, *, *, *, *}

GroupoidName[*G*]	return the name of the groupoid *G*
OperatorSymbol[*G*]	return the symbol used for the operator of *G*
SortGroupoid[*G*]	return the groupoid *G* with the elements sorted according to the Sort function
ReorderGroupoid[*G*, *newOrder*]	return the groupoid *G* with the elements in the order given by the list *newOrder*
ElementQ[*x*, *G*]	give True if *x* is an element of the groupoid *G*, and False otherwise
ElementsQ[*els*, *G*]	give True if all members of *els* are elements of the groupoid *G*, and False otherwise
GeneratingSet[*G*]	give a set of generators for determining *G*, if known, and {} otherwise

Functions related to the structure of a groupoid.

- While SubgroupGenerated returns its list of elements in the order they are generated, applying SortGroupoid to the result reorders them.

In[30]:= **{G = SubgroupGenerated[U[7], 3], SortGroupoid[G]}**

Out[30]= {Groupoid[{3, 2, 6, 4, 5, 1}, Mod[#1 #2, 7] &],
Groupoid[{1, 2, 3, 4, 5, 6}, Mod[#1 #2, 7] &]}

- If we want the elements to appear in some other order, it can always be arranged with the ReorderGroupoid function. Note that this function only changes the order in which the elements appear; there is no structural change to the mathematical object.

In[31]:= **{H = Randomize[{1, 2, 3, 4, 5, 6}], ReorderGroupoid[G, H]}**

Out[31]= {{1, 5, 3, 4, 2, 6},
Groupoid[{1, 5, 3, 4, 2, 6}, Mod[#1 #2, 7] &]}

- Given any object (or list of objects), we can determine if it belongs to some groupoid.

In[32]:= **{ElementQ[23, U[99]], ElementsQ[{23, 33}, U[99]]}**

Out[32]= {True, False}

- Here we see the generating sets for the group D_5 when it is formed in two different ways.

In[33]:= **Map[GeneratingSet,**
      {DihedralGroup[5, Form → Permutations], DihedralGroup[5]}]

Out[33]= {{{2, 3, 4, 5, 1}, {5, 4, 3, 2, 1}}, {Rot, Ref}}

▣ 2.3 Testing the defining properties of a group

■ 2.3.1 The four standard functions

The standard definition of a group is that it is a set of elements with an operation such that the set of elements is closed under this operation (commonly stated as being a *binary operation*), there is an identity element, every element has an inverse, and the operation is associative. There are Boolean functions corresponding to each of these requirements.

ClosedQ[*G*]	give True if the groupoid *G* is closed under its operation, and False otherwise
ClosedQ[{*G₁*, *G₂*, ..., *Gₙ*}]	list of values True or False depending on the values of ClosedQ[*Gᵢ*]
HasIdentityQ[*G*]	give True if *G* has an identity, and False otherwise
HasInversesQ[*G*]	give True if every element in *G* has an inverse, and False otherwise
AssociativeQ[*G*]	give True if the operation of *G* is associative, and False otherwise

Functions for testing the basic properties of a groupoid.

As with ClosedQ, the functions HasIdentityQ, HasInversesQ, and AssociativeQ can also take a list of groupoids and return a list of Boolean values corresponding to the values of the function on each groupoid.

The Textual and Visual modes are available for each of these four functions. Generally, we do not need to add the Structure option; while working with groups the Group value is automatically used, and when working with rings the Ring value is used. (This option affects only the Textual mode.)

option name	default value	
Mode	Computational	specifies what type of information, if any, should be given in addition to the result of the computation
Structure	Group	specifies if the underlying structure is considered as a groupoid or a ringoid

Options for the four property-defining functions when working with groupoids.

■ Consider the following groupoids.

```
In[34]:= Clear[G]
        {G₁ = FormGroupoid[{1, 2, 3, 4}, Plus, "+"],
         G₂ = FormGroupoidByTable[
            {α, β, γ}, {{γ, α, β}, {α, β, γ}, {α, γ, β}},
            CayleyForm → StandardForm]}

Out[35]= {Groupoid[{1, 2, 3, 4}, -Operation-],
         Groupoid[{α, β, γ}, -Operation-]}
```

■ The groupoid G_1 is not closed under the operation Plus. When using the Textual mode, an example illustrating nonclosure is given when the result is False.

```
In[36]:= ClosedQ[G₁, Mode → Textual]
```

> We say a set S is closed under an
>   operation op if whenever we have x and y
>   in S, we also have op[x,y] (or x~op~y) in S.

> In this case, the Groupoid TheGroup is NOT closed.
>   For example, since 1 + 4 = 5 (which is not in
>   the set), it is clear that it is not closed.

```
Out[36]= False
```

■ When using the Visual mode with ClosedQ, the products in the table are colored yellow if they belong to the groupoid, and red otherwise.

```
In[37]:= ClosedQ[G₁, Mode → Visual]
```

> All the elements marked with yellow are original
>   elements in the set. Those in red are from outside.

TheGroup   x * y

y \ x	1	2	3	4
1	2	3	4	5
2	3	4	5	6
3	4	5	6	7
4	5	6	7	8

Out[37]= False

- When the `Visual` mode is used with multiple groupoids, the graphics are arranged with `GraphicsArray`.

In[38]:= **ClosedQ[{G₁, G₂, U[15], Zx[8]}, Mode → Visual]**

TheGroup   x * y

y \ x	1	2	3	4
1	2	3	4	5
2	3	4	5	6
3	4	5	6	7
4	5	6	7	8

TheGroup   x * y

y \ x	α	β	γ
α	γ	α	β
β	α	β	γ
γ	α	γ	β

U[15]   x * y

x	1	2	4	7	8	11	13	14
1	1	2	4	7	8	11	13	14
2	2	4	8	14	1	7	11	13
4	4	8	1	13	2	14	7	11
7	7	14	13	4	11	2	1	8
8	8	1	2	11	4	13	14	7
11	11	7	14	2	13	1	8	4
13	13	11	7	1	14	8	4	2
14	14	13	11	8	7	4	2	1

Zx[8]   x * y

x	0	1	2	3	4	5	6	7
0	0	0	0	0	0	0	0	0
1	0	1	2	3	4	5	6	7
2	0	2	4	6	0	2	4	6
3	0	3	6	1	4	7	2	5
4	0	4	0	4	0	4	0	4
5	0	5	2	7	4	1	6	3
6	0	6	4	2	0	6	4	2
7	0	7	6	5	4	3	2	1

Out[38]= {False, True, True, True}

- A group identity must be both a left identity and a right identity. Warning messages may give additional information in case of failure.

In[39]:= **HasIdentityQ[{G₁, G₂}]**

Identity::lfail : TheGroup does not have a left identity.

Identity::rfail : TheGroup does not have a right identity.

Identity::fail : TheGroup does not have an identity.

Out[39]= {False, True}

- While the `Textual` mode of `HasIdentityQ` reveals the identity, if it exists, the identity can also be obtained with the `Identity` function.

In[40]:= **Identity[G₂, Mode → Textual]**

We say a Groupoid G has an identity e if
  for all other elements g in G we have e * g =
  g * e = g (where * indicates the operation).

In this case, TheGroup has the identity β.

Out[40]= β

- The `Visual` mode illustrates how an identity is both a left and right identity.

In[41]:= **HasIdentityQ[MinTaker[4], Mode → Visual]**

red->left identity   red->right identity

Out[41]= True

- `FirstTaker` is a groupoid with multiple right identities but no left identity.

In[42]:= **HasIdentityQ[FirstTaker[5]]**

Identity::lfail :
FirstTaker[5] does not have a left identity.

Identity::rmultiple : FirstTaker[5] has
  the following right identities: {1, 2, 3, 4, 5}.

Out[42]= False

- When a groupoid fails to have an inverse for all of its elements, the `Textual` mode of `HasInversesQ` reveals an example of such an element.

In[43]:= **HasInversesQ[G₂, Mode → Textual]**

```
Given a Groupoid G, we say an element g in G
    has an inverse h if G has an identity e and g *
    h = h * g = e (where * indicates the operation).
```

```
The Groupoid TheGroup contains some elements without
    inverses. For example, α does NOT have an inverse.
```

Out[43]= False

■ The Visual mode of HasInversesQ connects elements that are inverses of each other, using a loop for self-inversive elements. Which, if any, of these groups are isomorphic?

In[44]:= **HasInversesQ[{U[8], U[10], U[12], Z[4]}, Mode → Visual]**

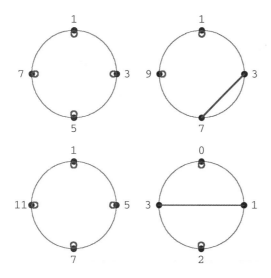

Out[44]= {True, True, True, True}

■ When a groupoid is not associative, the Textual mode illustrates triples that do not obey this property.

In[45]:= **AssociativeQ[G₂, Mode → Textual]**

```
Given a structured set S (Groupoid or
    Ringoid), we say the operation * is associative
    if for every g, h, and k in S we have (g*
    h)*k = g*(h*k), where * is the operation.
```

```
In this case, TheGroup is NOT associative since we
    have (α*α)*α = α, which is not equal to β = α*(α*α)!
```

```
Consider the following table illustrating triples that do
    not associate. Pay attention to the last two columns.
```

i	j	k	(i*j)*k	i*(j*k)
α	α	α	α	β
α	α	γ	β	α
α	γ	α	α	γ
α	γ	γ	γ	α
γ	α	α	γ	β
γ	α	γ	β	γ

Out[45]= False

- The Visual mode of AssociativeQ chooses a triple of elements at random and illustrates whether the operation is associative with this triple.

In[46]:= **AssociativeQ[U[8], Mode → Visual]**

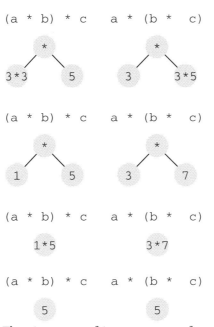

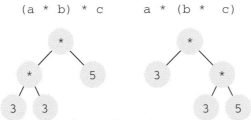

The two results are equal.
Associativity is possible.

Out[46]= True

■ 2.3.2 Related functions

There are a number of functions related to the four functions that handle the defining properties of a group. Many of these functions have `Textual` and/or `Visual` modes.

`ClosedQ[G, H]`	give `True` if *H* is a closed subset of the groupoid *G* (under the operation of *G*), and `False` otherwise

Function related to `ClosedQ`.

- The `Visual` mode of `ClosedQ`, when used with a subset, illustrates why the result is `True` or `False` by coloring the products (sums) that are in the subset yellow and those that are not red.

In[47]:= **ClosedQ[Z[6], {0, 2}, Mode → Visual]**

Z[6]   x * y

x \ y	0	2	1	3	4	5
0	0	2	1	3	4	5
2	2	4	3	5	0	1
1	1	3	2	4	5	0
3	3	5	4	0	1	2
4	4	0	5	1	2	3
5	5	1	0	2	3	4

Out[47]= False

- The `FirstTaker` function does not have a left identity.

In[48]:= **LeftIdentity[FirstTaker[5]]**

Out[48]= $Failed

- Although there are a number of right identities, the `RightIdentity` function takes the "first" one.

In[49]:= **{RightIdentity[FirstTaker[5]],**
        GroupIdentity[FirstTaker[5]]}

Out[49]= {1, $Failed}

The `HasIdentityQ` function checks for a two-sided identity. One-sided identity functions are also available.

HasLeftIdentityQ[*G*]	give True if the groupoid *G* contains a left identity, and False otherwise
HasRightIdentityQ[*G*]	give True if the groupoid *G* contains a right identity, and False otherwise
LeftIdentity[*G*]	give the left identity of the groupoid *G*, if it exists, and $Failed otherwise
RightIdentity[*G*]	give the right identity of the groupoid *G*, if it exists, and $Failed otherwise
GroupIdentity[*G*]	give the (two-sided) identity of the groupoid *G*, if it exists, and $Failed otherwise
Identity[*G*]	identical to GroupIdentity[*G*]

Functions related to HasIdentityQ.

- To determine if the element g^2 has a left inverse in the cyclic group of order 5, the following is used. (Note that the default generator *g* is a string and is thus enclosed in quotes.)

In[50]:= **LeftInvertibleQ[CyclicGroup[5], "g"^{2}]**

Out[50]= True

- Considering the element {{-1, 0}, {0, -1}} in the quaternion group (represented as matrices), we confirm that it is invertible.

In[51]:= **InvertibleQ[**
    QuaternionGroup[Form → AsMatrices], {{-1, 0}, {0, -1}}]

Out[51]= True

- Since the element Rot^2 ** Ref is a reflection in D_5, its right inverse is itself.

In[52]:= **RightInverse[DihedralGroup[5], Rot^2 ** Ref]**

Out[52]= Rot^2 ** Ref

- The Visual mode of the Inverses function uses loops for self-inversive elements and line segments to connect pairs of elements that are inverses to each other.

In[53]:= **Inverses[QuaternionGroup[Form → AsSymbols], Mode → Visual]**

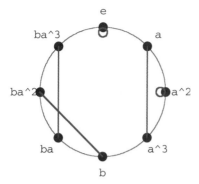

Out[53]= {{e, e}, {a, a^3}, {a^2, a^2}, {b, ba^2}, {ba, ba^3}}

The `HasInversesQ` function checks for a two-sided inverse; checking for the one-sided variety requires one of the following functions.

`LeftInvertibleQ[G, g]`	give `True` if the groupoid G contains the left inverse for the element g, and `False` otherwise
`LeftInverse[G, g]`	give the left inverse of the element g in the groupoid G, if it exists, and `$Failed` otherwise
`RightInvertibleQ[G, g]`	give `True` if the groupoid G contains the right inverse for the element g, and `False` otherwise
`RightInverse[G, g]`	give the right inverse of the element g in the groupoid G, if it exists, and `$Failed` otherwise
`InvertibleQ[G, g]`	give `True` if the groupoid G contains the (two-sided) inverse for the element g, and `False` otherwise
`GroupInverse[G, g]`	give the (two-sided) inverse of the element g in the groupoid G, if it exists, and `$Failed` otherwise
`Inverses[G]`	list pairs of the form $\{g, g^{-1}\}$, when g has an inverse, or $\{g, \text{no inverse}\}$ if g does not have an inverse
`Inverse[G, g]`	identical to `GroupInverse[G, g]`

Functions related to `HasInversesQ`.

■ When an element does not have an inverse, it is paired with "no inverse."

In[54]:= **Inverses[Zx[6]]**

Out[54]= {{0, no inverse}, {1, 1}, {2, no inverse},
         {3, no inverse}, {4, no inverse}, {5, 5}}

Since the `AssociativeQ` function checks all possible triples in the groupoid, the time it takes to check these grows in proportion to the cube of the order of the groupoid. There are faster methods, such as the Light test, that rely on using a generating set, but they are not currently implemented in these packages. In most cases, if an operation is not associative, the property is violated by a large percentage of the possible triples.

RandomAssociativeQ [*G*]	give True if the groupoid *G* satisfies the associative property after testing 25 triples at random, and False otherwise
RandomAssociativeQ [*G*, *n*]	give True if the groupoid *G* satisfies the associative property after testing *n* triples at random, and False otherwise
NonAssociatingTriples [*G*]	give a list of triples {*a*, *b*, *c*} such that $(a\,b)\,c \neq a\,(b\,c)$

Functions related to AssociativeQ.

- Lack of associativity is detected quite easily for the following groupoid, even by testing only one triple.

In[55]:= **RandomAssociativeQ[G = FormGroupoid[{2, 3, 4}, Subtract], 1]**

Out[55]= False

- The foregoing example works so well because all 27 possible triples fail associativity.

In[56]:= **{nat = NonAssociatingTriples[G], Length[nat]}**

Out[56]= {{{2, 2, 2}, {2, 2, 3}, {2, 2, 4}, {2, 3, 2},
        {2, 3, 3}, {2, 3, 4}, {2, 4, 2}, {2, 4, 3},
        {2, 4, 4}, {3, 2, 2}, {3, 2, 3}, {3, 2, 4}, {3, 3, 2},
        {3, 3, 3}, {3, 3, 4}, {3, 4, 2}, {3, 4, 3}, {3, 4, 4},
        {4, 2, 2}, {4, 2, 3}, {4, 2, 4}, {4, 3, 2}, {4, 3, 3},
        {4, 3, 4}, {4, 4, 2}, {4, 4, 3}, {4, 4, 4}}, 27}

We have seen how to test individual properties of being a group. With one test, we can check to see if a groupoid is indeed a group or not. If the groupoid is large, we can use the random approach using ProbableGroupQ, which uses RandomAssociativeQ. Additionally, there are structures called semigroups and monoids that have some of the group properties; each of them can also be tested.

When GroupQ is called, it first tests to see whether the set is closed under the operation, next whether there is an identity, then whether every element has an inverse, and finally whether the operation is associative. On the first failure of a test, it returns False.

- The GroupQ function tests the four basic properties for being a group and returns True or False accordingly.

In[57]:= **GroupQ[TwistedZ[13]]**

Out[57]= True

GroupQ [*G*]	give True if the groupoid *G* satisfies the properties of a group, and False otherwise
ProbableGroupQ [*G*]	give True if the groupoid *G* satisfies the properties of a group, using RandomAssociativeQ [*G*] for AssociativeQ [*G*], and False otherwise
SemiGroupQ [*G*]	give True if the groupoid *G* satisfies the properties of a semigroup (closed and associative), and False otherwise
MonoidQ [*G*]	give True if the groupoid *G* satisfies the properties of a monoid (closed, associative, and having an identity), and False otherwise
GroupInfo [*G*]	list the accumulated facts about the groupoid *G* (prefaced with name and description, if available), determined from various tests done on *G*

Related functions.

- Either by adding Mode → Textual to the GroupQ function, or by using GroupInfo afterwards, we can learn details about the result of testing the four group properties.

In[58]:= **GroupInfo[TwistedZ[13]]**

Out[58]= {TwistedZ[13], the set is closed under the operation, the left identity is 0, the right identity is 0, the identity is 0, every element has an inverse, the operation is associative with these elements, this is a group}

- Replacing AssociativeQ with RandomAssociativeQ, Probable-GroupQ is comparable to GroupQ.

In[59]:= **ProbableGroupQ[TwistedZ[12]]**

Out[59]= False

- Consider the following groupoid.

In[60]:= **G = FormGroupoid[{-1, 0, 1}, Times]**

Out[60]= Groupoid[{-1, 0, 1}, -Operation-]

- What kind of structure is it?

In[61]:= **{GroupQ[G], MonoidQ[G], SemiGroupQ[G]}**

Out[61]= {False, True, True}

▓ 2.4 Built-in groupoids

Although you are encouraged to try to build your own groupoids, there is a wide variety that are predefined and can be created with a few keystrokes.

■ 2.4.1 Groupoids based on the integers mod *n*

One of the simplest groups is $\mathbb{Z}_n$, the group of integers under addition mod *n*.

Z[*n*]	group (or ring) of integers $\{0, 1, 2, ..., n - 1\}$ under addition mod *n* (and multiplication mod *n*), when DefaultStructure is set to Group (Ring)
Z[*n*, Structure → Group]	group of integers $\{0, 1, 2, ..., n - 1\}$ under addition mod *n*
Z[*n*, Structure → Ring]	ring of integers $\{0, 1, 2, ..., n - 1\}$ under addition and multiplication mod *n*
ZG[*n*]	identical to Z[*n*, Structure → Group]
ZR[*n*]	identical to Z[*n*, Structure → Ring]
Z[{n_1, n_2, ..., n_k}]	return {Z[n_1], Z[n_2], ..., Z[n_k]}

The z function.

Since $\mathbb{Z}_n$ can be viewed either as a group or as a ring, the current value of the global variable DefaultStructure specifies the interpretation.

- Since the current value of DefaultStructure is Group, Z[5] returns the group rather than the ring.

In[62]:= **{DefaultStructure, Z[5]}**

Out[62]= {Group, Groupoid[{0, 1, 2, 3, 4}, Mod[#1 + #2, 5] &]}

- The current structure can be overridden by explicitly specifying the intended Structure.

In[63]:= **Z[5, Structure → Ring]**

Out[63]= Ringoid[{0, 1, 2, 3, 4}, Mod[#1 + #2, 5] &, Mod[#1 #2, 5] &]

- As shortcuts, ZG and ZR return $\mathbb{Z}_n$ as a group and a ring respectively, independent of the current value of DefaultStructure.

In[64]:= **{ZG[5], ZR[5]}**

Out[64]= {Groupoid[{0, 1, 2, 3, 4}, Mod[#1 + #2, 5] &],
        Ringoid[{0, 1, 2, 3, 4}, Mod[#1 + #2, 5] &, Mod[#1 #2, 5] &]}

There are two variations on the Z function, Z[n, k] and Z[n, I]. In both cases, the result depends on the value of DefaultStructure but can be overridden by the Structure option.

Z[n, k]	if k is a divisor of n, the group (ring) of integers $\{0, k, 2k, \ldots, (\frac{n}{k} - 1)k\}$ under addition mod n (and multiplication mod n)
Z[n, I]	Gaussian integers $\{a + b\,i \mid a, b \in \mathbb{Z}_n\}$ (group or ring) under addition mod n (and multiplication mod n)
GaussianIntegers[n]	identical to Z[n, I, Structure → Ring]
GaussianIntegersAdditive[n]	identical to Z[n, I, Structure → Group]

Variations of the Z function.

- We can view the group $\mathbb{Z}_n$ visually as the elements on an *n*-hour clock.

In[65]:= **Z[8, Mode → Visual]**

> Think of the elements as the numbers on a (modified) clock, where we view the last element, 8, as being equivalent to zero. Addition of two numbers is just like adding hours on the clock.

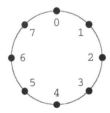

Out[65]= Groupoid[{0, 1, 2, 3, 4, 5, 6, 7}, Mod[#1 + #2, 8] &]

- A variation is to count by multiples of some divisor of *n*. Here we have the multiples of two.

In[66]:= **Z[8, 2]**

Out[66]= Groupoid[{0, 2, 4, 6}, Mod[#1 + #2, 8] &]

- Another variation is to consider the Gaussian integers, reduced by some modulus. In this example, they are reduced modulo six.

In[67]:= **Z[6, I]**

Out[67]= Groupoid[{0, i, 2 i, 3 i, 4 i, 5 i, 1, 1 + i, 1 + 2 i, 1 + 3 i,
  1 + 4 i, 1 + 5 i, 2, 2 + i, 2 + 2 i, 2 + 3 i, 2 + 4 i, 2 + 5 i, 3, 3 + i,
  3 + 2 i, 3 + 3 i, 3 + 4 i, 3 + 5 i, 4, 4 + i, 4 + 2 i, 4 + 3 i, 4 + 4 i,
  4 + 5 i, 5, 5 + i, 5 + 2 i, 5 + 3 i, 5 + 4 i, 5 + 5 i}, -Operation-]

- By adding the option Structure → Ring, we consider each of these as Ringoids.

In[68]:= **{Z[8, Structure → Ring],**
  Z[8, 2, Structure → Ring], Z[6, I, Structure → Ring]}

Out[68]= {Ringoid[{0, 1, 2, 3, 4, 5, 6, 7},
  Mod[#1 + #2, 8] &, Mod[#1 #2, 8] &],
  Ringoid[{0, 2, 4, 6}, Mod[#1 + #2, 8] &, Mod[#1 #2, 8] &],
  Ringoid[{0, i, 2 i, 3 i, 4 i, 5 i, 1, 1 + i, 1 + 2 i, 1 + 3 i,
  1 + 4 i, 1 + 5 i, 2, 2 + i, 2 + 2 i, 2 + 3 i, 2 + 4 i, 2 + 5 i, 3,
  3 + i, 3 + 2 i, 3 + 3 i, 3 + 4 i, 3 + 5 i, 4, 4 + i, 4 + 2 i, 4 + 3 i,
  4 + 4 i, 4 + 5 i, 5, 5 + i, 5 + 2 i, 5 + 3 i, 5 + 4 i, 5 + 5 i},
  -Addition-, -Multiplication-]}

- The functions Z, U, and many others can take a *list* of indices to produce a list of groupoids.

In[69]:= **{Z[Table[k, {k, 3, 4}]], U[{6, 7}]}**

Out[69]= {{Groupoid[{0, 1, 2}, Mod[#1 + #2, 3] &],
  Groupoid[{0, 1, 2, 3}, Mod[#1 + #2, 4] &]},
  {Groupoid[{1, 5}, Mod[#1 #2, 6] &],
  Groupoid[{1, 2, 3, 4, 5, 6}, Mod[#1 #2, 7] &]}}

There are several other groupoids related to the Z function.

Zx[n]	integers $\{0, 1, 2, ..., n - 1\}$ under multiplication mod n
Zx[n, k]	if k is a divisor of n, the groupoid of integers $\{0, k, 2k, ..., (\frac{n}{k} - 1)k\}$ under multiplication mod n
Zx[n, I]	Gaussian integers $\{a + bi \mid a, b \in \mathbb{Z}_n\}$ under multiplication mod n
GaussianIntegersMultiplicative[n]	identical to Zx[n, I]
U[n]	integers from $\{0, 1, 2, ..., n - 1\}$ that are relatively prime to n—the elements in Zx[n] that are invertible
TwistedZ[n]	integers $\{0, 1, 2, ..., n - 2\}$ with operation Mod[$x + y + xy$, n] for elements x and y

Functions related to the Z function.

■ The Groupoid Zx[n] is rarely a group since some elements are not invertible. (Is it *ever* a group?)

In[70]:= **{Zx[10], Inverses[Zx[10]]}**

Out[70]= {Groupoid[{0, 1, 2, 3, 4, 5, 6, 7, 8, 9}, Mod[#1 #2, 10] &],
  {{0, no inverse}, {1, 1}, {2, no inverse},
   {3, 7}, {4, no inverse}, {5, no inverse},
   {6, no inverse}, {8, no inverse}, {9, 9}}}

■ The elements in Zx[n] that have inverses are exactly the elements in U[n].

In[71]:= **U[10]**

Out[71]= Groupoid[{1, 3, 7, 9}, Mod[#1 #2, 10] &]

■ To obtain the elements in U_n, we can simply form the list of elements {0, 1, ..., $n-1$} and then select those whose gcd with n has the value 1. The following code accomplishes this (and is used in the definition of U).

In[72]:= **n = 10;**
    Range[0, n - 1] // Select[#, (GCD[#, n] == 1) &] &

Out[73]= {1, 3, 7, 9}

■ Z[n, k] and Z[n, I] have their parallels in the Zx function.

In[74]:= **{Zx[10, 2], Zx[4, I]}**

Out[74]= {Groupoid[{0, 2, 4, 6, 8}, Mod[#1 #2, 10] &],
  Groupoid[{0, i, 2 i, 3 i, 1, 1 + i, 1 + 2 i, 1 + 3 i, 2, 2 + i,
   2 + 2 i, 2 + 3 i, 3, 3 + i, 3 + 2 i, 3 + 3 i}, -Operation-]}

■ The TwistedZ groupoid has an operation that is quite different from the others.

In[75]:= **{G = TwistedZ[13], Operation[G]} // ColumnForm**

Out[75]= Groupoid[{0, 1, 2, 3, 4, 5, 6, 7, 8, 9, 10, 11}, -Operation-]
  Mod[#1 + #2 + #1 #2, 13] &

■ This groupoid is sometimes a group, and sometimes not. For what values of n is TwistedZ[n] a group?

In[76]:= **Table[{n, GroupQ[TwistedZ[n]]}, {n, 11, 15}]**

Out[76]= {{11, True}, {12, False},
  {13, True}, {14, False}, {15, False}}

■ 2.4.2 Other numeric-based groupoids

There is a number of groupoids whose elements are `Range[n]` or `Range[m, n]`.

`MaxTaker[n]`	elements are the integers $\{1, 2, ..., n\}$ and operation returns the maximum of the two inputs
`MaxTaker[m, n]`	elements are the integers $\{m, m+1, m+2, ..., n\}$ and operation returns the maximum of the two inputs
`MinTaker[n]`	elements are the integers $\{1, 2, ..., n\}$ and operation returns the minimum of the two inputs
`MinTaker[m, n]`	elements are the integers $\{m, m+1, m+2, ..., n\}$ and operation returns the minimum of the two inputs
`FirstTaker[n]`	elements are the integers $\{1, 2, ..., n\}$ and operation returns the first of the two inputs

The Taker family of functions.

■ Viewing a Cayley table of these `Taker` groupoids is an easy way of seeing their operations at work.

```
In[77]:= CayleyTable[{MaxTaker[5], MinTaker[5], FirstTaker[5]},
            ShowName → False, Mode → Visual];
```

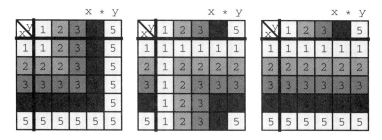

There are also several groupoids based on the divisors of the integer n.

`MeetDivisors[n]`	elements are the divisors of n and the operation returns the GCD of the two inputs
`JoinDivisors[n]`	elements are the divisors of n and the operation returns the LCM of the two inputs
`MixedDivisors[n]`	elements are the divisors of n and the operation returns the LCM/GCD of the two inputs

The Divisors family of functions.

- Viewing a Cayley table of these `Divisors` groupoids is again a good way of seeing their operations at work.

In[78]:= **CayleyTable[{MeetDivisors[15], JoinDivisors[15],**
MixedDivisors[15]}, ShowName → False, Mode → Visual];

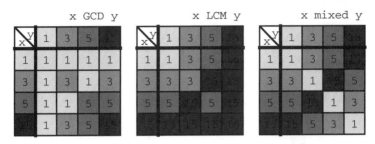

- While `MeetDivisors` and `JoinDivisors` do not appear to be groups (with $n = 15$), `MixedDivisors` does. What about other values of n for `MixedDivisors`?

In[79]:= **Table[{n, GroupQ[MixedDivisors[n]]}, {n, 10, 18}]**

Out[79]= {{10, True}, {11, True}, {12, False}, {13, True}, {14, True}, {15, True}, {16, False}, {17, True}, {18, False}}

`RootsOfUnity[n]`	solutions of $x^n = 1$, under multiplication
`IntegerUnits`	solutions of $x^2 = 1$, under multiplication
`GaussianUnits`	solutions of $x^4 = 1$, under multiplication

`RootsOfUnity` functions.

- Here we compare the elements in `IntegerUnits` with `RootsOf‐Unity[2]` and those in `GaussianUnits` with `RootsOfUnity[4]`. Then we ask if these are the same sets in both cases.

In[80]:= **{els = Map[Elements, {{IntegerUnits, RootsOfUnity[2]},**
{GaussianUnits, RootsOfUnity[4]}}, 1],
Apply[SameSetQ, els, {1}]}

Out[80]= {{{{1, -1}, {1, -1}}, {{1, -1, i, -i}, {1, i, -1, -i}}}, {True, True}}

- Observe that the elements of RootsOfUnity[5] are evenly distributed on the unit circle.

In[81]:= **RootsOfUnity[5, Mode → Visual]**

These are the (complex) numbers that are solutions to the equation x^5 - 1 = 0. The operation is ordinary (complex) multiplication, which in this case simplifies to adding the arguments (angles relative to the positive x-axis).

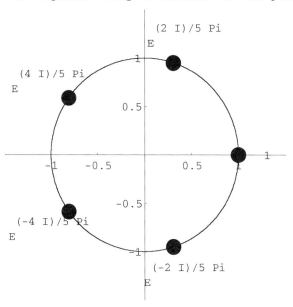

Out[81]= $\text{Groupoid}\left[\left\{1,\ e^{\frac{2i\pi}{5}},\ e^{\frac{4i\pi}{5}},\ e^{-\frac{4i\pi}{5}},\ e^{-\frac{2i\pi}{5}}\right\},\ e^{i\,(\text{Arg}[\#1]+\text{Arg}[\#2])}\ \&\right]$

■ 2.4.3 Groups of permutations

Given a list of integers $T = \{1, 2, 3, \ldots, n\}$, a permutation on T is a bijective function $\pi : T \to T$. The collection of all permutations on T is called the *symmetric group* on T. Each permutation can be more easily viewed by considering the ordered set of images of the set T under π.

- The group S_3 is obtained as the set of permutations of the elements $\{1, 2, 3\}$ under permutation multiplication.

In[82]:= **G1 = Symmetric[3]**

Out[82]= Groupoid[{{1, 2, 3}, {1, 3, 2}, {2, 1, 3}, {2, 3, 1}, {3, 1, 2}, {3, 2, 1}}, -Operation-]

`Symmetric[n]`	elements are all permutations of $\{1, 2, 3, ..., n\}$ and the operation is `MultiplyPermutations`, which assumes `ProductOrder → RightToLeft`
`Symmetric[n, opts]`	elements are all permutations of $\{1, 2, 3, ..., n\}$ and the operation is `MultiplyPermutations`, with the options given by *opts*
`S[n, opts]`	identical to `Symmetric[n, opts]`
`SymmetricGroup[n, opts]`	identical to `Symmetric[n, opts]`
`PermutationGroup[n, opts]`	identical to `Symmetric[n, opts]`
`SymmetricGroup[list, opts]`	elements are all permutations of the elements of *list* and the operation multiplies these permutations, with the options given by *opts*
`PermutationGroup[ list, opts]`	identical to `SymmetricGroup[list, opts]`

`Symmetric` and related functions.

There are several options specific to the permutation functions.

option name	default value	
`ProductOrder`	`RightToLeft`	specifies whether the permutations should be multiplied from the right to the left (default) or from the left to the right (`LeftToRight`)
`IndexLimit`	6	specifies the maximum value for the index n in the permutation functions

Options for `Symmetric`.

- The operation of the group S_3 is the function `MultiplyPermutations` using the option `ProductOrder` with the default value of `RightToLeft` (as opposed to `LeftToRight`).

In[83]:= **op = Operation[G1]**

Out[83]= `MultiplyPermutations[#1, #2, ProductOrder → DefaultOrder]` &

- The value for the global variable `DefaultOrder` is set directly or by the `ProductOrder` option.

In[84]:= **DefaultOrder**

Out[84]= `RightToLeft`

- Given permutations π_1 and π_2, the product $\pi_1 \circ \pi_2$ is obtained by doing π_2 first, followed by π_1 (when the value for `ProductOrder` is `RightToLeft`).

In[85]:= **op[{1, 3, 2}, {2, 3, 1}]**

Out[85]= {3, 2, 1}

- By changing the value of `ProductOrder` to `LeftToRight`, we can change how the group operation works. (Some mathematicians prefer this interpretation, so it is available.) Here we perform the operation from left to right.

In[86]:= **Operation[Symmetric[3, ProductOrder → LeftToRight]][**
    {1, 3, 2}, {2, 3, 1}]

Out[86]= {2, 1, 3}

- Since S_n becomes quite large with even small values of n, the user is warned when trying to create a group with an index greater than the value of `Index-Limit`.

In[87]:= **Symmetric[7]**

> *Group::size :*
> *Are you sure you want S[7]? This group has 5040 elements*
>   *in it. By default, the index must be less than*
>   *or equal to 6. If you wish to increase it, add*
>   *the option 'IndexLimit -> k' to this function,*
>   *where k is the desired maximum for the index.*

Out[87]= $Failed

- Adding a higher value for the option `IndexLimit` allows creation of such groups.

In[88]:= **Symmetric[7, IndexLimit → 7]**

Out[88]= Groupoid[{-Elements-}, -Operation-]

- We can also specify that the new value become the default value for the option.

In[89]:= **SetOptions[Symmetric, IndexLimit → 7]**

Out[89]= {Mode → Computational,
        ProductOrder → RightToLeft, IndexLimit → 7}

- We can obtain the permutation group on the set $\{\alpha, \beta, \gamma\}$ as follows.

In[90]:= **G2 = PermutationGroup[{α, β, γ}]**

Out[90]= Groupoid[{{α, β, γ}, {α, γ, β}, {β, α, γ},
        {β, γ, α}, {γ, α, β}, {γ, β, α}}, -Operation-]

- The default operation is again from right to left.

In[91]:= **Operation[G2][{α, γ, β}, {β, γ, α}]**

Out[91]= {γ, β, α}

- The left-to-right option is still available.

In[92]:= **Operation[PermutationGroup[{α, β, γ},**
        ProductOrder → LeftToRight]][{α, γ, β}, {β, γ, α}]

Out[92]= {β, α, γ}

Each permutation can be described as being either odd or even. (More details, and functions to determine whether a permutation is odd or even, are given in chapter 5.) The collection of even permutations is called the *alternating group*.

Alternating[*n*]	elements are all *even* permutations of {1, 2, 3, ..., *n*} and the operation is MultiplyPermutations, which assumes ProductOrder → RightToLeft
Alternating[*n*, *opts*]	elements are all *even* permutations of {1, 2, 3, ..., *n*} and the operation is MultiplyPermutations, using the options given by *opts*
A[*n*, *opts*]	identical to Alternating[*n*, *opts*]
AlternatingGroup[*n*, *opts*]	identical to Alternating[*n*, *opts*]

Alternating and related functions.

- The group A_4 is obtained as the set of even permutations of the elements {1, 2, 3, 4} under permutation multiplication.

In[93]:= **G = Alternating[4]**

Out[93]= Groupoid[{{1, 2, 3, 4}, {1, 3, 4, 2},
            {1, 4, 2, 3}, {2, 1, 4, 3}, {2, 3, 1, 4}, {2, 4, 3, 1},
            {3, 1, 2, 4}, {3, 2, 4, 1}, {3, 4, 1, 2}, {4, 1, 3, 2},
            {4, 2, 1, 3}, {4, 3, 2, 1}}, -Operation-]

- Note that these are all even permutations.

In[94]:= **Map[EvenPermutationQ, Elements[G]]**

Out[94]= {True, True, True, True, True,
            True, True, True, True, True, True, True}

■ 2.4.4 Dihedral and cyclic groups

The dihedral and cyclic groups are frequently used to describe the symmetries of various objects.

- One way of thinking of Dihedral[3] is to consider the symmetries of the equilateral triangle. There are three reflections (through a vertex and midpoint of the opposite side) and three rotations (of 120°, 240°, and 0°). Using the symbol Rot for the smallest (positive) degree rotation and Ref to represent one of the reflections, the rest of the elements can be represented in terms of these two.

In[95]:= **Dihedral[3, Mode → Visual]**

> The dihedral group D[3] consists of all the symmetries of the 3-sided regular polygon shown below. Included in the group are the 3 reflections through the lines connecting the vertices to the midpoints of opposite sides, as well as the 3 rotations through angles consisting of multiples of 120. degrees (360 divided by the index).

> The elements are considered as follows: The rotation of 120. degrees is called Rot and any one of the reflections is called Ref. The elements are then seen as powers of Rot and products of these powers and the reflection Ref.

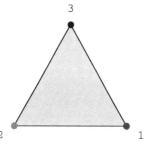

Out[95]= Groupoid[
{1, Rot, Rot², Ref, Rot ** Ref, Rot² ** Ref}, -Operation-]

- This use of Dihedral exercises the following default options.

In[96]:= **Options[Dihedral]**

Out[96]= {Mode → Computational,
Form → RotRef, RotSym → Rot, RefSym → Ref}

Dihedral[*n*]	dihedral group of order 2*n* whose elements are represented by *n* rotations and *n* reflections
Dihedral[*n*, Form → Permutations]	dihedral group of order 2*n* whose elements are represented by permutations of length *n*
Dihedral[*n*, RotSym → *rot*, RefSym → *ref*]	dihedral group of order 2*n* whose elements are represented by *n* rotations and *n* reflections, using the symbol *rot* for the lowest order rotation and *ref* for the reflection generator
D[*n*, *opts*]	identical to Dihedral[*n*, *opts*]
DihedralGroup[*n*, *opts*]	identical to Dihedral[*n*, *opts*]

Variations of the Dihedral function.

- The another acceptable value for Form is Permutations.

In[97]:= **Dihedral[3, Form → Permutations]**

Out[97]= Groupoid[{{1, 2, 3}, {1, 3, 2}, {2, 1, 3}, {2, 3, 1}, {3, 1, 2}, {3, 2, 1}}, –Operation–]

option name	default value	
Form	RotRef	specifies whether the elements should take on the form of permutations (Permutations) or be represented in terms of the lowest order rotation and a reflection (RotRef)
RotSym	Rot	when the Form → RotRef option is used, specifies the symbol used to represent the rotation
RefSym	Ref	when the Form → RotRef option is used, specifies the symbol used to represent the reflection

Options specific to Dihedral.

- Instead of using Rot to represent the lowest order rotation and Ref for a reflection, we can choose other symbols for these groups.

In[98]:= **CayleyTable[Dihedral[3, RotSym → "R", RefSym → "L"]]**

Out[98]= {{1, R, R^2, L, R**L, R^2**L}, {R, R^2, 1, R**L, R^2**L, L}, {R^2, 1, R, R^2**L, L, R**L}, {L, R^2**L, R**L, 1, R^2, R}, {R**L, L, R^2**L, R, 1, R^2}, {R^2**L, R**L, L, R^2, R, 1}}

- We can use the traditional name for the dihedral group, D, even though it is a reserved name in *Mathematica*.

In[99]:= **D[4]**

Out[99]= Groupoid[{1, Rot, Rot^2, Rot^3, Ref,
　　　　　Rot ** Ref, Rot^2 ** Ref, Rot^3 ** Ref}, -Operation-]

- The standard definition of D, in terms of differentiation, still works.

In[100]:= **{D[4, x], D[x^2, x]}**

Out[100]= {0, 2 x}

Although it is an easy result in group theory to show that there is essentially only one finite cyclic group, namely $\mathbb{Z}_n$, sometimes this group is expressed multiplicatively using a generic generator *g*. Cyclic is used to create this form.

Cyclic[*n*]	cyclic group of order *n*, expressed multiplicatively with generator "*g*"
Cyclic[*n*, *k*]	cyclic group of order *n/k*, expressed multiplicatively with generator "*g^k*"
Cyclic[*n*, Generator → *gen*]	cyclic group of order *n*, expressed multiplicatively with generator *gen*
CyclicGroup[*n*, *opts*]	identical to Cyclic[*n*, *opts*]

Variations of the Cyclic function.

- The default generator is used if no other is specified.

In[101]:= **Cyclic[8]**

Out[101]= Groupoid[{1, g, g^2, g^3, g^4, g^5, g^6, g^7}, -Operation-]

- If desired, a different symbol for the generator can be given.

In[102]:= **CyclicGroup[7, Generator → "α"]**

Out[102]= Groupoid[{1, α, α^2, α^3, α^4, α^5, α^6}, -Operation-]

- As with the other means of generating a cyclic group (using Z), we can also obtain variations by looking at multiples of a divisor of the index.

In[103]:= **Cyclic[8, 2, Generator → "β", Mode → Visual]**

```
Think of the elements as the numbers on a (modified)
clock. The differences, however, are that we view
the element 1 as being equivalent to twelve and the
other elements as powers of β^2. Multiplication of
two elements is done by adding the exponents on the
elements, just like adding the hours of a clock.
```

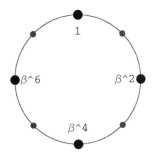

Out[103]= Groupoid[{1, β^2, β^4, β^6}, -Operation-]

■ 2.4.5 Other groupoids

In this section, we introduce the remaining built-in groupoids.

QuaternionGroup[]	quaternion group of order 8, given as matrices generated by {{0, 1}, {−1, 0}} and {{0, I}, {I, 0}}
QuaternionGroup[Form → AsMatrices]	identical to QuaternionGroup[]
Quaternion Group[Form → AsIJK]	quaternion group of order 8, given as elements {± 1, ± I, ± JJ, ± KK}
QuaternionGroup[Form → AsSymbols]	quaternion group of order 8, given as elements {e, a, a^2, a^3, b, b a, b a^2, b a^3}
Trivial	trivial group of order 1
Klein4	group whose elements are {e, a, b, c}, commonly called the Klein 4 group

Various other groupoids.

■ Klein4 is just a particular implementation of D_2.

In[104]:= **CayleyTable[Klein4, Mode → Visual]**

Klein4::warning :
    The elements e, a, b, c are considered strings and thus
        need to have double quotes around them when being used.

Klein4   x ⋆ y

Out[104]= {{e, a, b, c}, {a, e, c, b}, {b, c, e, a}, {c, b, a, e}}

- From the Cayley table, *b* appears to be one of the elements of the group.

In[105]:= **SubgroupGenerated[Klein4, b]**

> *Klein4::warning :*
> *The elements e, a, b, c are considered strings and thus*
> *need to have double quotes around them when being used.*

> *MemberQ::elmnt : b is not an element of Klein4.*

Out[105]= $Failed

- The elements in the Klein4 group are strings, so we need quotes around them when trying to work with them.

In[106]:= **SubgroupGenerated[Klein4, "b"]**

> *Klein4::warning :*
> *The elements e, a, b, c are considered strings and thus*
> *need to have double quotes around them when being used.*

Out[106]= Groupoid[{b, e}, -Operation-]

- The default form of the QuaternionGroup is to use the AsMatrices approach.

In[107]:= **QuaternionGroup[]**

Out[107]= Groupoid[
        {{{-1, 0}, {0, -1}}, {{0, -1}, {1, 0}}, {{0, -i}, {-i, 0}},
         {{0, i}, {i, 0}}, {{0, 1}, {-1, 0}}, {{-i, 0}, {0, i}},
         {{i, 0}, {0, -i}}, {{1, 0}, {0, 1}}}, -Operation-]

- There are two other approaches to this group (all of which are isomorphic to each other).

In[108]:= **{QuaternionGroup[Form → AsSymbols],**
        QuaternionGroup[Form → AsIJK]}

> *QuaternionGroup::JJKK :*
> *(Note that KK is used because K is a reserved symbol*
>   *in version 3 of Mathematica and JJ is used because*
>   *J is reserved as a generic name for an ideal.)*

Out[108]= {Groupoid[{e, a, a^2, a^3, b, ba, ba^2, ba^3}, -Operation-],
        Groupoid[{1, -1, i, -i, JJ, -JJ, KK, -KK}, -Operation-]}

2.5 Uses of the Cayley table

As has been shown, the Cayley table is a tool that can be valuable in working with groupoids. There is a large number of variations in implementing this function.

CayleyTable[*G*]	give the Cayley table of the groupoid *G* as a list of lists
CayleyTable[*G, opts*]	use the options in *opts* to form the table
CayleyTable[{$G_1, G_2, ..., G_n$} , *opts*]	give the Cayley tables of the groupoids $G_1, G_2,$ ..., G_n applying the options *opts* to each groupoid
CayleyTable[{$G_1, G_2, ..., G_n$}, {{$opts_1$}, {$opts_2$}, ..., {$opts_n$}}]	give the Cayley tables of the groupoids G_1, G_2 , ..., G_n applying the options $opts_i$ to groupoid G_i
CayleyTable[{$G_1, G_2, ..., G_n$}, {{$opts_1$}, {$opts_2$}, ..., {$opts_n$}}, Mode → *mode*]	give the Cayley tables of the groupoids $G_1, G_2, ...,$ G_n applying the options $opts_i$ to groupoid G_i, as well as applying Mode → *mode* to each groupoid
CayleyTable[{{G_1, $opts_1$}, { G_2, $opts_2$}, ..., { G_n, $opts_n$}}]	give the Cayley tables of the groupoids $G_1, G_2,$ ..., G_n applying the options $opts_i$ to groupoid G_i
CayleyTable[{{G_1, $opts_1$}, { G_2, $opts_2$}, ..., { G_n, $opts_n$}}, Mode → *mode*]	give the Cayley tables of the groupoids $G_1, G_2,$ ..., G_n applying the options $opts_i$ to groupoid G_i as well as applying Mode → *mode* to each groupoid

Variations for CayleyTable.

Because the Cayley table is such a rich tool for illustrating algebraic ideas, there are also many options available to control the display of these tables (all of which apply only when the Visual mode is being used).

option name	*default value*	
CayleyForm	OutputForm	specifies the form that should be applied to the elements in the body of the table
HeadingsColored	True	specifies whether the top and side headings of the table should be colored
KeyForm	InputForm	specifies the form that should be applied to the elements in the key of the table, when a key is used
ShowBodyText	True	specifies whether the text of the body of the table should be shown
ShowKey	True	specifies whether the key should be given, when a key is called for
ShowName	True	specifies whether the name should be shown above the table
ShowOperator	True	specifies whether the operator symbol should be shown in the corner of the table
ShowSidesText	True	specifies whether the text of the top row and left column of the table should be shown
TheSet	{}	specifies a reordering of the elements to be used in the table
VarToUse	"g"	specifies the symbol to use as the generic element name when a key is needed

Options specific to `CayleyTable`.

■ Consider the groupoids G_1 and G_2 defined as shown.

```
In[109]:= Clear[G]
          {G₁ = FormGroupoid[{2, 3, 1, 0}, Mod[#1 + #2, 4] &, "+"],
           G₂ = FormGroupoidByTable[{b, a, a ** b, aᵇ},
              {{a, a ** b, b, aᵇ}, {b, a, aᵇ, a ** b}, {a ** b, aᵇ, b, a},
               {aᵇ, a ** b, a, b}}, "*", WideElements → True]}
```

```
Out[110]= {Groupoid[{2, 3, 1, 0}, -Operation-],
           Groupoid[{b, a, a ** b, aᵇ}, -Operation-]}
```

■ The `Visual` mode of `CayleyTable` gives each element a unique color and then uses the colors throughout the body of the table.

```
In[111]:= CayleyTable[G₁, Mode → Visual]
```

TheGroup   x + y

y / x	2	3	1	0
2	0	1	3	2
3	1	2	0	3
1	3	0	2	1
0	2	3	1	0

Out[111]= {{0, 1, 3, 2}, {1, 2, 0, 3}, {3, 0, 2, 1}, {2, 3, 1, 0}}

- When a groupoid is formed and the option `WideElements → True` is added, a key is included in the table.

In[112]:= **CayleyTable[G₂, Mode → Visual]**

KEY for TheGroup: label used →
    element: {g1 → b, g2 → a, g3 → a**b, g4 → a^b}

TheGroup   x * y

y / x	g1	g2	g3	g4
g1	g2	g3	g1	g4
g2	g1	g2	g4	g3
g3	g3	g4	g1	g2
g4	g4	g3	g2	g1

Out[112]= {{a, a**b, b, a^b}, {b, a, a^b, a**b}, {a**b, a^b, b, a}, {a^b, a**b, a, b}}

- In addition to the options for `CayleyTable`, we can also pass graphics options.

In[113]:= **CayleyTable[G₁, Mode → Visual, TheSet → {0, 1, 2, 3},**
    HeadingsColored → False, ShowOperator → False,
    Background → Cyan, Frame → True, ShowSidesText → False]

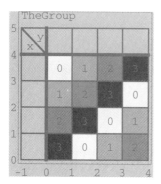

Out[113]= {{0, 1, 2, 3}, {1, 2, 3, 0}, {2, 3, 0, 1}, {3, 0, 1, 2}}

- As with many of the functions in these packages, multiple inputs are permissible. Here, the argument for Z is the list {3, 4}, which in turn, after evaluation, becomes a list of two groupoids for CayleyTable.

In[114]:= **CayleyTable[Z[{3, 4}], Mode → Visual]**

Z[3]   x + y   Z[4]   x + y

⟍y / x	0	1	2
0	0	1	2
1	1	2	0
2	2	0	1

⟍y / x	0	1	2	3
0	0	1	2	3
1	1	2	3	0
2	2	3	0	1
3	3	0	1	2

Out[114]= {{{0, 1, 2}, {1, 2, 0}, {2, 0, 1}},
           {{0, 1, 2, 3}, {1, 2, 3, 0}, {2, 3, 0, 1}, {3, 0, 1, 2}}}

- We can specify different options for each Groupoid.

In[115]:= **CayleyTable[{{G₁, ShowBodyText → False},**
          {G₂, ShowKey → False}}, Mode → Visual]

TheGroup   x + y   TheGroup   x * y

⟍y / x	2	3	1
2			
3			
1			

⟍y / x	g1	g2	g3	g4
g1	g2	g3	g1	g4
g2	g1	g2	g4	g3
g3	g3	g4	g1	g2
g4	g4	g3	g2	g1

```
Out[115]= {{{0, 1, 3, 2}, {1, 2, 0, 3}, {3, 0, 2, 1}, {2, 3, 1, 0}},
          {{a, a**b, b, aᵇ}, {b, a, aᵇ, a**b},
          {a**b, aᵇ, b, a}, {aᵇ, a**b, a, b}}}
```

TextCayley[*G*]	give a text-only version of a two-dimensional Cayley table

- While the graphical Cayley tables are usually more desirable, there is also a version for text-only systems.

```
In[118]:= TextCayley[G₂]
```

Out[118]//TableForm=

	b	a	a**b	aᵇ
–	–	–	–	–
b	a	a**b	b	aᵇ
a	b	a	aᵇ	a**b
a**b	a**b	aᵇ	b	a
aᵇ	aᵇ	a**b	a	b

2.6 Building other structures

When given an algebraic structure, such as a group, three questions naturally arise.

1. Can we take two structures, G_1 and G_2, and build a new structure based on them?

2. Given a structure G and a subset H of this structure, under what conditions is H itself a structure of the same type?

3. Given a structure G and a structure H that is also a subset of G, can we form a quotient structure G/H?

In this section we explore these three questions.

2.6.1 Direct products

For groups, the answer to the first question is to form the direct product or direct sum of the groups.

DirectProduct [G_1, G_2, ..., G_n]	give the direct product of the groupoids G_1, G_2, ..., G_n
DirectSum[G_1, G_2, ..., G_n]	identical to DirectProduct[G_1, G_2, ..., G_n]

How to form the direct product of several groupoids.

- Consider the following groupoids. Note that `DirectSum` and `DirectProduct` are functionally equivalent.

```
In[119]:= {G1 = DirectSum[Z[4], Z[4, 2]],
           G2 = DirectProduct[Z[3], Z[2, I]]}
```

```
Out[119]= {Groupoid[{{0, 0}, {0, 2}, {1, 0}, {1, 2},
              {2, 0}, {2, 2}, {3, 0}, {3, 2}}, -Operation-],
           Groupoid[{{0, 0}, {0, i}, {0, 1}, {0, 1 + i},
              {1, 0}, {1, i}, {1, 1}, {1, 1 + i}, {2, 0},
              {2, i}, {2, 1}, {2, 1 + i}}, -Operation-]}
```

- Let's pick two elements at random from `G1`.

```
In[120]:= {a, b} = RandomElements[G1, 2]
```

```
Out[120]= {{0, 0}, {3, 2}}
```

- The operation of a direct product uses the operation of groupoid G_i in the ith coordinate.

```
In[121]:= Operation[G1][a, b]
```

```
Out[121]= {3, 2}
```

- The `DirectSum` and `DirectProduct` functions can act on any number of arguments.

```
In[123]:= DirectSum[Z[2], Z[2], Z[2], Z[2]]
```

```
Out[123]= Groupoid[{{0, 0, 0, 0}, {0, 0, 0, 1},
              {0, 0, 1, 0}, {0, 0, 1, 1}, {0, 1, 0, 0}, {0, 1, 0, 1},
              {0, 1, 1, 0}, {0, 1, 1, 1}, {1, 0, 0, 0}, {1, 0, 0, 1},
              {1, 0, 1, 0}, {1, 0, 1, 1}, {1, 1, 0, 0}, {1, 1, 0, 1},
              {1, 1, 1, 0}, {1, 1, 1, 1}}, -Operation-]
```

■ 2.6.2 Subgroups

We now pursue the second question given at the outset of this section. Given a groupoid *G* and a subset *H*, we can form a new groupoid using the elements from *H* and the operation from *G*.

- Given a subset of a groupoid, we can form a new groupoid from the old.

```
In[124]:= G1 = Subgroupoid[Z[9], {0, 4}]
```

```
Out[124]= Groupoid[{0, 4}, Mod[#1 + #2, 9] &]
```

- We can now use `GroupQ` to determine if this is a subgroup.

```
In[125]:= GroupQ[G1]
```

Out[125]= False

- Alternatively, we can use SubgroupQ.

In[126]:= {**SubgroupQ[{0, 4}, Z[9]], SubgroupQ[G1, Z[9]]**}

Out[126]= {False, False}

Subgroupoid[G, H]	given H as a subset of elements of the groupoid G, return the groupoid with elements from H and operation from G
Subgroupoid[G, $\{H_1, H_2, ..., H_n\}$]	given a collection of subsets $\{H_1, H_2, ..., H_n\}$ of elements of the groupoid G, return the list of groupoids with elements from H_i and operation from G
Subgroupoid[$\{\{G_1, H_1\}, \{G_2, H_2\}, ..., \{G_n, H_n\}\}$]	given that the subset H_i is a subset of the elements of the groupoid G_i, return the list of groupoids with elements from H_i and operation from G_i

Variations for the function Subgroupoid.

- How often will a random set of elements from $\mathbb{Z}_9$ be a subgroup? Here we form 10 subgroupoids and then list the elements and whether this set forms a group or not. For $\mathbb{Z}_9$, what is the probability that three randomly chosen elements form a subgroup?

In[127]:= **subs = Subgroupoid[Z[9],**
       Table[RandomElements[Z[9], Random[Integer, {1, 9}],
         Replacement → False], {10}]];
       Map[{Elements[#], GroupQ[#]} &, subs] // ColumnForm

Out[128]= {{2, 6, 1, 0, 4}, False}
       {{8, 5, 7, 2, 6, 1, 0}, False}
       {{0, 1, 2, 3, 4, 5, 6, 7, 8}, True}
       {{3, 8, 2, 0, 7, 1, 6, 4}, False}
       {{6, 0, 2, 7}, False}
       {{2, 0, 8, 1, 3, 6, 7, 5}, False}
       {{5}, False}
       {{7, 6}, False}
       {{0, 1, 2, 3, 4, 5, 6, 7, 8}, True}
       {{0, 1, 2, 3, 4, 5, 6, 7, 8}, True}

- We can form several subgroupoids at once, as well as viewing visually what it means for the subset to be a subgroup.

In[129]:= **Subgroupoid[{{U[15], {1, 2}}, {Dihedral[3], {Rot, Ref}}},**
       Mode → Visual]

```
KEY for D[3]: label used → element: {g1 → Rot, g2 → Ref,
     g3 → 1, g4 → Rot^2, g5 → Rot**Ref, g6 → Rot^2**Ref}
```

U[15]   x * y   D[3]   x * y

X\Y	1	2	4	7	8	11	13	14
1	1	2	4	7	8	11	13	14
2	2	4	8	14	1	7	11	13
4	4	8	1	13	2	14	7	11
7	7	14	13	4	11	2	1	8
8	8	1	2	11	4	13	14	7
11	11	7	14	2	13	1	8	4
13	13	11	7	1	14	8	4	2
14	14	13	11	8	7	4	2	1

X\Y	g1	g2	g3	g4	g5	g6
g1	g4	g5	g1	g3	g6	g2
g2	g6	g3	g2	g5	g4	g1
g3	g1	g2	g3	g4	g5	g6
g4	g3	g6	g4	g1	g2	g5
g5	g2	g1	g5	g6	g3	g4
g6	g5	g4	g6	g2	g1	g3

```
Out[129]= {Groupoid[{1, 2}, Mod[#1 #2, 15] &],
           Groupoid[{Rot, Ref}, -Operation-]}
```

The syntax for the `SubgroupQ` function is similar to `Subgroupoid`, but the order of the arguments is reversed. This is because the notation $H < G$ is frequently used to indicate that H is a subgroup of G. Following are the basic forms; each of these can handle the standard modes, as well as `Visual2`.

`SubgroupQ[H, G]`	give `True` if the subset H is a subgroup of the groupoid G, and `False` otherwise
`SubgroupQ[` `{H_1, H_2, ...}, G]`	give a list of `True` or `False` depending on whether the subset H_i is a subgroup of the groupoid G
`SubgroupQ[{{H_1,` `G_1}, {H_2, G_2}, ...}]`	give a list of `True` or `False` depending on whether the subset H_i is a subgroup of groupoid G_i

Variations of the `SubgroupQ` function.

- The `TwistedZ` groupoid was introduced earlier. Is the set $\{0, 2, 8\}$ a subgroup of `TwistedZ[13]`?

```
In[130]:= SubgroupQ[ {0, 2, 8}, TwistedZ[13]]
```

```
Out[130]= True
```

- Before we can ask if $\{0, 4\}$ is a subgroup of `TwistedZ[6]`, we need this parent groupoid to be a group itself.

```
In[132]:= SubgroupQ[{0, 4}, TwistedZ[6]]
```

```
Group::fail :
   The Groupoid TwistedZ[6] fails at least one of the tests
      for being a group, which is needed for this function.
```

```
Out[132]= $Failed
```

- Let's pick some random sets of elements from TwistedZ[7] and see which of these are subgroups.

```
In[131]:= SubgroupQ[ Table[RandomElements[TwistedZ[7],
             Random[Integer, {1, Size[TwistedZ[7]]}],
             Replacement → False], {4}], TwistedZ[7], Mode → Visual]
```

TwistedZ[7]   x * y

x\y	3	0	1	2	4	5
3	1	3	0	4	5	2
0	3	0	1	2	4	5
1	0	1	3	5	2	4
2	4	2	5	1	0	3
4	5	4	2	0	3	1
5	2	5	4	3	1	0

TwistedZ[7]   x * y

x\y	2	0	1	3	4	5
2	1	2	5	4	0	3
0	2	0	1	3	4	5
1	5	1	3	0	2	4
3	4	3	0	1	5	2
4	0	4	2	5	3	1
5	3	5	4	2	1	0

TwistedZ[7]   x * y

x\y	3	2	1	4	0	5
3	1	4	0	5	3	2
2	4	1	5	0	2	3
1	0	5	3	2	1	4
4	5	0	2	3	4	1
0	3	2	1	4	0	5
5	2	3	4	1	5	0

TwistedZ[7]   x * y

x\y	5	2	0	4	3	1
5	0	3	5	1	2	4
2	3	1	2	0	4	5
0	5	2	0	4	3	1
4	1	0	4	3	5	2
3	2	4	3	5	1	0
1	4	5	1	2	0	3

Out[131]= {False, False, False, True}

- We can check several distinct groups at once for subgroups.

```
In[132]:= SubgroupQ[{{ {0, 3}, Z[5]}, { {1, 4}, U[9]}}, Mode → Visual]
```

Z[5]   x + y

x\y	0	3	1	2	4
0	0	3	1	2	4
3	3	1	4	0	2
1	1	4	2	3	0
2	2	0	3	4	1
4	4	2	0	1	3

U[9]   x * y

x\y	1	4	2	5	7	8
1	1	4	2	5	7	8
4	4	7	8	2	1	5
2	2	8	4	1	5	7
5	5	2	1	7	8	4
7	7	1	5	8	4	2
8	8	5	7	4	2	1

Out[132]= {False, False}

- The Visual2 mode requires that the subset be a subgroup to illustrate its visualization.

In[133]:= **SubgroupQ[**
        {{{0, 3}, Z[5]}, {{1, 4, 7}, U[9]}}, Mode → Visual2]

MemberQ::sbgrp : {0, 3} is not a subgroup of Z[5].

All the elements marked with yellow are elements in the
   subgroup. The others are colored according to the
   various left cosets of the subgroup in the group.

U[9]				x * y		
x\y	1	4	7	2	8	5
1	1	4	7	2	8	5
4	4	7	1	8	5	2
7	7	1	4	5	2	8
2	2	8	5	4	7	1
8	8	5	2	7	1	4
5	5	2	8	1	4	7

Graphics Error

Out[133]= {False, True}

Although not every subset *H* of a group *G* is a subgroup, the *closure* of every subset is a subgroup of *G*.

Closure[*G*, *H*]	given *H* as a subset of the elements of the groupoid *G*, return the groupoid generated by the elements in *H*
Closure[*G*, *H*, *opts*]	given *H* as a subset of the elements of the groupoid *G*, return the groupoid generated by the elements in *H* according to the options specified in *opts*
SubgroupClosure[*G*, *H*, *opts*]	identical to Closure[*G*, *H*, *opts*]

Use of Closure.

- Starting with the elements 1 and 4, we see that the element 7 is needed to complete the list before we have a subgroup.

In[134]:= **K = Closure[U[9], {1, 4}]**

Out[134]= Groupoid[{1, 4, 7}, Mod[#1 #2, 9] &]

- We can confirm that this is indeed a subgroup of U_9.

In[135]:= **SubgroupQ[K, U[9]]**

Out[135]= True

In addition to the standard Mode option, several other options for this function are available.

option name	default value	
ReportIterations	False	specifies whether the results of each iteration should be reported, in addition to the final closure
Sort	False	specifies whether the elements of the closure should be sorted in the output
Staged	False	specifies, when the mode is Visual, whether the results of each iteration should be reported visually one step at a time or all at once

Options for Closure.

- When ReportIterations is set to True, a list of the form {closure, number of iterations, result at each iteration} is returned.

In[137]:= **Closure[Z[9], {1, 4}, ReportIterations → True]**

Out[137]= {Groupoid[{1, 4, 2, 5, 8, 3, 6, 0, 7}, Mod[#1 + #2, 9] &],
{3, {{1, 4}, {1, 4, 2, 5, 8}, {1, 4, 2, 5, 8, 3, 6, 0, 7}}}}

- When the option Sort is set to True, the elements are sorted by the Sort function before being returned.

In[138]:= **Closure[Z[9], {1, 4}, Sort → True]**

Out[138]= Groupoid[{0, 1, 2, 3, 4, 5, 6, 7, 8}, Mod[#1 + #2, 9] &]

- When the Visual mode is used without Staged set to True, a visualization for each iteration is returned. Without the Output → GraphicsArray option, three full-size graphics are created, ready for animation, if desired.

In[139]:= **SubgroupClosure[Symmetric[3], {{2, 3, 1}},**
Mode → Visual, Output → GraphicsArray]

KEY for S[3]: label used → element: {g1 →
{2, 3, 1}, g2 → {1, 2, 3}, g3 → {1, 3, 2},
g4 → {2, 1, 3}, g5 → {3, 1, 2}, g6 → {3, 2, 1}}

Out[139]= - GraphicsArray -

- The `Staged` option to `Closure` causes all the graphics to be generated, but only the first iteration is displayed (at first).

In[140]:= **SubgroupClosure[Symmetric[3],**
      {{2, 3, 1}}, Mode → Visual, Staged → True]

      KEY for S[3]: label used → element: {g1 →
        {2, 3, 1}, g2 → {1, 2, 3}, g3 → {1, 3, 2},
        g4 → {2, 1, 3}, g5 → {3, 1, 2}, g6 → {3, 2, 1}}

      S[3]   x * y

Y\X	g1	g2	g3	g4	g5	g6
g1	g5	g1	g4	g6	g2	g3
g2	g1	g2	g3	g4	g5	g6
g3	g6	g3	g2	g5	g4	g1
g4	g3	g4	g1	g2	g6	g5
g5	g2	g5	g6	g3	g1	g4
g6	g4	g6	g5	g1	g3	g2

Out[140]= Groupoid[{{2, 3, 1}, {3, 1, 2}, {1, 2, 3}}, -Operation-]

The `NextStage` and `PreviousStage` functions are used to rotate through the cycle of iterations when the `Staged` option is used. (Note: A second argument, consisting of an integer, can be added to specify how many steps forward or backwards to take.)

`NextStage[Closure]`	when used with the `Staged` option, show the visualization of the next iteration of `Closure`
`PreviousStage[Closure]`	when used with the `Staged` option, show the visualization of the previous iteration of `Closure`

Functions to be used with the `Staged` option of `Closure`.

Another function for working with individual subgroups is the `SubgroupGenerated` function.

`SubgroupGenerated[G, g]`	given a group G and an element g from G, give the subgroup generated by g, $\langle g \rangle$

- The `SubgroupGenerated` function also reveals the order of an element, indicated by the number of elements in the subgroup that is generated.

In[141]:= **SubgroupGenerated[Alternating[4], {2, 3, 1, 4}]**

Out[141]= Groupoid[
        {{2, 3, 1, 4}, {3, 1, 2, 4}, {1, 2, 3, 4}}, -Operation-]

- Use of the Textual mode gives additional information about how a subgroup is generated.

In[142]:= **SubgroupGenerated[U[25], 4, Mode → Textual]**

    The group U[25] has the subgroup <4> given as

index	powers	simplified
1	(4)^1	4
2	(4)^2	16
3	(4)^3	14
4	(4)^4	6
5	(4)^5	24
6	(4)^6	21
7	(4)^7	9
8	(4)^8	11
9	(4)^9	19
10	(4)^10	1

    as generated: 4 16 14   6 24 21   9 11 19   1
          sorted: 1   4   6   9 11 14 16 19 21 24

Out[142]= Groupoid[{4, 16, 14, 6, 24, 21, 9, 11, 19, 1}, Mod[#1 #2, 25] &]

- There are two visual modes for this function. This one is best viewed as an animation after it is generated, or, as is done below, evaluated with the Output → GraphicsArray option.

In[143]:= **SubgroupGenerated[Z[8], 6,**
        Mode → Visual, Output → GraphicsArray]

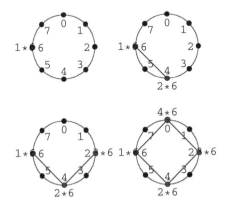

- This second visual mode uses the context of the Cayley table to illustrate the subgroup that is generated.

In[144]:= **SubgroupGenerated[U[11], 4, Mode → Visual2]**

U[11]										x * y	
⊠	1	2	3	4	5	6	7	8	9	10	
1	1	2	3	4	5	6	7	8	9	10	
2	2	4	6	8	10	1	3	5	7	9	
3	3	6	9	1	4	7	10	2	5	8	
4	4	8	1	5	9	2	6	10	3	7	
5	5	10	4	9	3	8	2	7	1	6	
6	6	1	7	2	8	3	9	4	10	5	
7	7	3	10	6	2	9	5	1	8	4	
8	8	5	2	10	7	4	1	9	6	3	
9	9	7	5	3	1	10	8	6	4	2	
10	10	9	8	7	6	5	4	3	2	1	

n 1 2 3 4 5
g^n 4 5 9 3 1

Out[144]= Groupoid[{4, 5, 9, 3, 1}, Mod[#1 #2, 11] &]

Sometimes we are interested in finding all the subgroups of a group.

CyclicSubgroups [*G*]	give the list of all cyclic subgroups of the groupoid *G*
Subgroups [*G*]	give the list of all subgroups of the groupoid *G*

Functions to obtain subgroups.

■ Here we generate all the cyclic subgroups of D_3.

In[145]:= **CyclicSubgroups[D[3]]**

Out[145]= {Groupoid[{1}, -Operation-],
 Groupoid[{1, Ref}, -Operation-],
 Groupoid[{1, Rot ** Ref}, -Operation-],
 Groupoid[{1, Rot^2 ** Ref}, -Operation-],
 Groupoid[{1, Rot, Rot^2}, -Operation-]}

■ Note the addition when we ask for *all* the subgroups of D_3.

In[146]:= **Subgroups[D[3]]**

Out[146]= {Groupoid[{1}, -Operation-],
 Groupoid[{1, Ref}, -Operation-],
 Groupoid[{1, Rot ** Ref}, -Operation-],
 Groupoid[{1, Rot^2 ** Ref}, -Operation-],
 Groupoid[{1, Rot, Rot^2}, -Operation-], Groupoid[
  {1, Rot, Rot^2, Ref, Rot ** Ref, Rot^2 ** Ref}, -Operation-]}

There are several other functions for working with subgroups.

SubgroupIntersection [G, H_1, H_2]	given a group G and subgroups H_1 and H_2, give the subgroupoid of G determined by the intersection of H_1 and H_2
SubgroupJoin [G, H_1, H_2]	given a group G and subgroups H_1 and H_2, give the subgroupoid of G generated by the elements in H_1 and H_2
SubgroupProduct [G, H_1, H_2]	given a group G and subgroups H_1 and H_2, give the subgroupoid $\{h_1 \, h_2 \mid h_1 \in H_1, h_2 \in H_2\}$ of G
SubgroupUnion [G, H_1, H_2]	given a group G and subgroups H_1 and H_2, give the subgroupoid of G determined by the union of H_1 and H_2

Various functions to combine two subgroups of a group.

- Consider the subgroups $\langle 2 \rangle$ and $\langle 3 \rangle$ in $\mathbb{Z}_{18}$. Here is the groupoid formed by their intersection.

```
In[147]:= G1 = SubgroupIntersection[
              Z[18], H1 = SubgroupGenerated[Z[18], 2],
              H2 = SubgroupGenerated[Z[18], 3]]

Out[147]= Groupoid[{0, 6, 12}, Mod[#1 + #2, 18] &]
```

- Using the same group and subgroups, here is the groupoid formed by their union.

```
In[148]:= G2 = SubgroupUnion[Z[18], H1, H2]

Out[148]= Groupoid[{0, 2, 3, 4, 6, 8, 9, 10, 12, 14, 15, 16},
              Mod[#1 + #2, 18] &]
```

- We can see which of these two groupoids are subgroups of $\mathbb{Z}_{18}$.

```
In[149]:= SubgroupQ[{G1, G2}, Z[18]]

Out[149]= {True, False}
```

- Here we form the SubgroupJoin of two subgroups of U_{24} and also the join of two subgroups of S_3.

```
In[150]:= {G1, G2} = {SubgroupJoin[U[24], {1, 17}, {1, 13}],
              SubgroupJoin[Symmetric[3],
                {{1, 2, 3}, {2, 1, 3}}, {{1, 2, 3}, {3, 2, 1}}]]

Out[150]= {Groupoid[{1, 5, 13, 17}, Mod[#1 #2, 24] &],
              Groupoid[{{1, 2, 3}, {1, 3, 2}, {2, 1, 3},
                {2, 3, 1}, {3, 1, 2}, {3, 2, 1}}, -Operation-]}
```

- Note that both these SubgroupJoins are subgroups (the first being a proper subgroup).

In[151]:= **SubgroupQ[{{ G1, U[24]}, { G2, Symmetric[3]}}]**

Out[151]= {True, True}

- Now we form the `SubgroupProduct` of the same subgroups in U_{24} and also the same subgroups of S_3.

In[152]:= **{G1, G2} = {SubgroupProduct[U[24], {1, 17}, {1, 13}],**
        SubgroupProduct[Symmetric[3],
          {{1, 2, 3}, {2, 1, 3}}, {{1, 2, 3}, {3, 2, 1}}]}

Out[152]= {Groupoid[{1, 5, 13, 17}, Mod[#1 #2, 24] &], Groupoid[
        {{1, 2, 3}, {2, 1, 3}, {2, 3, 1}, {3, 2, 1}}, -Operation-]}

- Note that these are not both subgroups (and that the product is different from the join in the S_3 case).

In[153]:= **SubgroupQ[{{ G1, U[24]}, { G2, Symmetric[3]}}]**

Out[153]= {True, False}

- Here is one way to generate a noncyclic subgroup of D_4.

In[154]:= **{G = SubgroupJoin[Dihedral[4],**
      SubgroupGenerated[Dihedral[4], Ref],
      SubgroupGenerated[Dihedral[4], Rot^2]], CyclicQ[G]}

Out[154]= {Groupoid[{1, Ref, Rot^2, Rot^2 ** Ref}, -Operation-], False}

■ 2.6.3 Quotient groups

The third question deals with quotient structures. The fundamental function related to this for groupoids is `QuotientGroup`. There is a number of related functions, as well as a number of options to consider for this function. Since a quotient group requires a normal subgroup H (from some group G) for its construction, and a normal subgroup is one in which every left coset $g H$ is equal to the right coset $H g$ (where $g \in G$), we start this section by considering cosets.

LeftCoset[G, H, g]	given a group G, subgroup H, and element g, give the left coset of H in G containing g
RightCoset[G, H, g]	given a group G, subgroup H, and element g, give the right coset of H in G containing g

Functions to create cosets.

- The left coset of the subgroup {0, 4} in the group $\mathbb{Z}_8$ containing 7 is the set 7 + {0, 4}.

In[155]:= **LeftCoset[Z[8], {0, 4}, 7]**

Out[155]= {7, 3}

■ The Visual mode is intended to illustrate how this coset is derived.

In[156]:= **LeftCoset[Z[8], {0, 4}, 7, Mode → Visual]**

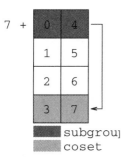

Out[156]= {7, 3}

■ The Textual mode walks through the calculation.

In[157]:= **LeftCoset[Z[8], {0, 4}, 7, Mode → Textual]**

Given an element g in a Groupoid G and a subgroup
   H of G, the left coset of H in G containing g
   is the set of all elements gH = {g h | h in H}.
In this case, the left coset of {0, 4} in Z[8] containing
   7 is given by {7, 3}. This can be seen as follows:
7 + {0, 4}
= {7 + 0, 7 + 4}
= {7, 3}

Out[157]= {7, 3}

■ Note that the subgroup for the second argument can be given as a list or as a
   groupoid. The coset *H* + *g* may happen to be the subgroup *H* itself.

In[158]:= **RightCoset[U[11],**
        SubgroupGenerated[U[11], 3], 3, Mode → Visual]

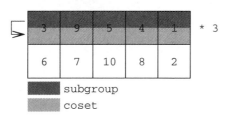

Out[158]= {9, 5, 4, 1, 3}

The examples shown thus far have been Abelian, so the distinction between left and right cosets is blurred.

■ For this nonabelian group with this subgroup and element, we get different results depending on whether we use left or right cosets.

In[159]:= **H = SubgroupGenerated[Symmetric[3], {3, 2, 1}];**
{LeftCoset[Symmetric[3], H, {3, 1, 2}],
RightCoset[Symmetric[3], H, {3, 1, 2}]}

Out[159]= {{{1, 3, 2}, {3, 1, 2}}, {{2, 1, 3}, {3, 1, 2}}}

LeftCosets[*G*, *H*]	given a group *G* and subgroup *H*, give the left cosets of *H* in *G*
RightCosets[*G*, *H*]	given a group *G* and subgroup *H*, give the right cosets of *H* in *G*

Functions to list all of the cosets.

■ Given a group *G* and a subgroup *H*, Lagrange's Theorem indicates that the number of cosets is $|G| / |H|$.

In[160]:= **LeftCosets[Z[8], {0, 4}]**

Out[160]= {{0, 4}, {1, 5}, {2, 6}, {3, 7}}

■ The Visual mode attempts to make a Cayley table of these cosets. Note: Output → Graphics causes the normal output to be suppressed so that the graphic image is the output. This is saved here and shown again.

In[161]:= **gr1 = RightCosets[Z[8],**
{0, 4}, Mode → Visual, Output → Graphics];

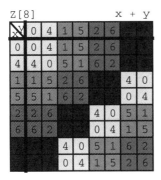

■ Compare the previous image to the following one.

In[162]:= **gr2 = CayleyTable[Z[4], Mode → Visual, Output → Graphics];**

- Putting the two graphics side by side adds great insight into the nature of the quotient group. (The first image was initially inspired by Ladnor Geissinger's work in his pioneering program *Exploring Small Groups*.)

In[163]:= **Show[GraphicsArray[{gr1, gr2}]];**

As indicated at the beginning of this section, a prerequisite to forming a quotient group is that the chosen subgroup be normal. The NormalQ function is used to test this. Additionally, the Index function measures the number of cosets, which in turn is also the size of the quotient group. Note that since we sometimes denote that the subgroup N is normal in G by $N \lhd G$, the order of the arguments in this function reflect this notation.

NormalQ[N, G]	given a group G and subgroup N, give True if N is normal in G, and False otherwise
Index[G, N]	given a group G and subgroup N, give the number of cosets of N in G

Related coset functions.

- The subgroup $H = \langle \{3, 2, 1\} \rangle$ in S_3 is not a normal subgroup.

In[164]:= **NormalQ[H = SubgroupGenerated[Symmetric[3], {3, 2, 1}],
Symmetric[3]]**

Out[164]= False

- We can still compute the index for this subgroup, however.

In[165]:= **Index[Symmetric[3], H]**

Out[165]= 3

- A quotient group is a group with a valid binary operation. The following shows why the lack of normality in *H* prevents a quotient structure from being defined.

In[166]:= **LeftCosets[Symmetric[3], H, Mode → Visual];**

```
KEY for S[3]: label used → element: {g1 →
    {3, 2, 1}, g2 → {1, 2, 3}, g3 → {2, 3, 1},
    g4 → {1, 3, 2}, g5 → {3, 1, 2}, g6 → {2, 1, 3}}
```

We are now ready to consider forming a quotient group.

QuotientGroup [*G, N*]	given a group *G* and normal subgroup *N*, give the quotient group *G/N*
QuotientGroup [*G, N, opts*]	give the quotient group *G/N* according to the options given in *opts*
FactorGroup [*G, N, opts*]	equivalent to QuotientGroup [*G, N, opts*]

Use of QuotientGroup.

- The default form of the elements in a quotient group is to use the symbol *NS* for the normal subgroup used in the construction and then list all the elements as left cosets of the form *g* + *NS* (or *g* NS if a multiplicative group).

In[167]:= **QuotientGroup[Z[8], {0, 4}]**

> *QuotientGroup::NS :*
> *This quotient group uses NS to represent*
>   *the normal subgroup {0, 4} that you*
>   *specified. Use CosetToList to convert this*
>   *coset representation to a list of elements.*

Out[167]= Groupoid[{NS, 1 + NS, 2 + NS, 3 + NS}, –Operation–]

■ The subgroup needs to be normal before the construction can take place.

In[168]:= **FactorGroup[Symmetric[3], H]**

> *Group::notnorm :*
> *{{3, 2, 1}, {1, 2, 3}} is not a normal subgroup of S[3].*

Out[168]= $Failed

A number of options are available for the QuotientGroup function.

option name	value	
Form	Cosets	return cosets in the form g + NS or g NS (default value)
Form	CosetLists	return each coset in the form of a list of the actual elements
Form	Representatives	return cosets by simply listing a representative from each coset
Representat ives	Canonical	use the list of representatives consisting of the "smallest" element in each coset (default value)
Representat ives	{g_1 , g_2 , ...}	use the list of representatives given by {g_1 , g_2 , ...} (if properly chosen)
Representat ives	Random	use a list of representatives chosen randomly

Options for QuotientGroup.

■ When a valid list of representatives is given, the list is used in other computations as well.

In[169]:= **CayleyTable[**
   QuotientGroup[Z[8], {0, 4}, Form → Representatives,
   Representatives → {4, 5, 6, 7}], Mode → Visual]

Z[8]/NS   x + y

Out[169]= {{4, 5, 6, 7}, {5, 6, 7, 4}, {6, 7, 4, 5}, {7, 4, 5, 6}}

- The CosetLists value of Form returns the elements as the actual cosets in the form of a list of elements.

In[170]:= **CayleyTable[**
          QuotientGroup[Z[8], {0, 4}, Form → CosetLists]]

Out[170]= {{{0, 4}, {1, 5}, {2, 6}, {3, 7}},
           {{1, 5}, {2, 6}, {3, 7}, {0, 4}},
           {{2, 6}, {3, 7}, {0, 4}, {1, 5}},
           {{3, 7}, {0, 4}, {1, 5}, {2, 6}}}

CosetToList[Q, *coset*]	return *coset* represented as a list of elements from G where *coset* is an element of $Q = G/N$
ElementToCoset[Q, *el*]	return the coset found in the quotient group $Q = G/N$ that contains the element *el* from G

Additional coset functions.

- Suppose we define G to be the following quotient group, using the default settings.

In[171]:= **Q = QuotientGroup[Z[16], {0, 4, 8, 12}]**

          QuotientGroup::NS :
          This quotient group uses NS to represent
          the normal subgroup {0, 4, 8, 12} that you
          specified. Use CosetToList to convert this
          coset representation to a list of elements.

Out[171]= Groupoid[{NS, 1 + NS, 2 + NS, 3 + NS}, -Operation-]

- Although the quotient group lists the elements in the form $g + NS$, we can recover the complete list of elements in any coset with the CosetToList function.

In[172]:= **CosetToList[Q, 2 + NS]**

Out[172]= {2, 6, 10, 14}

- Given an element in a group *G*, one can determine the coset to which it belongs in the quotient group *G* / *N* by using the ElementToCoset function.

In[173]:= **ElementToCoset[Q, 13]**

Out[173]= 1 + NS

▦ 2.7 Other group properties

This section collects together a number of functions that illustrate some property of a group, a subgroup, or an element in a group.

RandomElement[*G*]	give a random element from the groupoid *G*
RandomElements[*G*, *n*]	give *n* random elements from the groupoid *G*

Generating random elements.

- If a stop sign had a random element from D_8 applied to it as a transformation, the probability that a motorist might be confused is 15/16.

In[174]:= **RandomElement[Dihedral[8]]**

Out[174]= Rot^5 ** Ref

option name	value	
SelectFrom	NonIdentity	choose any element except the identity (default for RandomElement)
SelectFrom	Any	choose any element (default for RandomElements)
Replacement	True	assume replacement when choosing the elements (default for RandomElements)

Options for generating random elements.

- Although you can *ask* for something that you can't get, you won't get it.

In[175]:= **RandomElements[U[12], 5, Replacement → False]**

> *RandomElements::toomany :*
> *You can't ask for 5 random elements*
> *when there are only 4 available.*

Out[175]= {}

- The `SelectFrom` option allows us to avoid the identity element.

In[176]:= **RandomElements[U[12], 5, SelectFrom → NonIdentity]**

Out[176]= {5, 7, 11, 5, 11}

- `RandomElement` and `RandomElements` work on ringoids, polynomials, and other structures. Here we pick we a random element in $\mathbb{Z}\left[\sqrt{5}\,\right]$ of the form $a + b\sqrt{5}$ where $|a| \le 50$ and $|b| \le 50$.

In[177]:= **RandomElement$\left[\mathbf{Z}\left[\sqrt{5}\,\right], 50\right]$**

Out[177]= $22 - 12\sqrt{5}$

- `RandomElement` and `RandomElements` can operate on any list of elements, not just groupoids or ringoids.

In[178]:= **RandomElements[{α, β, γ, δ, ε, ζ, η, θ, ι,**
  κ, λ, μ, ν, ξ, ο, π, ρ, σ, τ, υ, φ, χ, ψ, ω}, 4]

Out[178]= {ι, υ, ρ, λ}

`Size[G]`	give the order of the groupoid G
`Order[G]`	equivalent to `Size[G]`
`OrderOfElement[G, g]`	give the order of the element g in the groupoid G
`Order[G, g]`	equivalent to `OrderOfElement[G, g]`
`OrderOfAllElements[G]`	give a list of the orders of all the elements in the groupoid G in the form $\{g, |g|\}$
`Orders[G]`	equivalent to `OrderOfAllElements[G]`
`Orders[G, list]`	equivalent to `OrderOfElement[G, list]`

Functions related to the order of a groupoid or its elements.

- The `Visual` mode of `OrderOfAllElements` reveals a number of facts, including the order of the groupoid, the order of each element and whether the groupoid is cyclic.

In[179]:= **OrderOfAllElements[U[8], Mode → Visual]**

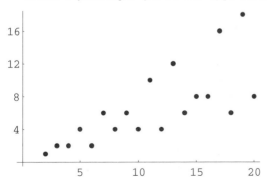

Out[179]= {{1, 1}, {3, 2}, {5, 2}, {7, 2}}

- While `OrderOfAllElements` returns the orders for all elements in pairs {*g*, |*g*|}, `OrderOfElement` can be given a list of elements and returns just the orders of those elements.

In[180]:= **OrderOfElement[U[8], {3, 5}]**

Out[180]= {2, 2}

- The `Orders` function has output similar to `OrderOfAllElements`.

In[181]:= **Orders[U[8], {3, 5}]**

Out[181]= {{3, 2}, {5, 2}}

- The order of the group U_n does not increase monotonically.

In[182]:= **ListPlot[Table[{n, Size[U[n]]}, {n, 2, 20}],**
              PlotStyle → {PointSize[0.02], RGBColor[0, 0, 1]},
              Ticks → {Table[i, {i, 5, 20, 5}], Table[i, {i, 4, 20, 4}]}];

- The group S_3 is nonabelian, as illustrated by the lack of symmetry across the main diagonal.

In[183]:= **AbelianQ[Symmetric[3], Mode → Visual]**

```
KEY for S[3]: label used → element: {g1 →
    {1, 2, 3}, g2 → {1, 3, 2}, g3 → {2, 1, 3},
    g4 → {2, 3, 1}, g5 → {3, 1, 2}, g6 → {3, 2, 1}}
```

S[3]   x * y

x\y	g1	g2	g3	g4	g5	g6
g1	g1	g2	g3	g4	g5	g6
g2	g2	g1	g5	g6	g3	g4
g3	g3	g4	g1	g2	g6	g5
g4	g4	g3	g6	g5	g1	g2
g5	g5	g6	g2	g1	g4	g3
g6	g6	g5	g4	g3	g2	g1

Out[183]= False

■ These are the pairs that do not commute.

In[184]:= **NonCommutingPairs[Symmetric[3]]**

Out[184]= {{{1, 3, 2}, {2, 1, 3}},
            {{1, 3, 2}, {2, 3, 1}}, {{1, 3, 2}, {3, 1, 2}},
            {{1, 3, 2}, {3, 2, 1}}, {{2, 1, 3}, {2, 3, 1}},
            {{2, 1, 3}, {3, 1, 2}}, {{2, 1, 3}, {3, 2, 1}},
            {{2, 3, 1}, {3, 2, 1}}, {{3, 1, 2}, {3, 2, 1}}}

Although commutativity is not a required property for a structure to be a group, it is an important property.

AbelianQ[G]	give True if the groupoid G is Abelian, and False otherwise
CommutativeQ[G]	equivalent to AbelianQ[G]
NonCommutingPairs[G]	give a list of the pairs of elements in the groupoid G that do not commute
Commutator[G, x, y]	give the commutator $x\,y\,x^{-1}\,y^{-1}$ in the groupoid G
Commutators[G]	give the set of commutators in the groupoid G
CommutatorSubgroup[G]	give the commutator subgroup in the groupoid G

Functions related to commutativity.

■ This shows the three commutators for D_3 and how they arise from various pairs of elements.

In[185]:= **Commutators[Dihedral[3], Mode → Visual]**

```
KEY for D[3]: label used → element: {g1 → 1, g2 → Rot,
    g3 → Rot^2, g4 → Ref, g5 → Rot**Ref, g6 → Rot^2**Ref}
```

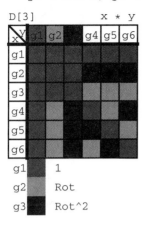

Out[185]= $\{1, \text{Rot}, \text{Rot}^2\}$

- From the foregoing table, we see that the commutator $x\,y\,x^{-1}\,y^{-1}$ for $x = \text{Rot}^2$ and $y = \text{Ref}$ is Rot, confirmed as follows.

In[186]:= **Commutator[Dihedral[3], Rot^2, Ref]**

Out[186]= Rot

The notion of the *center* of a group is related to commutativity. By definition, the center of a group G is $\{x \in G \mid g\,x = x\,g \ \forall\, g \in G\}$. Given an element g in G, the *centralizer* of g is $\{x \in G \mid g\,x = x\,g\}$. Related to this is the centralizer of a subgroup H of G: $\{x \in G \mid g\,x = x\,g \ \forall g \in H\}$.

GroupCenter[G]	give the center of the groupoid G
Center[G]	identical to GroupCenter[G]
Centralizer[G, g]	give the centralizer of the element g in the groupoid G
Centralizer[G, H]	give the centralizer of the subgroup H in the groupoid G

Functions related to the group center.

- The center of the group D_4 consists of two elements.

In[187]:= **GroupCenter[DihedralGroup[4]]**

Out[187]= $\{1, \text{Rot}^2\}$

- There are four elements in D_4 that commute with the element Ref.

In[188]:= **Centralizer[DihedralGroup[4], Ref]**

Out[188]= Groupoid[{1, Rot^2, Ref, Rot^2 ** Ref}, -Operation-]

- The only element in D_4 that commutes with all the elements in the subgroup {1, Ref} is the identity element, 1.

In[189]:= **Centralizer[DihedralGroup[4], {1, Ref}]**

Out[189]= Groupoid[{1}, -Operation-]

Given a group G and elements h and x, we say $x\,h\,x^{-1}$ is the *conjugate* of h by x. The *conjugacy class* of h in G is defined to be $\{x\,h\,x^{-1} \mid x \in G\}$. Given a subgroup H, the conjugate of H by x is $\{x\,h\,x^{-1} \mid h \in H\}$ and is also a subgroup. Related to this is the *normalizer* of H in G: $\{x \in G \mid x\,H\,x^{-1} = H\}$; this too is a subgroup.

ElementConjugate[G, h, x]	give the element $x\ h\ x^{-1}$ in the groupoid G
Conjugate[G, h, x]	identical to ElementConjugate[G, h, x]
SubgroupConjugate[G, H, x]	give the groupoid that is the conjugate of the subgroup H by x
Conjugate[G, H, x]	identical to SubgroupConjugate[G, H, x]
ConjugacyClass[G, h]	give the conjugacy class of h in the group G
Normalizer[G, H]	give the normalizer of the subgroup H in G

Functions related to conjugation.

- Since $\mathbb{Z}_5$ is Abelian, conjugating 2 by 4 simply returns 2.

In[190]:= **ElementConjugate[Z[5], 2, 4]**

Out[190]= 2

- Conjugating the subgroup {1, Ref} in D_3 by Rot returns the subgroup {1, Rot^2 ** Ref}.

In[191]:= **SubgroupConjugate[Dihedral[3], {1, Ref}, Rot]**

Out[191]= Groupoid[{1, Rot^2 ** Ref}, -Operation-]

- Here we find the conjugacy class of the element Ref in D_3.

In[192]:= **CR = ConjugacyClass[Dihedral[3], Ref]**

Out[192]= {Ref, Rot ** Ref, Rot^2 ** Ref}

- We see that this conjugacy class is not a subgroup.

In[193]:= **SubgroupQ[CR, Dihedral[3]]**

Out[193]= False

CyclicQ[*G*]	give True if the groupoid *G* is cyclic, and False otherwise
CyclicGenerators[*G*]	give a list of elements in the groupoid *G* that are cyclic generators of *G*

Functions related to cyclicity.

- Since there is no element of order 8, the group U_{15} is not cyclic.

In[194]:= **CyclicQ[U[15], Mode → Visual]**

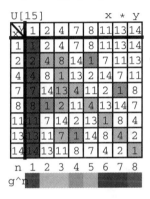

Out[194]= False

- One third of the elements in U_{13} can act as a generator for the group.

In[195]:= **{CyclicGenerators[U[13]], Size[U[13]]}**

Out[195]= {{2, 6, 7, 11}, 12}

- The following exponents (including the 1 implicitly given on g) are intimately related to 12. How?

In[196]:= **CyclicGenerators[Cyclic[12]]**

Out[196]= {g, g^5, g^7, g^{11}}

ElementToPower[*G*, *g*, *n*]	give the element g^n in the groupoid *G*
GroupExponent[*G*]	give the smallest positive integer *n* such $g^n = \mathrm{id}_G$ for all elements *g* in the groupoid *G*

Miscellaneous functions.

- Using the Table function with ElementToPower, we can easily see various powers of a given element.

In[197]:= **Table[{n, ElementToPower[U[15], 2, n]}, {n, -2, 4}] //**
         TableForm[#, TableHeadings → {None, {"n", "2ⁿ\n"}}] &

Out[197]//TableForm=

n	2^n
-2	4
-1	8
0	1
1	2
2	4
3	8
4	1

- We see that the exponent of the group U_{15} is 4.

In[198]:= **GroupExponent[U[15], Mode → Visual]**

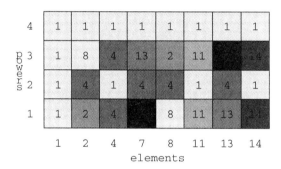

Out[198]= 4

Chapter 3

Ringoids

🖿 3.1 Forming Ringoids

The principal means by which ringoids are formed is with `FormRingoid`. Additionally, there is `FormRingoidByTable`, which parallels the function `FormGroupoidByTable`.

■ 3.1.1 FormRingoid

`FormRingoid[` `{els}, add, mult]`	form a `Ringoid` with elements {*els*} and operations *add* and *mult* using + and * as operation symbols
`FormRingoid[{els},` `add, mult, {asym, msym}]`	form a `Ringoid` with elements {*els*} and operations *add* and *mult* using *asym* and *msym* as operation symbols
`FormRingoid[` `{els}, add, mult, opts]`	form a `Ringoid` with elements {*els*}, operations *add* and *mult* using + and * as operation symbols and options *opts*
`FormRingoid[{els}, add,` `mult, {asym, msym}, opts]`	form a `Ringoid` with elements {*els*} and operations *add* and *mult* using *asym* and *msym* as operation symbols and options *opts*

Variations in using `FormRingoid`.

The syntax of the `FormRingoid` function requires the first parameter to be a list of elements and the next two parameters to be binary operations that serve as the addition and multiplication of the ringoid that is to be created. These binary operations are usually pure functions with two `Slot` variables, #1 and #2. An optional fourth parameter can be provided, which is a list of two symbols to be used for addition and multiplication. Finally, any option of `FormRingoid` can be included as a final argument.

The choice of operation symbols does not change the ringoid's properties; it is simply a cosmetic change that makes some outputs more appealing.

- The following are essentially the same ringoid.

In[1]:= `{R1 = FormRingoid[{True, False}, Xor, And],`
        `R2 = FormRingoid[{True, False}, Xor, And, {"⊕", "∧"}]}`

Out[1]= `{Ringoid[{True, False}, -Addition-, -Multiplication-],`
          `Ringoid[{True, False}, -Addition-, -Multiplication-]}`

option name	default value	
`CayleyForm`	`OutputForm`	specifies how to format elements appearing in Cayley tables
`FormatElements`	`False`	specifies whether elements should be displayed or formatted to `-Elements-`
`FormatOperator`	`True`	specifies whether addition and multiplication should be displayed or formatted to `-Addition-` and `-Multiplication-`
`IsARing`	`False`	specifies whether we can assume that the ringoid is a ring
`KeyForm`	`InputForm`	specifies how to format elements appearing in a key when `CayleyTables` or `MultiplicationTable` is called with a ringoid having `WideElements` set to `True`
`MaxElementsTo List`	`50`	specifies the number of elements to list before the list of elements is automatically formatted.
`RingoidDescription`	`" "`	brief description of the ringoid
`RingoidName`	`"TheRing"`	name of the ringoid
`WideElements`	`False`	specifies whether the elements are considered too wide to fit into a Cayley table, initiating a key to be set up

Options for `FormRingoid`.

The operations on some larger ringoids can be time-consuming. Some time can be saved if you know that a ringoid is indeed a ring by declaring it as such with the `IsARing` option. Caution: Your declaration is not checked, so you should use this option only if you are sure you have a ring.

- Here is a big ring.

In[2]:= **BigRing = FormRingoid[Range[0, 1000],**
   Mod[#1 + #2, 1001] &, Mod[#1 #2, 1001] &, IsARing → True]

Out[2]= Ringoid[{-Elements-}, -Addition-, -Multiplication-]

- The following timing has been greatly reduced by the use of IsARing.

In[3]:= **Timing[RingQ[BigRing]]**

Out[3]= {0.0166667 Second, True}

- Normally the elements of a small ringoid are listed; they can, however, be suppressed.

In[4]:= **FormRingoid[Range[0, 1],**
   Mod[#1 + #2, 2] &, Times, MaxElementsToList → 1]

Out[4]= Ringoid[{-Elements-}, -Addition-, -Multiplication-]

- In most cases, the usual addition and multiplication symbols are adequate. Here, as the fourth parameter, alternate symbols are specified.

In[5]:= **UR = FormRingoid[{"a", "b", "c"}, #1 &,**
   #2 &, {"Fst", "Lst"}, RingoidName → "UnRing",
   RingoidDescription → "A nonsense Ringoid"]

Out[5]= Ringoid[{a, b, c}, -Addition-, -Multiplication-]

- The alternate symbols appear in the Cayley tables.

In[6]:= **CayleyTables[UR, Mode → Visual];**

  For each element, a different color is used.
   The entries in the table corresponding to the
   elements are then colored and labeled accordingly.

Add.	x Fst y			Mult.	x Lst y		
y\x	a	b	c	y\x	a	b	c
a	a	a	a	a	a	b	c
b	b	b	b	b	a	b	c
c	c	c	c	c	a	b	c

- Here is a ring of multiples of π mod 3π with CayleyForm set to TraditionalForm.

In[7]:= **piRing = FormRingoid$\left[\text{Range}[0,\ 2\ \pi,\ \pi],\ \pi\ \text{Mod}\left[\dfrac{\#1 + \#2}{\pi},\ 3\right]\ \&,\right.$**

$\left.\pi\ \text{Mod}\left[\dfrac{\#1\ \#2}{\pi^2},\ 3\right]\ \&,\ \text{CayleyForm} \rightarrow \text{TraditionalForm}\right]$**

Out[7]= Ringoid[{0, π, 2 π}, -Addition-, -Multiplication-]

- The Cayley tables are displayed with the TraditionalForm of π.

In[8]:= **CayleyTables[piRing, Mode → Visual];**

- Now we form the same ring with CayleyForm set to InputForm.

In[9]:= **piRing2 = FormRingoid$\left[\text{Range}[0,\ 2\ \pi,\ \pi],\ \pi\ \text{Mod}\left[\dfrac{\#1 + \#2}{\pi},\ 3\right]\ \&,\right.$**

$\left.\pi\ \text{Mod}\left[\dfrac{\#1\ \#2}{\pi^2},\ 3\right]\ \&,\ \text{CayleyForm} \rightarrow \text{InputForm}\right]$**

Out[9]= Ringoid[{0, π, 2 π}, -Addition-, -Multiplication-]

- Now π is displayed as Pi in the Cayley tables.

In[10]:= **CayleyTables[piRing2, Mode → Visual];**

- By using the option CayleyForm with the CayleyTables function, we can always override any preset value. Here we override piRing2's CayleyForm value of InputForm.

```
In[11]:= CayleyTables[piRing2,
             Mode → Visual, CayleyForm → TraditionalForm];
```

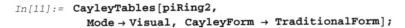

Add.   x + y   Mult.   x * y

■ 3.1.2 FormRingoidByTable

A second general means of forming ringoids is with FormRingoidByTable.

FormRingoidByTable[{els}, atable, mtable]	form a Ringoid with elements { els} and operation tables *atable* and *mtable* using + and * as operation symbols
FormRingoidByTable[{els}, atable, mtable, {asym, msym}, opts]	form a Ringoid with elements { els} and operation tables *atable* and *mtable* using *asym* and *msym* as operation symbols and options *opts*

Variations in using FormRingoidByTable.

■ Here are two tables of integers between 0 and 3.

```
In[12]:= {atable = (RotateLeft[Range[0, 3], #1] &) /@Range[0, 3],
          mtable = (RotateLeft[Range[0, 3], #1] &) /@Range[1, 7, 2]} //
          ColumnForm

Out[12]= {{0, 1, 2, 3}, {1, 2, 3, 0}, {2, 3, 0, 1}, {3, 0, 1, 2}}
         {{1, 2, 3, 0}, {3, 0, 1, 2}, {1, 2, 3, 0}, {3, 0, 1, 2}}
```

■ These tables can be used to define a ringoid. FormRingoidByTable accepts the same options as FormRingoid.

```
In[13]:= FormRingoidByTable[Range[0, 3], atable, mtable,
          RingoidDescription → "Experimental Ringoid",
          FormatElements → True, FormatOperator → False]

Out[13]= Ringoid[{-Elements-},
          First[Flatten[{{0, 1, 2, 3}, {1, 2, 3, 0}, {2, 3, 0, 1}, {3,
                0, 1, 2}}[[First[(Position[{0, 1, 2, 3}, #1] &)[#1]],
             First[(Position[{0, 1, 2, 3}, #1] &)[#2]]]]]] &,
```

```
First[Flatten[{{1, 2, 3, 0}, {3, 0, 1, 2}, {1, 2, 3, 0}, {3,
      0, 1, 2}}[[First[(Position[{0, 1, 2, 3}, #1] &)[#1]],
   First[(Position[{0, 1, 2, 3}, #1] &)[#2]]]]]] &]
```

▨ 3.2 Structure of Ringoids

▪ 3.2.1 Basic functions

A ringoid is a structure that has head `Ringoid` and four arguments, the fourth being suppressed and used for internal purposes. The first argument is the domain of the ringoid, which can be referenced with `Domain`. The second argument is the addition operation on the domain, referenced with `Addition`. The third argument is the multiplication operation on the domain, referenced with `Multiplication`.

Domain[R]	return the domain (elements) of R
Elements[R]	identical to Domain[R]
Addition[R]	return the addition operation of R (as a pure function)
Multiplication[R]	return the multiplication operation of R (as a pure function)

Functions for working with the structure of a ringoid.

▪ The expression ZR[5] constructs the ringoid $\mathbb{Z}_5$.

In[14]:= **ZR[5]**

Out[14]= Ringoid[{0, 1, 2, 3, 4}, Mod[#1 + #2, 5] &, Mod[#1 #2, 5] &]

▪ Here are the elements in the ringoid.

In[15]:= **Domain[ZR[5]]**

Out[15]= {0, 1, 2, 3, 4}

▪ This is how to obtain the addition and multiplication operations.

In[16]:= **{Addition[ZR[5]], Multiplication[ZR[5]]}**

Out[16]= {Mod[#1 + #2, 5] &, Mod[#1 #2, 5] &}

Behind the scenes, when ZR[5] is evaluated, `FormRingoid` is called in order to construct it.

■ 3.2.2 Related functions

`RingoidName[`*R*`]`	return the name of the ringoid *R*
`PlusSymbol[`*R*`]`	return the symbol used for the addition of the ringoid *R*
`TimesSymbol[`*R*`]`	return the symbol used for the multiplication of the ringoid *R*
`ElementQ[`*x*`,` *R*`]`	give `True` if *x* is an element of *R*, and `False` otherwise
`ElementsQ[`*els*`,` *R*`]`	give `True` if all members of *els* are elements of the ringoid *R*, and `False` otherwise

Functions related to the structure of a ringoid.

■ Here is an example of a nonstandard ring with information provided as options to `FormRingoid`.

```
In[17]:= R = FormRingoid[Range[1, 4], Min[#1, #2] &,
            Max[#1, #2] &, {"Min", "Max"}, RingoidDescription →
            "An example of a Ringoid that is not a ring.",
            RingoidName → "Rex"]

Out[17]= Ringoid[{1, 2, 3, 4}, -Addition-, -Multiplication-]
```

■ Here is how we can learn the name of this ringoid and also the symbols used for its operators.

```
In[18]:= {RingoidName[R], PlusSymbol[R], TimesSymbol[R]}

Out[18]= {Rex, Min, Max}
```

■ The `ElementQ` and `ElementsQ` functions work just as they do for groupoids.

```
In[19]:= {ElementQ[3, R], ElementsQ[{3, 5}, R]}

Out[19]= {True, False}
```

■ In this ringoid, the "sum" of two numbers is the minimum of the two. Before doing any testing for ring properties, here is what is known about this ringoid.

```
In[20]:= RingInfo[R]

Out[20]= {Rex, An example of a Ringoid that is not a ring.}
```

■ As tests are performed on a ringoid, information is collected and retrievable by the `RingInfo` function.

```
In[21]:= {Zero[R], RingQ[R], RingInfo[R]}

Out[21]= {4, False,
          {Rex, An example of a Ringoid that is not a ring.,
```

```
the Ringoid has as a zero the element 4,
the set is closed under addition,
the set is closed under multiplication,
there are elements that do not have an additive inverse,
this is NOT a ring}}
```

- Most would agree that a ring where all products are zero is "freaky." Why not use a "freaky" multiplication symbol?

In[22]:= **FreakyRing = FormRingoid[**
    Range[0, 2], Mod[#1 + #2, 3] &, 0 &, {"+", "☺"}];
MultiplicationTable[FreakyRing, Mode → Visual];

(TheRing, x)   x ☺ y

y / x	0	1	2
0	0	0	0
1	0	0	0
2	0	0	0

Sometimes it is useful to choose a random ring element or a list of random elements.

RandomElement[R, *opts*]	return a random element from the ringoid R according to the options *opts*
RandomElements[R, n, *opts*]	return n random elements from the ringoid R according to the options *opts*

Choosing random elements.

- To obtain nine random elements (allowing replacement) from $\mathbb{Z}_{11}$, we can use either of the following methods.

In[24]:= **{Table[RandomElement[Z[11]], {9}],**
    RandomElements[Z[11], 9]}

Out[24]= {{7, 1, 4, 2, 8, 2, 6, 7, 3}, {8, 0, 1, 0, 2, 7, 6, 2, 10}}

- If you are interested in nine distinct elements and avoiding both of the identities, we can use the following options to accomplish this.

In[25]:= **RandomElements[Z[11], 9,**
    Replacement → False, SelectFrom → NonIdentity]

Out[25]= {7, 5, 8, 10, 9, 4, 6, 2, 3}

option name	value	
SelectFrom	NonZero	choose any element except the zero (default for RandomElement)
SelectFrom	NonUnity	choose any element except the unity
SelectFrom	NonIdentity	choose any element except the unity or zero
SelectFrom	Any	choose any element (default for RandomElements)
Replacement	True	assume replacement when choosing the elements (default for RandomElements)

Options for generating random elements.

■ 3.2.3 Groupoids from ringoids

Since a ringoid is a collection of elements with two binary operations, it makes sense to consider splitting off either of the operations with the elements and forming a groupoid. There are several directions to go here: the additive groupoid, the multiplicative groupoid, and the multiplicative groupoid consisting of just the nonzero elements.

AdditiveGroupoid[R]	groupoid associated with ringoid R consisting of the elements of R and its addition operation
AGroupoid[R]	identical to AdditiveGroupoid[R]
MultiplicativeGroupoid[R]	groupoid associated with ringoid R consisting of the elements of R and its multiplication operation
MGroupoid[R]	identical to MultiplicativeGroupoid[R]
NonZeroMGroupoid[R]	groupoid of nonzero elements of ringoid R with the multiplication of R

Groupoids associated with a ringoid.

■ The additive groupoid of the ring $\mathbb{Z}_n$ is essentially the group $\mathbb{Z}_n$.

```
In[26]:= AdditiveGroupoid[ZR[11]]
```

```
Out[26]= Groupoid[{0, 1, 2, 3, 4, 5, 6, 7, 8, 9, 10}, Mod[#1 + #2, 11] &]
```

■ The additive groupoid of a ring is an Abelian group.

```
In[27]:= G = AdditiveGroupoid[LatticeRing[30]];
         {GroupQ[G], AbelianQ[G]}
```

```
Out[28]= {True, True}
```

- The multiplicative groupoid of a ring is never a group.

In[29]:= `GroupQ[MultiplicativeGroupoid[Z[9]]]`

Out[29]= `False`

- In some cases, the multiplicative groupoid of nonzero elements is a group.

In[30]:= `{G = NonZeroMGroupoid[Z[7]], GroupQ[G]}`

Out[30]= `{Groupoid[{1, 2, 3, 4, 5, 6}, Mod[#1 #2, 7] &], True}`

3.3 Testing properties of a ring

3.3.1 Additive properties

The additive properties of a ring require that the domain be closed with respect to addition, an identity (the zero) for addition must exist in the domain, each domain element must have an additive inverse in the domain, and addition must be both commutative and associative.

`ClosedQ[R, Operation → Addition]`	give `True` if the elements of the ringoid R are closed with respect to the addition, and `False` otherwise
`HasZeroQ[R]`	give `True` if R has a zero, and `False` otherwise
`HasInversesQ[R, Operation → Addition]`	give `True` if every element in R has an additive inverse in R, and `False` otherwise
`AssociativeQ[R, Operation → Addition]`	give `True` if the addition of R is associative, and `False` otherwise
`CommutativeQ[R, Operation → Addition]`	give `True` if the addition of R is commutative, and `False` otherwise
`HasIdentityQ[R, Operation → Addition]`	identical to `HasZeroQ[R]`

Functions for testing the additive properties of a ring.

- Although it is common knowledge that $\mathbb{Z}_5$ satisfies all the additive ring properties, here the *Mathematica* commands confirm it.

In[31]:= `{ClosedQ[Z[5], Operation → Addition], HasZeroQ[Z[5]],`
`        HasInversesQ[Z[5], Operation → Addition],`
`        AssociativeQ[Z[5], Operation → Addition],`
`        CommutativeQ[Z[5], Operation → Addition]}`

Out[31]= `{True, True, True, True, True}`

■ The `Visual` mode illustrates the symmetry involved in the commutative property.

In[32]:= **CommutativeQ[Z[5], Operation → Addition, Mode → Visual]**

Add(Z[5])   x + y

y \ x	0	1	2	3	4
0	0	1	2	3	4
1	1	2	3	4	0
2	2	3	4	0	1
3	3	4	0	1	2
4	4	0	1	2	3

Out[32]= True

`Zero[R]`	return the zero of R, if one exists, and `$Failed` otherwise
`HasNegativeQ[R, r]`	give `True` if r has a negative in R, and `False` otherwise
`NegationOf[R, r]`	return the additive inverse of r in R, if it exists, and `$Failed` otherwise
`RandomAssociativeQ[ R, Operation → Addition]`	give `True` if addition on R appears to be associative (based on testing 25 random triples from R), and `False` otherwise
`InvertibleQ[R, r, Operation → Addition]`	identical to `HasNegativeQ[R, r]`
`Inverses[R, Operation → Addition]`	return a list of pairs of the form $\{x, -x\}$, or $\{x,$ no inverse$\}$ if x has no negation

Related functions for testing the additive properties of a ring.

In[33]:= **Inverses[FormRingoid[{2, 3, -2, 0}, Plus, Times], Operation → Addition]**

Out[33]= {{2, -2}, {0, 0}, {3, no inverse}}

■ By the following, we know that 1 has a negative in $\mathbb{Z}_5$.

In[34]:= **HasNegativeQ[Z[5], 1]**

Out[34]= True

■ Here is why 0 is the zero of $\mathbb{Z}_5$.

In[35]:= **Zero[Z[5], Mode → Visual]**

```
Add(Z[5])        x + y       Add(Z[5])        x + y
```

y\x	0	1	2	3	4		y\x	0	1	2	3	4
0	0	1	2	3	4		0	0	1	2	3	4
1	1	2	3	4	0		1	1	2	3	4	0
2	2	3	4	0	1		2	2	3	4	0	1
3	3	4	0	1	2		3	3	4	0	1	2
4	4	0	1	2	3		4	4	0	1	2	3

```
red->left identity          red->right identity
```

Out[35]= 0

- This explains why the negation of 1 in $\mathbb{Z}_5$ is 4.

In[36]:= **NegationOf[Z[5], 1, Mode → Textual]**

> Given a Ringoid R, we say an element r in R has an
>   additive inverse s if R has an additive identity 0 and
>   r + s = s + r = 0 (where + indicates the operation).
>
> In this case, 1 has 4 as the inverse.

Out[36]= 4

Associativity in a ringoid with n elements involves n^3 tests, so RandomAssociativeQ is suggested for larger ringoids.

- If we used AssociativeQ here, we would wait for 64,000 tests.

In[37]:= **RandomAssociativeQ[Z[40], Operation → Addition]**

Out[37]= True

option name	default value	
Mode	Computational	specifies what type of information, if any, should be given in addition to the result of the computation
Operation	Both	for functions that apply to both addition and multiplication, specifies which of the operations should be considered; Both, Addition, or Multiplication are values for this option

Options for functions that test ring properties.

- Most of the functions that test the various properties of a ring allow the Operation option. Here we illustrate the four possibilities; note that (in this case—not true for all functions) not specifying the option assumes that both operations are to be checked.

```
In[38]:= R = FormRingoid[{-1, 0, 1}, Plus, Times];
         {ClosedQ[R, Operation → Addition],
          ClosedQ[R, Operation → Multiplication],
          ClosedQ[R, Operation → Both], ClosedQ[R]}

Out[39]= {False, True, False, False}
```

■ 3.3.2 Multiplicative properties

The only purely multiplicative properties a ring must satisfy are closure and associativity, but there are several other additional properties that are of interest and can be tested.

ClosedQ[*R*, Operation → Multiplication]	give True if the elements of the ringoid *R* are closed with respect to the multiplication, and False otherwise
AssociativeQ[*R*, Operation → Multiplication]	give True if the multiplication on *R* is associative, and False otherwise

Functions for the required multiplicative ring properties.

As with the additive properties, there are not many surprises when you consider standard rings such as the $\mathbb{Z}_n$ family.

■ The unity is 1.

```
In[40]:= {WithUnityQ[Z[15]], Unity[Z[15]]}

Out[40]= {True, 1}
```

■ For the units of $\mathbb{Z}_{15}$, here is how elements are paired with their multiplicative inverses.

```
In[41]:= Map[{#, MultiplicativeInverse[Z[15], #]} &, Units[Z[15]]]

Out[41]= {{1, 1}, {2, 8}, {4, 4}, {7, 13},
          {8, 2}, {11, 11}, {13, 7}, {14, 14}}
```

■ Here is an alternate method that considers all the elements in the ring, providing appropriate information for those lacking multiplicative inverses.

```
In[42]:= Inverses[Z[15], Operation → Multiplication]

Out[42]= {{0, no inverse}, {1, 1}, {2, 8}, {3, no inverse}, {4, 4},
          {5, no inverse}, {6, no inverse}, {7, 13}, {9, no inverse},
          {10, no inverse}, {11, 11}, {12, no inverse}, {14, 14}}
```

WithUnityQ[*R*]	give True if a multiplicative identity (unity) exists in *R*, and False otherwise
Unity[*R*]	return the unity of *R*, if one exists, and $Failed otherwise
UnitQ[*R*, *r*]	give True if *r* has a multiplicative inverse in *R*, and False otherwise
MultiplicativeInverse[*R*, *r*]	return the multiplicative inverse of *r* in *R*, if it exists, and $Failed otherwise
CommutativeQ[*R*, Operation → Multiplication]	give True if multiplication is commutative in *R*, and False otherwise
RandomAssociativeQ[*R*, Operation → Multiplication]	give True if multiplication in *R* appears to be associative (based on testing 25 random triples), and False otherwise
HasUnityQ[*R*]	identical to WithUnityQ[*R*]
HasIdentityQ[*R*, Operation → Multiplication]	identical to WithUnityQ[*R*]
InvertibleQ[*R*, *r*, Operation → Multiplication]	identical to UnitQ[*R*, *r*]
Inverses[*R*, Operation → Multiplication]	return a list of pairs of the form $\{x, x^{-1}\}$, or $\{x, \text{no inverse}\}$ if x is not a unit

Related multiplicative functions of ring properties.

- The nonstandard rings can provide some interesting surprises.

In[43]:= **Unity[Z[10, 2]]**

Out[43]= 6

- Multiplication in a ring does not need to be commutative. The ring of two-by-two matrices over $\mathbb{Z}_2$ is not commutative.

In[44]:= **CommutativeQ[ToRingoid[Mat[Z[2], 2]], Operation → Multiplication]**

Out[44]= False

Units[*R*]	return the list of units of *R*
ZeroDivisorQ[*R*, *r*]	give True if *r* is a zero divisor in *R*, and False otherwise
ZeroDivisors[*R*]	return the list of all zero divisors of *R*

Additional functions related to the multiplication in a ringoid.

■ Of the elements $\{1, 2, 3\}$ in $\mathbb{Z}_{15}$, which are zero divisors?

```
In[45]:= Map[ZeroDivisorQ[Z[15], #] &, {1, 2, 3}]

Out[45]= {False, False, True}
```

■ Here are all the zero divisors with an explanation of why they are so designated.

```
In[46]:= zd = ZeroDivisors[Z[15], Mode → Textual]

         One reason why 3 is a zero divisor is that 3 * 10 = 0.
         One reason why 5 is a zero divisor is that 5 * 6 = 0.
         One reason why 6 is a zero divisor is that 6 * 10 = 0.
         One reason why 9 is a zero divisor is that 9 * 5 = 0.
         One reason why 10 is a zero divisor is that 10 * 9 = 0.
         One reason why 12 is a zero divisor is that 12 * 10 = 0.

Out[46]= {3, 5, 6, 9, 10, 12}
```

■ Here are the units of the same ring.

```
In[47]:= u = Units[Z[15]]

Out[47]= {1, 2, 4, 7, 8, 11, 13, 14}
```

■ Here we see that every element in the ring is either a zero divisor, a unit, or the zero. Is it true for any ring that an element is a zero divisor, a unit, or the zero?

```
In[48]:= Union[zd, u, {0}] == Elements[Z[15]]

Out[48]= True
```

■ 3.3.3 Distributive property

The distributive property can be tested in four ways: `DistributiveQ`, `LeftDistributiveQ`, `RightDistributiveQ`, and `RandomDistributiveQ`. As with associativity, the full test is always correct but can be time-consuming. The random test tends to be shorter, but not guaranteed. Distributivity must be satisfied both on the left and the right: $a(b + c) = a\,b + a\,c$ and $(b + c)\,a = b\,a + c\,a$. Testing can be done from both sides or one of the sides.

■ The ring of integers modulo n has the distributive property.

```
In[49]:= DistributiveQ[Z[12]]

Out[49]= True
```

There are $2\,n^3$ tests performed by `DistributiveQ` on a ringoid with n elements. For large ringoids, `RandomDistributiveQ` is recommended

■ To avoid asking for 65,536,000 tests to be performed, we use the random version and opt for ten tests here.

In[50]:= **RandomDistributiveQ[Z[320], 10]**

Out[50]= True

DistributiveQ [*R*]	give True if multiplication is distributive (from both sides) over addition in the ringoid *R*, checking all possible triples, and False otherwise
LeftDistributiveQ [*R*]	give True if multiplication is left distributive over addition, and False otherwise
RightDistributiveQ [*R*]	give True if multiplication is right distributive over addition, and False otherwise
RandomDistributiveQ [*R*]	give True if multiplication is likely to be distributive over addition, checking 25 triples, and False otherwise
RandomDistributiveQ [*R, n*]	give True if multiplication is likely to be distributive over addition, checking *n* triples, and False otherwise

Test for distributivity in a ringoid.

■ 3.3.4 RingQ test

All of the basic ring properties can be tested using one function, RingQ. Probable-RingQ, approaching associativity and distributivity randomly, is recommended for larger ringoids. The results of ring tests are recorded and can be retrieved using RingInfo.

RingQ [*R*]	give True if *R* is a ring, and False otherwise
ProbableRingQ [*R*]	give True if *R* is probably a ring (using random associativity and distributivity tests), and False otherwise
RingInfo [*R*]	return a list of facts about *R* that are generated by various tests of *R*

Testing for all ring properties and retrieval of results.

■ Here we check $\mathbb{Z}_7$ to verify that it is a ring.

In[51]:= **RingQ[Z[7]]**

Out[51]= True

■ Since BooleanRing[6] has 64 elements, we use ProbableRingQ.

In[52]:= **ProbableRingQ[BooleanRing[6]]**

Out[52]= True

■ Let's see what is known about this ringoid now.

In[53]:= **RingInfo[BooleanRing[6]]**

Out[53]= {Bool[6], the boolean Ring on {1,...,6},
       the set is closed under addition,
       the set is closed under multiplication,
       the Ringoid has as a zero the element {},
       every element has an additive inverse,
       the addition operation is commutative,
       multiplication is probably distributive over addition,
       the addition operation is probably associative,
       the multiplication operation is probably associative,
       this is probably a ring}

■ 3.3.5 Specialized rings

FieldQ[R]	give True if R is a field, and False otherwise
IntegralDomainQ[R]	give True if R is an integral domain, and False otherwise

Specialized rings can be tested with these functions.

■ The ring $\mathbb{Z}_n$ is a field if and only if n is prime.

In[54]:= **Map[{#, FieldQ[Z[#]]} &, Range[3, 9]]**

Out[54]= {{3, True}, {4, False}, {5, True},
     {6, False}, {7, True}, {8, False}, {9, False}}

■ A quotient ring of polynomials over a field F is a field if and only if the modulus polynomial is irreducible over F.

In[55]:= **FieldQ[QuotientRing[Z[3], Poly[Z[3], $x^2 + x + 2$]]]**

Out[55]= True

■ 3.3.6 Closure of subsets

For a ringoid to be a ring, its domain must be closed with respect to addition and multiplication. We can use ClosedQ to test for closure not only of the whole domain but also of subsets.

ClosedQ[*R, W*]	give True if *W* is closed with respect to addition and multiplication in *R*, and False otherwise
ClosedQ[*R, W,* Operation → *op*]	give True if *W* is closed with respect to the operation *op* in *R*, and False otherwise
ClosedPlusQ[*R, W*]	give True if *W* is closed with respect to addition in *R*, and False otherwise
ClosedDiffQ[*R, W*]	give True if *W* is closed with respect to subtraction in *R*, and False if not, or if any negation fails to exist
ClosedTimesQ[*R, W*]	give True if *W* is closed with respect to multiplication in *R*, and False otherwise

Other closure functions on ringoids.

Closure properties are most often considered when determining whether a subset is a subsystem.

- Here is a set that is not closed with respect to both ring operations, but it is closed with respect to multiplication, however.

In[56]:= **{ClosedQ[Z[8], {1, 3, 5, 7}],**
ClosedQ[Z[8], {1, 3, 5, 7}, Operation → Multiplication]}

Out[56]= {False, True}

- Differences between even elements of $\mathbb{Z}_n$ are again even. Likewise, they are closed under multiplication.

In[57]:= **{ClosedDiffQ[Z[8], {0, 2, 4, 6}],**
ClosedTimesQ[Z[8], {0, 2, 4, 6}]}

Out[57]= {True, True}

- The second argument of ClosedQ can be a set, or a structure.

In[58]:= **{ClosedQ[LatticeRing[30], Elements[LatticeRing[15]]],**
ClosedQ[LatticeRing[30], LatticeRing[15]]}

Out[58]= {True, True}

- The Visual mode gives you a sense of "how closed" a set is.

In[59]:= **ClosedQ[Z[9], {0, 2, 4, 6, 8},**
Mode → Visual, Operation → Addition]

All the elements marked with yellow are original
elements in the set. Those in red are from outside.

TheGroup   x * y

x\y	0	2	4	6	8
0	0	2	4	6	8
2	2	4	6	8	1
4	4	6	8	1	3
6	6	8	1	3	5
8	8	1	3	5	7

Out[59]= False

■ 3.3.7 Testing other properties

Given any finite ring R, the characteristic of the ring is the least positive integer n such that $n\,x = 0$ for all x in R. We can use the `MultipleOfElement` function to determine values of $n\,x$. An analogous function, `ElementToPower`, returns the multiplicative equivalent, x^n.

`ElementToPower [R, a, n]`	return the power a^n in R
`MultipleOfElement [R, a, n]`	return the multiple na in R
`Characteristic [R]`	return the characteristic of R

Functions related to powers and multiples of elements.

- The characteristic of $\mathbb{Z}_n$ is n, the characteristic of any Boolean ring is 2, and the characteristic of a direct product is related to the characteristics of the factors.

```
In[61]:= Map[Characteristic,
            { BooleanRing[3], Z[6], Z[9], DirectProduct[Z[6], Z[9]]}]
```

Out[61]= {2, 6, 9, 18}

- Here is a series of multiples and powers of (3, 6) in $\mathbb{Z}_6 \times \mathbb{Z}_9$.

```
In[60]:= TableForm[
            Map[{#, MultipleOfElement[DirectProduct[Z[6], Z[9]],
                {3, 6}, #], ElementToPower[DirectProduct[Z[6], Z[9]],
                {3, 6}, #]} &, Range[-1, 4]],
            TableHeadings → {None, {"n", "n (3,6)\n", "(3,6)ⁿ\n"}},
            TableDepth → 2]

        Inverse::fail :
          {3, 6} does not have an inverse in Mult(Z[6] x Z[9]).
```

```
Out[60]//TableForm=
        n          n (3,6)        (3,6)ⁿ

       -1          {3, 3}         $Failed
        0          {0, 0}         {1, 1}
        1          {3, 6}         {3, 6}
        2          {0, 3}         {3, 0}
        3          {3, 0}         {3, 0}
        4          {0, 6}         {3, 0}
```

IdempotentQ[R, r]	give True if r is an idempotent element of R (i.e., $r^2 = r$), and False otherwise
Idempotents[R]	return the list of idempotent elements of R
NilpotentQ[R, r]	give True if r is a nilpotent element of R (i.e., $r^k = 0$ for some positive k), and False otherwise
Nilpotents[R]	return the list of nilpotent elements of R
NilpotentDegree[R, r]	return the least positive integer k for which $r^k = 0$ in R if r is nilpotent, and 0 if r is not nilpotent in R

Other functions related to powers of an element.

■ Here we test the elements of $\mathbb{Z}_8$ to see which are idempotent.

```
In[62]:= Map[{#, IdempotentQ[Z[8], #]} &, Elements[Z[8]]]

Out[62]= {{0, True}, {1, True}, {2, False}, {3, False},
          {4, False}, {5, False}, {6, False}, {7, False}}
```

■ All the elements of a Boolean ring are idempotent.

```
In[63]:= Idempotents[BooleanRing[3]]

Out[63]= {{}, {3}, {2}, {2, 3}, {1}, {1, 3}, {1, 2}, {1, 2, 3}}
```

■ The element 6 is nilpotent in $\mathbb{Z}_{24}$ because $6^3 \equiv 0 \pmod{24}$.

```
In[64]:= NilpotentQ[Z[24], 6]

Out[64]= True
```

■ Here are the nilpotents of $\mathbb{Z}_{16}$.

```
In[65]:= Nilpotents[Z[16]]

Out[65]= {0, 2, 4, 6, 8, 10, 12, 14}
```

■ Here are the elements of $\mathbb{Z}_{16}$ with their nilpotent degrees listed below.

```
In[66]:= {Elements[Z[16]],
           NilpotentDegree[Z[16], #] & /@Elements[Z[16]]} //
          TableForm[#, TableSpacing → {1, 1.2}] &
```

Out[66]//TableForm=

0	1	2	3	4	5	6	7	8	9	10	11	12	13	14	15
1	0	4	0	2	0	4	0	2	0	4	0	2	0	4	0

- In a commutative ring, the nilpotents form an ideal.

```
In[67]:= IdealQ[Nilpotents[Z[24]], Z[24]]
```

Out[67]= True

3.4 Built-in Ringoids

There are several ringoids that are easily constructed by functions in the packages. By far the most important of the rings in elementary abstract algebra is the ring of integers mod n, $\mathbb{Z}_n$.

■ 3.4.1 Numeric rings

ZR[n]	ring of integers mod n with mod n addition and multiplication
ZR[n, k]	ring of multiples of k mod n, if k is a divisor of n
Z[n]	identical to ZR[n] if DefaultStructure is Ring
Z[n, k]	identical to ZR[n, k] if DefaultStructure is Ring
Z[n, I]	ring of Gaussian integers with mod n addition and multiplication, if the DefaultStructure is Ring
TrivialZR[n]	ring of integers mod n with mod n addition but the multiplication is zero for all products
LatticeRing[n]	divisors of n with LCM/GCD for addition and GCD for multiplication

Some built-in numeric ringoids.

Here we illustrate several ringoids with eight elements.

- We've seen the Z family many times already. We have mostly used the symbol Z, but when using ZR we know that we will get a ringoid in all cases.

```
In[68]:= ZR[8]
```

Out[68]= Ringoid[{0, 1, 2, 3, 4, 5, 6, 7},
        Mod[#1 + #2, 8] &, Mod[#1 #2, 8] &]

- The following ring happens to be an ideal of $\mathbb{Z}_{16}$.

In[69]:= **{R = Z[16, 2], IdealQ[R, Z[16]]}**

Out[69]= {Ringoid[{0, 2, 4, 6, 8, 10, 12, 14},
        Mod[#1 + #2, 16] &, Mod[#1 #2, 16] &], True}

- This is a standard counterexample in elementary algebra courses; any product is
  equal to zero.

In[71]:= **CayleyTables[TrivialZR[8], Mode → Visual];**

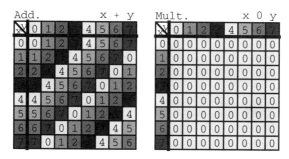

- Here is another interesting ring with eight elements. The operations in *L* are not
  displayed, so they are also listed.

In[71]:= **{L = LatticeRing[30], Addition[L], Multiplication[L]}**

Out[71]= $\Big\{$Ringoid[{1, 2, 3, 5, 6, 10, 15, 30}, -Addition-,
        -Multiplication-], $\dfrac{\text{LCM}[\#1, \#2]}{\text{GCD}[\#1, \#2]}$ &, GCD[#1, #2] &$\Big\}$

■ 3.4.2 Other rings

BooleanRing[*n*]	Boolean ring consisting of the set of subsets of {1, 2, ..., *n*} with symmetric difference for addition and intersection for multiplication
BooleanRing[*list*]	Boolean ring with subsets of *list*

Nonnumeric ringoids.

- Here is another ring with eight elements.

In[72]:= **CayleyTables[BooleanRing[{α, β, γ}],**
        Mode → Visual, KeyForm → StandardForm];

KEY for Add(Bool[{α, β, γ}]): label used → element:
{g1 → {}, g2 → {γ}, g3 → {β}, g4 → {β, γ}, g5 →
{α}, g6 → {α, γ}, g7 → {α, β}, g8 → {α, β, γ}}

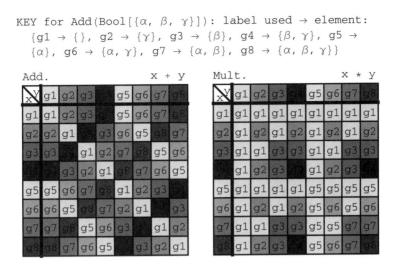

3.5 Using Cayley tables

Both the additive and multiplicative Cayley tables are produced by the function `Cayley-Tables` and individually with `CayleyTables` or `CayleyTable` (using the `Operation` option). `MultiplicationTable` can also be used to generate a multiplication table.

`CayleyTables[R]`	return the Cayley tables (in double array form) for Addition and Multiplication on *R*
`CayleyTables[R, Mode → Visual]`	return a PostScript rendition of the Addition and Multiplication tables of *R* in addition to the double array form
`CayleyTables[R, Operation → op]`	return the Cayley table of *R* for the operation *op* (Addition or Multiplication)
`MultiplicationTable[R]`	return the Multiplication table of *R*

Methods for generating Cayley tables of ringoids.

- Here is the computational output for the Cayley tables of the ring of integers mod 3.

In[73]:= **CayleyTables[Z[3]]**

Out[73]= {{{0, 1, 2}, {1, 2, 0}, {2, 0, 1}},
{{0, 0, 0}, {0, 1, 2}, {0, 2, 1}}}

- An alternative to using `MultiplicationTable` is to use `CayleyTables` with the option `Operation → Multiplication`. To get just the addition table, use the value `Addition` for this option. (Note: `CayleyTable`, without the ending `s`, can also be used when a single operation is chosen.)

```
In[74]:= CayleyTables[ZR[3], Operation → Addition,
            Mode → Visual, ShowName → False];
```

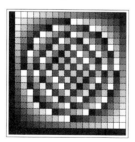

- Although the computational output is useful for further calculations, the `Visual` mode provides a result that is much easier to read.

```
In[75]:= CayleyTables[ZR[3], Mode → Visual]
```

```
Out[75]= {{{0, 1, 2}, {1, 2, 0}, {2, 0, 1}},
           {{0, 0, 0}, {0, 1, 2}, {0, 2, 1}}}
```

- Sometimes a density plot is an effective alternative to the visual mode.

```
In[76]:= ListDensityPlot[
            MultiplicationTable[ZR[19]], FrameTicks → False];
```

The options available to `CayleyTables` are generally the same as those for `CayleyTable`. See section 2.5 in chapter 2 for details.

The Cayley tables of a ringoid are often not very readable if the set of elements is large (20 or more). Occasionally insight can be gained by looking at just the color patterns of the tables, without any labels. We can use the options ShowKey, ShowBodyText, ShowOperator, ShowName, and ShowSidesText to suppress the printing of textual information.

- There are some interesting patterns that provide information even without seeing a key

In[77]:= **CayleyTables[BooleanRing[5], Mode → Visual, ShowKey → False, ShowBodyText → False, ShowSidesText → False, ShowOperator → False, ShowName → False];**

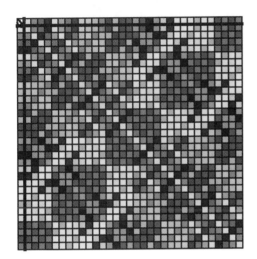 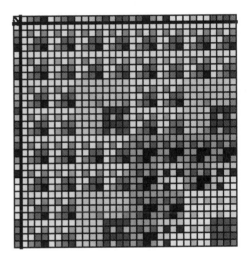

3.6 Building other structures

As with groups, the use of direct products, substructures, and quotient structures is an effective means of obtaining new ringoids.

3.6.1 Direct products

The direct product of two or more ringoids can be constructed with DirectProduct.

- Here is yet another ring with eight elements.

In[78]:= **T = DirectProduct[ZR[2], ZR[4]]**

Out[78]= Ringoid[{{0, 0}, {0, 1}, {0, 2}, {0, 3}, {1, 0}, {1, 1}, {1, 2}, {1, 3}}, -Addition-, -Multiplication-]

- Here is its unity—not exactly the number 1, but close.

In[79]:= **Unity[T]**

Out[79]= {1, 1}

DirectProduct[R_1, R_2]	return the ringoid with domain the Cartesian product of the domains of R_1 and R_2 and coordinate-wise operations
DirectProduct[R_1, R_2, ...]	return the direct products of ringoids R_1, R_2, ...
DirectSum[R_1, R_2, ...]	identical in functionality to DirectProduct[R_1, R_2, ...]

Functions that generate direct products/sums.

- The units in a direct product must have units in each coordinate.

In[80]:= **Units[T, Mode → Textual]**

```
{1, 1} is a unit because {1, 1} . {1, 1} = {1, 1}.
{1, 3} is a unit because {1, 3} . {1, 3} = {1, 1}.
```

Out[80]= {{1, 1}, {1, 3}}

- More than two ringoids can be "multiplied." In this case, there are three copies of $\mathbb{Z}_2$, giving one more example of a ring with eight elements.

In[81]:= **Elements[DirectProduct[ZR[2], ZR[2], ZR[2]]]**

Out[81]= {{0, 0, 0}, {0, 0, 1}, {0, 1, 0}, {0, 1, 1},
        {1, 0, 0}, {1, 0, 1}, {1, 1, 0}, {1, 1, 1}}

■ 3.6.2 Subrings and ideals

When the ringoid R is a ring, the expression SubringQ[S, R] tests whether a list of elements S is a subring of R. This is done by testing whether S is nonempty and closed with respect to addition and multiplication in R. Subrings must also contain negations of its elements, but in a finite ring, it is sufficient to verify closure to conclude that this property is true.

Given a subring S, we call it a two-sided *ideal* of the ring R if for every $r \in R$ and every $s \in S$ both rs and sr are in S. Left ideals and right ideals require only one of these products to be in S.

- If n is even, the even elements of $\mathbb{Z}_n$ are a subring. Here are two ways of testing: using a list of elements or a structure.

In[82]:= **{SubringQ[{0, 2, 4, 6, 8}, Z[10]], SubringQ[Z[10, 2], Z[10]]}**

Out[82]= {True, True}

- This is Z[10, 2] with the addition and the multiplication exchanged, showing that the operations, not just the elements, need to match.

```
In[83]:= {Rrev = FormRingoid[Range[0, 8, 2], Z[10][[3]], Z[10][[2]]],
           SubringQ[Rrev, Z[10]]}
```

> *Operation::fail :*
> *The operation of the substructure(s) does*
> *not match that of the parent structure.*

```
Out[83]= {Ringoid[{0, 2, 4, 6, 8},
           -Addition-, -Multiplication-], False}
```

SubringQ[*S, R*]	return True if *S* (given as a ringoid with operations matching those in *R* or as a list) is a subring of the ringoid *R*, and False otherwise
IdealQ[*S, R*]	return True if S is a (two-sided) ideal of the ringoid *R*, and False otherwise
LeftIdealQ[*S, R*]	return True if *S* is a left ideal of the ringoid *R*, and False otherwise
RightIdealQ[*S, R*]	return True if *S* is a right ideal of the ringoid *R*, and False otherwise

Subring and ideal testing.

- When a subring is an ideal, we are able to build a new ring with Quotient-Ring.

```
In[84]:= {ideal = Range[0, 8, 2], IdealQ[ideal, Z[10]],
           QuotientRing[Z[10], ideal]} // ColumnForm
```

> *QuotientRing::NS :*
> *This quotient ring uses NS to represent the*
> *ideal (normal subgroup) {0, 2, 4, 6, 8} you*
> *specified. Use CosetToList to convert this*
> *coset representation to a list of the elements.*

```
Out[84]= {0, 2, 4, 6, 8}
         True
         Ringoid[{NS, 1 + NS}, -Addition-, -Multiplication-]
```

- Not every subring is an ideal.

```
In[85]:= {SubringQ[{{}, {1, 2, 3}}, BooleanRing[3]],
           IdealQ[{{}, {1, 2, 3}}, BooleanRing[3]]}
```

```
Out[85]= {True, False}
```

- Also, not every ideal is both a left and a right ideal. In this example, we consider the two-by-two matrices over $\mathbb{Z}_2$ and consider the set of elements (calling it LI) whose first column is all zeros. (These happen to be in positions 1, 2, 5 and 6 in the ring.)

In[86]:= **M = ToRingoid[Mat[Z[2], 2]];**
        Map[MatrixForm, (LI = Elements[M][[{1, 2, 5, 6}]])]

Out[87]= $\left\{ \begin{pmatrix} 0 & 0 \\ 0 & 0 \end{pmatrix}, \begin{pmatrix} 0 & 0 \\ 0 & 1 \end{pmatrix}, \begin{pmatrix} 0 & 1 \\ 0 & 0 \end{pmatrix}, \begin{pmatrix} 0 & 1 \\ 0 & 1 \end{pmatrix} \right\}$

- Now we verify that LI is indeed a subring and also a left ideal, but not a right ideal.

In[88]:= **{SubringQ[LI, M], LeftIdealQ[LI, M], RightIdealQ[LI, M]}**

Out[88]= {True, True, False}

- The Textual mode is supported by SubringQ.

In[89]:= **SubringQ[{{}, {1, 2, 3}}, BooleanRing[3], Mode → Textual]**

        Prospective subring is not empty; passes the first test.

        Prospective subring is closed with
          respect to addition; passes the second test.

        Prospective subring is closed with respect
          to multiplication; passes the third test.

Out[89]= True

The *principal ideal generated by* r in a commutative ring R is defined to be $\langle r \rangle = \{s\,r \mid s \in R\}$. Given a subset S of the commutative ring R, we define the *annihilator* of S by $\mathrm{Ann}(S) = \{r \in R \mid r\,s = 0 \text{ for all } s \in S\}$; this is an ideal.

PrincipalIdeal[R, r]	return the principal ideal generated by r in R
Annihilator[R, S]	return the annihilator of S in R

Methods to create specialized ideals.

- We get a principal ideal by multiplying the generator by the ring elements, so here we get the multiples of 8.

In[90]:= **PrincipalIdeal[Z[20], 8]**

Out[90]= Ringoid[{0, 4, 8, 12, 16}, Mod[#1 + #2, 20] &, Mod[#1 #2, 20] &]

- Multiplication in Boolean rings is set intersection, so the ideal generated by {1, 2} is the intersection of {1, 2} with other sets in the ring.

In[91]:= **PrincipalIdeal[BooleanRing[4], {1, 2}]**

Out[91]= Ringoid[{{}, {1}, {2}, {1, 2}},
        -Addition-, -Multiplication-]

- Here are all the elements in $\mathbb{Z}_8$ that annihilate 4.

In[92]:= **Annihilator[Z[8], {4}]**

Out[92]= Ringoid[{0, 2, 4, 6}, Mod[#1 + #2, 8] &, Mod[#1 #2, 8] &]

In addition to principal ideals, prime and maximal ideals are also of importance. A proper ideal P of a commutative ring R is said to be a *prime ideal* of R if whenever $a, b \in R$ and $a b \in P$, we have either $a \in P$ or $b \in P$. A proper ideal M of a commutative ring R is said to be a *maximal ideal* of R if whenever N is an ideal of R and $M \subseteq N \subseteq R$, then either $M = N$ or $N = R$.

MaximalIdealQ[S, R]	give True if S is a maximal ideal of R, and False otherwise
PrimeIdealQ[S, R]	give True if S is a prime ideal of R, and False otherwise

Testing for maximal and prime ideals.

- Here are the elements of $\mathbb{Z}_{12}$ that generate a maximal principal ideal. Do you see a pattern?

In[93]:= **maxGen = Select[Range[1, 11],**
  MaximalIdealQ[PrincipalIdeal[Z[12], #], Z[12]] &]

Out[93]= {2, 3, 9, 10}

- Perhaps looking at the actual principal ideals (preceded by their generators) may be helpful.

In[94]:= **Map[{#, Elements[PrincipalIdeal[Z[12], #]]} &, maxGen]**

Out[94]= {{2, {0, 2, 4, 6, 8, 10}}, {3, {0, 3, 6, 9}},
  {9, {0, 3, 6, 9}}, {10, {0, 2, 4, 6, 8, 10}}}

- In a commutative ring with unity, every maximal ideal is prime. In finite rings, we also have every prime ideal being maximal (in a commutative ring with unity). Here are the generators of the prime (or maximal) principal ideals in $\mathbb{Z}_{14}$.

In[95]:= **Select[Range[1, 13],**
  PrimeIdealQ[PrincipalIdeal[Z[14], #], Z[14]] &]

Out[95]= {2, 4, 6, 7, 8, 10, 12}

■ 3.6.3 Quotient rings

The cosets of an ideal can form a ring using QuotientRing.

- Earlier we identified {0, 2, 4, 6, 8} as an ideal of $\mathbb{Z}_{10}$. Thus, we can form the quotient ring.

In[96]:= **QuotientRing[Z[10], {0, 2, 4, 6, 8}]**

Out[96]= Ringoid[{NS, 1 + NS}, -Addition-, -Multiplication-]

QuotientRing [*R*, *T*]	return the quotient ring formed by the ring *R* and the ideal *T*
LeftCosets [*R*, *T*]	return the set of left cosets of the subring *T* in the ringoid *R*
RightCosets [*R*, *T*]	return the set of right cosets of the subring *T* in the ringoid *R*
LeftCoset [*R*, *T*, *r*]	return the left coset *r* + *T* when given the element *r* from *R* and the subring *T* of *R*
RightCoset [*R*, *T*, *r*]	return the right coset *T* + *r*
CosetToList [*Q*, *coset*]	return *coset* represented as a list of elements from *R* where *coset* is an element of *Q* = *R*/*T*
FactorRing [*R*, *T*]	identical to QuotientRing [*R*, *T*]

Functions for quotient rings and related structures.

■ We are notified if a subset is not an ideal.

In[97]:= **FactorRing[DirectProduct[Z[4], Z[4]], {{0, 0}, {2, 2}}]**

> *QuotientRing::notideal :*
> *The set {{0, 0}, {2, 2}} is not an*
> *ideal of the Ringoid Z[4] x Z[4].*

Out[97]= $Failed

■ There is a variety of forms in which the cosets can appear. Here representatives are used to describe the cosets.

In[98]:= **QuotientRing[DirectProduct[Z[4], Z[4]],**
{{0, 0}, {1, 0}, {2, 0}, {3, 0}}, Form → Representatives]

Out[98]= Ringoid[{{0, 0}, {0, 1}, {0, 2}, {0, 3}},
-Addition-, -Multiplication-]

■ Here is a quotient ring with the coset representatives selected randomly, but the cosets are listed in the form *r* + NS.

In[99]:= **Q = QuotientRing[Z[21], {0, 7, 14}, Representatives → Random]**

Out[99]= Ringoid[{14 + NS, 1 + NS, 2 + NS, 3 + NS, 4 + NS, 5 + NS, 6 + NS},
-Addition-, -Multiplication-]

■ To retrieve the complete list of elements of a coset within a quotient ring, use CosetToList. The elements of one of the cosets are listed here.

In[100]:= **CosetToList[Q, 5 + NS]**

Out[100]= {5, 12, 19}

In cases where a subring is not an ideal we can still form cosets. However, in this case multiplication of cosets using coset representatives is not well defined. This is why `QuotientRing` requires an ideal as its second argument.

- We can illustrate the problem of an ill-defined multiplication with the following list of cosets.

```
In[101]:= (cosets = LeftCosets[BooleanRing[3], {{}, {1, 2, 3}}]) //
            ColumnForm

Out[101]= {{}, {1, 2, 3}}
          {{3}, {1, 2}}
          {{2}, {1, 3}}
          {{2, 3}, {1}}
```

- Using the first elements of the second and third cosets, the product is in the first coset. Using the last elements of the same cosets, the product is in the fourth coset. Thus, multiplication of cosets is not well-defined in this case.

```
In[102]:= {Multiplication[BooleanRing[3]][cosets[[2, 1]],
            cosets[[3, 1]]], Multiplication[BooleanRing[3]][
            cosets[[2, 2]], cosets[[3, 2]]]}

Out[102]= {{}, {1}}
```

- By studying the following Cayley table showing how the left cosets multiply, we can visually see the same results.

```
In[103]:= LeftCosets[BooleanRing[3], {{}, {1, 2, 3}},
            Mode → Visual, Operation → Multiplication];
```

```
KEY for Mult(Bool[3]): label used → element:
  {g1 → {}, g2 → {1, 2, 3}, g3 → {3}, g4 → {1,
  2}, g5 → {2}, g6 → {1, 3}, g7 → {2, 3}, g8 → {1}}
```

Mult(Bool[3])   x * y

x\y	g1	g2	g3	g4	g5	g6	g7	g8
g1	g1	g1	g1	g1	g1	g1	g1	g1
g2	g1	g2	g3	g4	g5	g6		
g3	g1	g3	g3	g1	g1	g3	g3	g1
g4	g1	g4	g1	g4	g5		g5	
g5	g1	g5	g1	g5	g5	g1	g5	g1
g6	g1	g6	g3		g1	g6	g3	
g7		g1		g3	g5	g5	g3	g1
g8		g1		g1		g1		g1

▦ 3.7 Extension ringoids

The time and memory needed to evaluate interesting expressions on large ringoids is unacceptably high. In order to make it feasible to work with polynomials, matrices, and functions over a ringoid, an extension ring structure has been created. In the case of matrices and functions, these extension rings are still finite, but they grow in proportion to n^{n^2} and n^n (where n is the order of the base ring). All but the most trivial cases become awkward to implement as `Ringoids`. Of course, polynomial rings are infinite and representation as a finite `Ringoid` is out of the question.

`PolynomialsOver[R]`	generate the extension ring of polynomials over R
`MatricesOver[R, {m, n}]`	generate the extension ring of m–by–n matrices over R
`MatricesOver[R, n]`	generate the extension ring of n–by–n matrices over R
`FunctionsOver[R]`	generate the extension ring of functions on R

Ring extensions.

- Here is a polynomial extension, a matrix extension, and a function extension over the base ring $\mathbb{Z}_3$.

```
In[104]:= {P = PolynomialsOver[Z[3]],
           M = MatricesOver[Z[3], 2], F = FunctionsOver[Z[3]]}
```

```
Out[104]= {-Ring of Polynomials over Z[3]-,
           -Mat₂(Z[3])-, -Ring of Functions over Z[3]-}
```

`ExtensionType[ext]`	return the type of extension of *ext*
`BaseRing[ext]`	return the base ring of *ext*
`ElementQ[x, ext]`	give `True` if x is an element of *ext*, and `False` otherwise
`Addition[ext]`	return the addition operation on *ext*
`Multiplication[ext]`	return the multiplication operation on *ext*

Defining parameters of an extension.

- Once an extension is created, its *type* can be retrieved.

```
In[105]:= Map[ExtensionType, {P, M, F}]
```

```
Out[105]= {PolyRing, Matrices, FuncRing}
```

- Its base ring is also accessible.

```
In[106]:= Map[BaseRing, {P, M, F}]
```

Out[106]= {Ringoid[{0, 1, 2}, Mod[#1 + #2, 3] &, Mod[#1 #2, 3] &],
　　　　Ringoid[{0, 1, 2}, Mod[#1 + #2, 3] &, Mod[#1 #2, 3] &],
　　　　Ringoid[{0, 1, 2}, Mod[#1 + #2, 3] &, Mod[#1 #2, 3] &]}

- To be a member of the polynomial ring *P* (defined earlier), not only does the element have to be a polynomial formed with the special function `Poly`, but the correct base ring must also be used in its formation.

In[107]:= **Map[ElementQ[#, P] &, {p = Poly[Z[3], x^2 + 2 x - 1],**
　　　　Poly[Z[4], x^2 + 2 x - 1], x^2 + 2 x - 1}]

Out[107]= {True, False, False}

3.8 Polynomials over a ringoid

The `PolynomialsOver` function returns an extension ringoid representing the ring of polynomials over the ringoid *R*. A current restriction on `Ringoids` is that they must be finite, but over any ring with two or more elements there are an infinite number of polynomials. Therefore, the resulting object is a `RingExtension`.

3.8.1 Forming polynomials

An individual polynomial over a ringoid *R* is created by providing input of the form `Poly[R, *expression*]` or `Poly[R, *coefficients*]`.

- Here is a polynomial over $\mathbb{Z}_5$ entered as an expression.

In[108]:= **p = Poly[Z[5], t^2 + 2 t + 3]**

Out[108]= 3 + 2 t + t^2

- Here is another formed by listing just the coefficients.

In[109]:= **q = Poly[Z[5], 4, 3, 2, 1]**

Out[109]= 4 + 3 x + 2 x^2 + x^3

The entries in a coefficient sequence are assumed, by default, to start with the constant term and increase in degree. The indeterminate is explicit when an expression is entered, but when only the coefficients are provided, the default is *x*. The option `Indeterminate` can be used to specify an alternate indeterminate. The option `PowersIncrease` reflects the order in which the coefficients should be entered, as well as specifying the output format.

- Here is how to specify an alternate indeterminate when listing coefficients.

In[110]:= **r = Poly[Z[5], 1, 0, 2, Indeterminate → λ]**

Out[110]= $1 + 2 \lambda^2$

Notice that any terms that are missing from a polynomial, because the coefficient is zero, must still be accounted for when giving the list of coefficients.

Poly[*R*, *expr*]	create the polynomial over the ringoid *R* given by *expr*
Poly[*R*, *expr*, *opts*]	create the polynomial over *R* given by *expr* using the options given by *opts*
Poly[*R*, *coeffs*, *opts*]	create the polynomial over *R* with coefficients *coeffs*, using options *opts*
Poly[Polynomials-Over[*R*], *args*]	identical to Poly[*R*, *args*]

Entering polynomials using Poly.

- If you are more comfortable entering coefficients starting with the term with the highest degree, you can consider the PowersIncrease option. Use this as an option to Poly if you want to order the coefficients this way only occasionally.

In[111]:= **Poly[Z[5], 4, 3, 2, 1, PowersIncrease → RightToLeft]**

Out[111]= $4\ x^3\ +\ 3\ x^2\ +\ 2\ x\ +\ 1$

- If you want to consistently start with the coefficient of the term with the highest degree, change the default.

In[112]:= **SetOptions[Poly, PowersIncrease → RightToLeft]**

Out[112]= {PowersIncrease → RightToLeft,
    Indeterminate → x, FlexibleEntering → True}

- Now this order is the default.

In[113]:= **Poly[Z[5], 4, 3, 2, 1]**

Out[113]= $4\ x^3\ +\ 3\ x^2\ +\ 2\ x\ +\ 1$

- You can always reverse the change and the original ordering is back in effect. (Note that we get a completely different polynomial; it is not just the output that is affected.)

In[114]:= **SetOptions[Poly, PowersIncrease → LeftToRight];**
    Poly[Z[5], 4, 3, 2, 1]

Out[115]= $4 + 3\ x + 2\ x^2 + x^3$

There are several options to consider for Poly.

option name	default value	
PowersIncrease	LeftToRight	for input, determines whether sequences of coefficients start with the constant or leading coefficient; for output, specifies whether the powers in the polynomials increase from left to right or right to left (RightToLeft)
Indeterminate	x	indeterminate to be used in a polynomial
FlexibleEntering	True	if set to True, allows negations of ring elements in polynomial expressions and arbitrary integers for polynomials over $\mathbb{Z}_n$

Options for Poly.

- With FlexibleEntering set to True, some natural extensions to notation are allowed. Over the integers mod n, any integers can be entered and they are reduced mod n.

In[116]:= **Poly[Z[5], x² - x + 11]**

Out[116]= $1 + 4 x + x^2$

- This is not valid if FlexibleEntering is set to False

In[117]:= **Poly[Z[5], x² - x + 11, FlexibleEntering → False]**

> *MemberQ::elmnts :*
> *At least one of the coefficients {11, -1, 1} is*
> *not an element of the base ring.*

Out[117]= $Failed

■ 3.8.2 Random polynomials

A request for a random polynomial is meaningless unless an upper limit on the degree is specified, so a degree must be specified as a second argument. If degree n is specified, then $n + 1$ random elements from the base ring are selected for coefficients. By default, the leading coefficient is usually not allowed to be the zero of the ring, so the degree will be exactly n. Options can be used to control the selection of coefficients to allow a degree less than or equal to n.

- Here is a random cubic polynomial over $\mathbb{Z}_6$.

In[118]:= **RandomElement[PolynomialsOver[Z[6]], 3]**

Out[118]= $5 + 5 x + 4 x^2 + 3 x^3$

`RandomElement[` `PolynomialsOver[R],n]`	return a random polynomial of degree n over R
`RandomElement[` `PolynomialsOver[R],n,opts]`	return a random polynomial of degree n (or less) over R according to options *opts*
`RandomElements[` `PolynomialsOver[R],n,k,opts]`	return k random polynomials of degree n (or less) over R according to options *opts*

Obtaining random polynomials.

option name	default value	
`SelectFrom`	`NonZero`	determines whether certain polynomials are to be excluded in selecting random polynomials; other possible values are `Any`, `NonUnity`, `NonIdentity`
`LowerDegreeOK`	`False`	specifies whether the requested degree must be exact or whether polynomials of lesser degree may be selected
`Monic`	`False`	specifies whether the random polynomial should be monic (leading coefficient equal to the unity of the base ringoid)

Options of `RandomElement` for polynomial extensions.

- The unity of `BooleanRing[3]` is `{1, 2, 3}`.

In[119]:= **RandomElement[**
        PolynomialsOver[BooleanRing[3]], 2, Monic → True]

Out[119]= $\{1, 2\} + \{2\} x + \{1, 2, 3\} x^2$

- With `SelectFrom` set to `Any`, we may expect a zero polynomial here.

In[120]:= **Table[RandomElement[PolynomialsOver[Z[2]],**
        2, LowerDegreeOK → True, SelectFrom → Any], {8}]

Out[120]= $\{1, x + x^2, 1, x + x^2, 1 + x + x^2, 0, 1, 1\}$

- Here are five random polynomials of degree 2, not requiring them to be monic and not allowing replacement.

In[121]:= **RandomElements[PolynomialsOver[Z[3]],**
        2, 5, Monic → False, Replacement → False]

Out[121]= $\{x + 2 x^2, 2 + x + 2 x^2, 1 + x + x^2, 1 + 2 x + x^2, 2 + x^2\}$

■ 3.8.3 Polynomial arithmetic

Addition[P][a, b]	return the sum of polynomials a and b in the polynomial extension P
Addition[P, a, b]	identical to Addition[P][a, b]
Addition[a, b]	if a and b are both over some polynomial extension P, return Addition[P][a, b], otherwise return $Failed
$a + b$	identical to Addition[a, b]
Multiplication[P][a, b]	return the product of polynomials a and b in the polynomial extension P

Arithmetic in a polynomial extension.

Addition and multiplication of polynomials are defined in the usual way, with the coefficients of the sums and products based on calculations in the base ring. If the base rings of the polynomial extension and polynomials do not match, the result is $Failed. The multiplication has available all the alternate methods shown for the addition.

■ Here we form polynomials over $\mathbb{Z}_3$.

In[122]:= **P = PolynomialsOver[Z[3]]**

Out[122]= -Ring of Polynomials over Z[3]-

■ We demonstrate polynomial arithmetic with these two polynomials.

In[123]:= **{a = Poly[Z[3], 1 + 2 x^3 + x^4], b = Poly[Z[3], x + x^2 + x^3]}**

Out[123]= $\{1 + 2 x^3 + x^4, x + x^2 + x^3\}$

■ Here are the four ways of finding the sum.

In[124]:= **{Addition[P][a, b], Addition[P, a, b], Addition[a, b], a + b}**

Out[124]= $\{1 + x + x^2 + x^4, 1 + x + x^2 + x^4, 1 + x + x^2 + x^4, 1 + x + x^2 + x^4\}$

■ There are analogous ways of finding the product.

In[125]:= **{Multiplication[P][a, b],**
Multiplication[P, a, b], Multiplication[a, b], a b, a * b}

Out[125]= $\{x + x^2 + x^3 + 2 x^4 + x^7, x + x^2 + x^3 + 2 x^4 + x^7,$
$x + x^2 + x^3 + 2 x^4 + x^7, x + x^2 + x^3 + 2 x^4 + x^7, x + x^2 + x^3 + 2 x^4 + x^7\}$

Many of the common functions that apply to ringoids also apply to polynomial extensions. Here are some of the more important ones.

HasZeroQ[*P*]	give True if polynomial extension *P* has a zero, and False otherwise
Zero[*P*]	return the zero of *P*, if one exists, and $Failed otherwise
NegationOf[*P*, *a*]	return the negation of polynomial *a* in *P* if all the coefficients of *a* have a negation in the base ring of *P*, and $Failed otherwise
WithUnityQ[*P*]	give True if *P* has a unity, and False otherwise
Unity[*P*]	return the unity of *P*, if one exists, and $Failed otherwise

Examples of general ringoid functions working in a polynomial extension.

- Recall *P*; its zero and unity are not too surprising.

In[126]:= **{P, HasZeroQ[P], Zero[P], WithUnityQ[P], un = Unity[P]}**

Out[126]= {-Ring of Polynomials over Z[3]-, True, 0, True, 1}

- What is behind a simple 1 may be a surprise, however. Since it is still a polynomial, it uses the (complicated) internal form of a general polynomial.

In[128]:= **un // InputForm**

Out[127]//InputForm=
```
        AbstractAlgebra`RingExtensions`Private`poly[
          {AbstractAlgebra`Core`Private`ringoid[{0, 1, 2}, Mod[#1 + #2,
          3] & , Mod[#1*#2, 3] & , {{}, {}, {}, {}, {}, {RingoidName ->
          "Z[3]", RingoidDescription -> "the ring of integers mod 3",
            FormatOperator -> False}}], LeftToRight, x, True}, {1}]
```

- All polynomials over $\mathbb{Z}_n$ have negations.

In[426]:= **NegationOf[P,**
      Poly[Z[3], 2 x^2 - x + 1, PowersIncrease → RightToLeft]]

Out[128]= x^2 + x + 2

A distinctive aspect of polynomials is the division property. If *a* and *b* are polynomials and the leading coefficient of *b* is a unit of the base ring, then there are two unique polynomials *q* and *r* such that *a* = *b q* + *r* with *r* = 0 or the degree of *r* is less than the degree of *b*.

PolynomialDivision[*P*, *a*, *b*]	return the pair {*q*, *r*} consisting of the quotient and remainder in the division of *a* by *b*
PolynomialRemainder[*P*, *a*, *b*]	return the remainder in the division of *a* by *b*
PolynomialQuotient[*P*, *a*, *b*]	return the quotient in the division of *a* by *b*

Extensions of built-in polynomial functions for working in polynomial extensions.

Note that each of these functions can also be used by giving just the two polynomials, assuming that they both have the same base ring. Additionally, note that these are extensions of built-in functions, which are still available for ordinary polynomials.

- Here we divide a fourth-degree polynomial by a cubic polynomial.

In[129]:= **{q, r} = PolynomialDivision[P,**
   a = Poly[Z[3], 1 + 2 x³ + x⁴], b = Poly[Z[3], x + x² + x³]]

Out[129]= $\{1 + x, \; 1 + 2x + x^2\}$

- We can verify that the quotient and remainder are correct.

In[130]:= **{a, Addition[P][Multiplication[P][b, q], r]}**

Out[130]= $\{1 + 2x^3 + x^4, \; 1 + 2x^3 + x^4\}$

Frequently only the remainder, *r*, is needed, such as in finite field calculations.

- Let's consider the polynomials over $\mathbb{Z}_2$. Let *m* be a modulus that is irreducible over $\mathbb{Z}_2$.

In[131]:= **P2 = PolynomialsOver[Z[2]];**
   m = Poly[Z[2], α² + α + 1]

Out[132]= $1 + \alpha + \alpha^2$

- Consider an element *t* in the Galois field of polynomials mod *m*.

In[133]:= **t = Poly[Z[2], α + 1]**

Out[133]= $1 + \alpha$

- To square *t*, we multiply it by itself and then divide by the modulus. The result is the remainder. (Using functions in `AbstractAlgebra`FiniteFields` makes this process easier.)

In[134]:= **PolynomialRemainder[P2, Multiplication[P2][t, t], m]**

Out[134]= α

- The base ring does not need to be a field, but the leading coefficient of the divisor must be a unit. Here the leading coefficient of the divisor is 3, which is a unit of $\mathbb{Z}_4$.

In[135]:= **PolynomialDivision[PolynomialsOver[Z[4]],**
   Poly[Z[4], x³ + 2 x + 1], Poly[Z[4], 3 x + 3]]

Out[135]= $\{1 + x + 3x^2, \; 2\}$

- This time, the leading coefficient of the divisor is 2, which is not a unit of $\mathbb{Z}_4$.

In[136]:= **PolynomialDivision[PolynomialsOver[Z[4]],**
        Poly[Z[4], x³ + 2 x + 1], Poly[Z[4], 2 x + 3]]

   PolynomialDivision::undef :
    Since the leading coefficient, 2, is
      not a unit in Z[4], division is undefined.

Out[136]= $Failed

- The Textual mode is supported by PolynomialDivision.

In[137]:= **PolynomialDivision[Poly[ZR[3], x³ + 2 x + 1],**
        Poly[ZR[3], 2 x + 1], Mode → Textual]

   a(x) = b(x) q(x) + r(x) where
   a(x) = 1 + 2 x + x³,
   b(x) = 1 + 2 x,
   q(x) = 2 x + 2 x² and
   r(x) = 1.
   Notice that either r(x) = 0 or deg r < deg b.

Out[137]= {2 x + 2 x², 1}

- Here are two new polynomials over $\mathbb{Z}_3$.

In[138]:= **{a = Poly[Z[3], x⁵ + x⁴ + 2 x³ + x + 1], b = Poly[Z[3], x⁴ + 2]}**

Out[138]= {1 + x + 2 x³ + x⁴ + x⁵, 2 + x⁴}

- Here is their greatest common divisor and least common multiple.

In[139]:= **{h = PolynomialGCD[a, b], g = PolynomialLCM[a, b]} //**
        ColumnForm

Out[139]= 1 + 2 x
        2 + x + x² + 2 x³ + 2 x⁴ + 2 x⁵ + 2 x⁶ + x⁷ + 2 x⁸

- The product of the LCM and GCD should equal the product of the original polynomials.

In[140]:= **Multiplication[PolynomialsOver[Z[3]], h, g] ==**
        Multiplication[PolynomialsOver[Z[3]], a, b]

Out[140]= True

The Euclidean algorithm for computing the greatest common factor of two polynomials has been implemented over a ringoid. The algorithm fails if any of the polynomial divisions are undefined.

- The problem here is that the leading coefficient of the divisor is not a unit, so the first division in the Euclidean algorithm cannot be performed.

```
In[141]:= PolynomialGCD[PolynomialsOver[Z[4]],
          Poly[Z[4], x³ + x + 1], Poly[Z[4], 2 x² + 1]]
```

PolynomialGCD::undefined :
Result of PolynomialGCD is undefined due
*   to a nonring base or an undefined division.*

Out[141]= $Failed

The standard (built-in) uses for `PolynomialGCD` and `PolynomialLCM` still work for ordinary polynomials.

`PolynomialGCD[`*P, a, b*`]`	return the greatest common divisor of the polynomials *a* and *b* in the polynomial extension *P*
`PolynomialGCD[`*a, b*`]`	if both *a* and *b* are in extension *P*, return `PolynomialGCD[`*P, a, b*`]`
`PolynomialLCM[`*P, a, b*`]`	return the least common multiple of the polynomials *a* and *b* in *P*
`PolynomialLCM[`*a, b*`]`	if both *a* and *b* are in extension *P*, return `PolynomialLCM[`*P, a, b*`]`

GCD and LCM on polynomial extensions.

■ 3.8.4 Quotient rings of polynomials

A transition back to ringoids is made by `QuotientRing`. Given a polynomial *p* of degree *k*, `QuotientRing[`*R, p*`]` generates a `Ringoid` with all polynomials of degree less than *k* over *R*. The operations are polynomial addition and multiplication modulo the polynomial *p*.

`QuotientRing[` `PolynomialsOver[`*R*`],` *p*`]`	return the ringoid of polynomials mod *p*, if *R* is a ringoid with unity, *p* is a polynomial over *R* and the leading coefficient of *p* is a unit of *R*
`QuotientRing[`*R, p*`]`	identical to `QuotientRing[PolynomialsOver[`*R*`],` *p*`]`
`ModulusPolynomial[`*Q*`]`	return the polynomial from which the quotient ring *Q* was created

Polynomial quotient rings.

■ With a cubic polynomial over $\mathbb{Z}_2$, the domain of the `QuotientRing` contains eight polynomials.

In[142]:= **Q = QuotientRing[Z[2], Poly[Z[2], x^3 + x + 1]]**

Out[142]= Ringoid[{0, x^2, x, x + x^2, 1, 1 + x^2, 1 + x, 1 + x + x^2},
        -Addition-, -Multiplication-]

- The first argument may also be a polynomial extension instead of the base ring. The result is the same.

In[143]:= **Q =**
        QuotientRing[PolynomialsOver[Z[2]], Poly[Z[2], x^3 + x + 1]]

Out[143]= Ringoid[{0, x^2, x, x + x^2, 1, 1 + x^2, 1 + x, 1 + x + x^2},
        -Addition-, -Multiplication-]

- The modulus can be extracted from the result of QuotientRing.

In[144]:= **ModulusPolynomial[Q]**

Out[144]= 1 + x + x^3

- The polynomial $x^3 + x + 1$ is irreducible over $\mathbb{Z}_2$, so Q is a finite field with eight elements, *GF*(8). (Q can also be constructed by GF[8, $x^3 + x + 1$].) Here is the multiplication table for this field.

In[145]:= **MultiplicationTable[Q, Mode → Visual,**
        KeyForm → StandardForm, ShowName → False];

        KEY for   3
        Mult(Quotient Ring mod 1 + x + x): label used →
          element: {g1 → 0, g2 → x^2, g3 → x, g4 → x + x^2,
          g5 → 1, g6 → 1 + x^2, g7 → 1 + x, g8 → 1 + x + x^2}

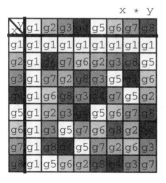

A limit to the size of a polynomial quotient ring is controlled by the option SizeLimit. The default limit is 50.

- This ringoid, with $11^4 = 14641$ elements, is far too large to generate (in this version of AbstractAlgebra).

In[146]:= **QuotientRing[Z[11], Poly[Z[11], x^4 + x + 1]]**

QuotientRing::toobig :
Requested quotient ring is likely to be
too large. Option SizeLimit can be reset.

Out[146]= $Failed

- The modulus polynomial need not be irreducible, as in the following case.

In[147]:= **Q2 = QuotientRing[Z[3], Poly[Z[3], x² + 2]]**

Out[147]= Ringoid[{0, x, 2 x, 1, 1 + x, 1 + 2 x, 2, 2 + x, 2 + 2 x},
         -Addition-, -Multiplication-]

- Fields do not have zero divisors, but Q2 clearly does.

In[148]:= **Multiplication[Q2][Poly[Z[3], x + 1], Poly[Z[3], x + 2]]**

Out[148]= 0

- Consider the element $2 x$ in Q2. Note that $(2 x)^2 + 2 = 4 x^2 + 2 = x^2 + 2$.
  Therefore $2 x$ is a root (zero) of the polynomial $x^2 + 2$ over $\mathbb{Z}_3$. The following
  function confirms this.

In[149]:= **EvaluationInExtension[Q2,**
        ModulusPolynomial[Q2], Poly[Z[3], 2 x]]

Out[149]= 0

EvaluationInExtension[E, p, q]	given E as the quotient ring of a field F over the irreducible polynomial p, this evaluates the induced coset polynomial equivalent of p at the element q in E

Doing evaluation in field extensions.

■ 3.8.5 Irreducibility of integer-based polynomials

Assuming p is a polynomial over the integers, $p(0) \neq 0$, then if $\frac{r}{s}$ is a rational root of p, r must be a divisor of the constant coefficient and s must divide the leading coefficient. This is the essence of the Rational Root Theorem. Eisenstein's Criterion and the Mod p Irreducibility Test are two other methods for determining the irreducibility of polynomials over the integers. Some or all of these approaches are in most standard abstract algebra books, as well as in *Exploring Abstract Algebra with Mathematica*. Note that these functions work with polynomials that are *not* constructed with the Poly function; ordinary *Mathematica* polynomials are used.

- The possible rational roots here are determined by the numbers -7 and 4:
  divisors of -7 divided by divisors of 4.

In[150]:= **RationalRootCandidates[4 x⁵ + 5 x³ - 2 x² - 7]**

Out[150]= $\left\{-7, -\frac{7}{2}, -\frac{7}{4}, -1, -\frac{1}{2}, -\frac{1}{4}, \frac{1}{4}, \frac{1}{2}, 1, \frac{7}{4}, \frac{7}{2}, 7\right\}$

`RationalRootCandidates[`*zpoly*`]`	list of candidates of rational roots of polynomial *zpoly*
`RationalRootTheorem[`*zpoly*`]`	return a pair of lists consisting of the rational roots of *zpoly*, followed by the candidates
`ModpIrreducibilityQ[`*p, zpoly*`]`	give `True` if the polynomial *zpoly* is irreducible according to the Mod *p* Irreducibility Test using the prime *p*, and `False` otherwise
`ModpIrreducibilityQ[`*zpoly, n*`]`	give `True` if the polynomial *zpoly* is determined irreducible according to the Mod *p* Irreducibility Test using the first *n* primes (defaulting to 25), and `False` otherwise
`ModpIrreducibilityQ[`*zpoly*`, Mode → Textual]`	give a textual commentary of the result of applying the Mod *p* Irreducibility Test to *zpoly*, also giving the computational result
`EisensteinsCriterionQ[`*zpoly*`]`	give `True` if the polynomial *zpoly* is irreducible according to Eisenstein's Criterion, and `False` otherwise
`EisensteinsCriterionQ[`*zpoly*`, Mode → Textual]`	give a textual commentary of the result of applying Eisenstein's Criterion to *zpoly*, also giving the computational result

Functions for testing irreducibility of integer-based polynomials.

- Now we consider the irreducibility of the polynomial $x^6 + 7x^2 - 6$ over the integers. Reducing the polynomial mod 2, we see that this reduced polynomial is not irreducible.

In[151]:= **ModpIrreducibilityQ[2, x^6 + 7 x^2 - 6]**

Out[151]= `False`

- Using the option `Modulus` in the `Factor` function, we can obtain similar results.

In[152]:= **Map[Factor[x^6 + 7 x^2 - 6, Modulus → Prime[#]] &,**
     Range[1, 6]] // ColumnForm

Out[152]= $x^2 (1 + x)^4$
     $x^2 (2 + x + x^2) (2 + 2x + x^2)$
     $(1 + 3x + x^2 + x^3) (4 + 3x + 4x^2 + x^3)$
     $(1 + x^2) (2 + x^2) (4 + x^2)$
     $(4 + x) (7 + x) (10 + 5x^2 + x^4)$
     $7 + 7x^2 + x^6$

- For certain polynomials, Eisenstein's Criterion ably determines irreducibility.

In[153]:= **EisensteinsCriterionQ[5 x^4 - 27 x^2 + 6 x + 12, Mode → Textual]**

> The coefficients to consider (from low
> degree to high degree) are: {12, 6, -27, 0, 5}
>
> 3 is a prime that divides
> all of the first n-1 coefficients.
>
> Is it true that 3 does not divide 5? → True
>
> Is it true that 9 does not divide 12? → True
>
> Therefore, 3 is a prime that
> illustrates the polynomial is irreducible.

Out[153]= True

■ 3.8.6 Functions related to solving equations or evaluation

PolynomialEvaluation[p, α]	given the polynomial p, evaluate p(α) for α in the base ring of p
PolynomialEvaluation[P, p, α]	given the polynomial p in the ring of polynomials P, evaluate p(α) for α in the base ring of p
PolynomialEvaluation[P, p, *rules*]	use *rules* to specify the value(s) at which the polynomial p should be evaluated
Solve[$p == \alpha$]	given a polynomial p and a value α in the base ring of p, solve the equation $p = \alpha$
Zeros[p]	give a list of all the zeros of the polynomial p

Functions for solving equations or evaluating polynomials in a polynomial extension.

- If we are interested in the zeros of the polynomial $x^2 + 2x - 3$ over the ring $\mathbb{Z}_5$, we can map the PolynomialEvaluation function over the elements of the ring and look for a zero.

In[154]:= **Map[{#, PolynomialEvaluation[**
        p = Poly[Z[5], x^2 + 2 x - 3], #]} &, Elements[Z[5]]] //
    TableForm[#, TableHeadings → {None, {"x", "p(x) \n"}}] &

Out[154]//TableForm=

x	p(x)
0	2
1	0
2	0
3	2
4	1

■ This result can also be obtained using the `Zeros` function.

In[155]:= **Zeros[p]**

Out[155]= {1, 2}

■ The `Solve` function is one more means of finding the zeros of a polynomial function.

In[156]:= **zeros = Solve[p == 0]**

Out[156]= {{x → 1}, {x → 2}}

■ The output of the `Solve` command is in the same form used by the standard `Solve` command. Furthermore, this output can be used by the `ReplaceAll` function (alias /.).

In[157]:= **p /. zeros**

Out[157]= {0, 0}

■ The `Solve` function is useful for more than locating zeros.

In[158]:= **sols = Solve[p == 2]**

Out[158]= {{x → 0}, {x → 3}}

■ By evaluating the polynomial *p* at the solutions given by the `Solve` command, we can confirm that indeed these yield the value 2 in each case.

In[159]:= **PolynomialEvaluation[PolynomialsOver[Z[5]], p, sols]**

Out[159]= {2, 2}

■ A single rule can be given, such as $x \rightarrow 3$, or a list of rules can be given as the third argument.

In[160]:= **{PolynomialEvaluation[Poly[Z[7], x^2 + 2 x - 3], x → 3],**
        PolynomialEvaluation[
          Poly[Z[7], x^2 + 2 x - 3], {x → 3, x → 2}]}

Out[160]= {5, {5, 5}}

■ 3.8.7 Extensions of ordinary *Mathematica* functions

We have already seen the functions `PolynomialQuotient`, `PolynomialDivision`, `PolynomialRemainder`, `PolynomialGCD`, and `PolynomialLCM` having natural extensions in an extension ringoid of polynomials. Here we consider some other extensions of built-in *Mathematica* functions.

`Equal[PolynomialsOver[` `R], p, q]`	give `True` if the polynomials p and q are equal, and `False` otherwise
`Equal[p, q]`	assuming that the polynomials p and q are both over the same ring, give `True` if they are equal, and `False` otherwise
`Equal[p, q,` `IgnoreIndeterminate →` `False]`	give `True` if the polynomials p and q have the same list of coefficients and the same indeterminate, and `False` otherwise
`Exponent[p]`	return the degree of the polynomial p
`Degree[p]`	identical to `Exponent[p]`
`Variables[p]`	return the variable used in the polynomial p
`Coefficient[p, ind, n]`	given a polynomial p in the indeterminate *ind*, return the coefficient of ind^n
`CoefficientList[p]`	return the list of coefficients for the polynomial p in the order given by `PowersIncrease → LeftToRight`

Extensions of ordinary *Mathematica* functions.

- Here are three polynomials to consider, as well as an illustration that the `AbstractAlgebra` packages do not currently support polynomials with more than one indeterminate.

$In[161]:=$ `{p = Poly[Z[3], `x^2` + 2 x], q = Poly[Z[4], `x^2` + 2 x],`
`r = Poly[Z[3], `y^2` + 2 y], s = Poly[Z[5], `$x^2 y^4$` - x y + 1]}`

> `Poly::mixvars :`
>   `In your polynomial 1 - x y + `$x^2 y^4$`, you should`
>   `be using only a single variable (such as`
>   `x), but the variables {x, y} were used.`

$Out[161]=$ `{2 x + `x^2`, 2 x + `x^2`, 2 y + `y^2`, $Failed}`

- Polynomials p and q are not the same since they have different base rings. Polynomials p and r are the same if the indeterminate is ignored, but otherwise they are different.

$In[162]:=$ `{Equal[p, q], Equal[p, r],`
`Equal[p, r, IgnoreIndeterminate → False]}`

$Out[162]=$ `{False, True, False}`

- The following illustrates some of the other functions.

$In[163]:=$ `{Exponent[p], Coefficient[p, x, 2], Coefficient[p, y, 2],`
`Coefficient[p, 2], CoefficientList[p]}`

Coefficient::ind :
Since $2x + x^2$ *uses the indeterminate* x, y *should not*
be specified as the indeterminate. Mention of
the indeterminate is optional; it can be omitted.

Out[163]= {2, 1, $Failed, 1, {0, 2, 1}}

■ 3.8.8 Miscellaneous functions

■ Here we form two polynomials.

In[164]:= {p = Poly[Z[7], 9 x⁴ - 12 x² + 3 x - 2],
        q =
          Poly[BooleanRing[{"a", "b", "c"}], {"a", "b"}, {}, {"a"}]}

Out[164]= {5 + 3 x + 2 x² + 2 x⁴, {a, b} + {} x + {a} x²}

■ The following shows that the Poly function was used to construct *p* and *q*, but not *r* (which is defined in the following cell).

In[165]:= Map[PolyQ, {p, q, r = x² + 3 x - 5}]

Out[165]= {True, True, False}

ToOrdinaryPolynomial[*p*]	given a polynomial produced by the Poly function, return a polynomial in *Mathematica*'s ordinary usage, where this makes sense
PolyQ[*p*]	give True if the polynomial *p* was created in the AbstractAlgebra packages with the Poly function, and False otherwise
Monomial[*R*, *c*, *n*]	return Poly[PolynomialsOver[*R*], *c* x^n]
BaseRing[*p*]	return the underlying ringoid of the polynomial *p*

Some miscellaneous functions.

■ Even if the dot product could be completed, the polynomial *q* has no meaning in *Mathematica* apart from the AbstractAlgebra packages, while the polynomial *p* does.

In[166]:= Map[ToOrdinaryPolynomial, {p, q, r}]

Dot::dotsh : Tensors {{a, b}, {}, {a}}
and {1, x, x²} *have incompatible shapes.*

Out[166]= {5 + 3 x + 2 x² + 2 x⁴, {{a, b}, {}, {a}}.{1, x, x²}, -5 + 3 x + x²}

■ The zero of LatticeRing[10] is 1, which explains why the quadratic term is "lost" here.

In[167]:= **ToOrdinaryPolynomial[Poly[LatticeRing[10], 5, 10, 1]]**

Out[167]= $5 + 10\,x$

- Every polynomial constructed using the `Poly` function has an underlying base ring.

In[168]:= **Map[BaseRing, {p, q}]**

Out[168]= {Ringoid[{0, 1, 2, 3, 4, 5, 6},
        Mod[#1 + #2, 7] &, Mod[#1 #2, 7] &], Ringoid[
        {{}, {c}, {b}, {b, c}, {a}, {a, c}, {a, b}, {a, b, c}},
        -Addition-, -Multiplication-]}

- Here we build a polynomial from some monomials.

In[169]:= **p = Sum[Monomial[Z[5], k, k], {k, 1, 3}]**

Out[169]= $x + 2\,x^2 + 3\,x^3$

- This polynomial *p* is the same as constructing it as a single polynomial.

In[170]:= **p == Poly[Z[5], x + 2 x² + 3 x³]**

Out[170]= True

PolynomialsOfDegreeN[*R, n, opts*]	return all polynomials of degree *n* over the ringoid *R*, according to any restrictions in the options *opts*
PolynomialsUpToDegreeN[*R, n, opts*]	return all polynomials up to (and including) degree *n* over the ringoid *R*, using the options *opts*

Generating complete lists of polynomials.

option name	default value	
SizeLimit	125	specifies maximum size of the list that is created
Indeterminate	x	specifies indeterminate to be used for the polynomials

Options for `PolynomialsOfDegreeN` and `PolynomialsUpToDegreeN`.

- Here are all the cubic polynomials over $\mathbb{Z}_2$.

In[171]:= **PolynomialsOfDegreeN[Z[2], 3]**

Out[171]= $\{x^3,\ x^2 + x^3,\ x + x^3,\ x + x^2 + x^3,$
        $1 + x^3,\ 1 + x^2 + x^3,\ 1 + x + x^3,\ 1 + x + x^2 + x^3\}$

- Here are all polynomials of degree 2 or less over $\mathbb{Z}_5$, which is as big a set as usually possible without using the `SizeLimit` option.

In[172]:= **PolynomialsUpToDegreeN[Z[5], 2] // Short**

Out[172]//Short=

$$\{x^2, 2\,x^2, 3\,x^2, 4\,x^2, x + x^2, x + 2\,x^2,$$
$$x + 3\,x^2, \ll 111 \gg, 4 + 3\,x, 4 + 4\,x, 0, 1, 2, 3, 4\}$$

- There are 1024 polynomials of degree 9 or less over $\mathbb{Z}_2$. Here they are in the indeterminate α.

In[173]:= **Short[PolynomialsUpToDegreeN[ZR[2],**
        9, Indeterminate → α, SizeLimit → 1024], 2]

Out[173]//Short=

$$\{\alpha^9, \alpha^8 + \alpha^9, \alpha^7 + \alpha^9, \alpha^7 + \alpha^8 + \alpha^9, \alpha^6 + \alpha^9,$$
$$\alpha^6 + \alpha^8 + \alpha^9, \ll 1012 \gg, 1 + \alpha^2, 1 + \alpha + \alpha^2, \alpha, 1 + \alpha, 0, 1\}$$

▓ 3.9 Matrices over a ringoid

For any ringoid R, the expression `MatricesOver[R, n]` generates an extension ringoid that represents the ringoid of all *n*-by-*n* matrices over R. The domain of this system is finite but tends to be quite large, so the normal `Ringoid` structure is not used. Alternatively, one can obtain *m*-by-*n* matrices over some ring R by using `MatricesOver[R, {m, n}]`.

- This represents a matrix extension with 625 matrices as elements.

In[174]:= **M = MatricesOver[Z[5], 2]**

Out[174]= $-\mathrm{Mat}_2(\mathrm{Z}[5])-$

There are several shortcuts to creating matrix extensions. The `Mat` function is equivalent to `MatricesOver`. The functions `MatA` and `MatM` are useful if one wishes to later convert these extensions to a groupoid; in the former case, the operation is addition, while multiplication is used in the latter case.

`Mat[R, n]`	identical to `MatricesOver[R, n]`
`Mat[R, {m, n}]`	identical to `MatricesOver[R, {m, n}]`
`MatA[R, n]`	identical to `MatricesOver[R, n, Operation → Addition]`
`MatA[R, {m, n}]`	identical to `MatricesOver[R, {m, n}, Operation → Addition]`
`MatM[R, n]`	identical to `MatricesOver[R, n, Operation → Multiplication]`

Alternate forms of creating matrix extensions.

- Using `Mat` to create an extension, we pick a random two-by-two matrix over $\mathbb{Z}_5$.

In[175]:= **RandomElement[Mat[Z[5], 2]] // MatrixForm**

Out[175]//MatrixForm=
$$\begin{pmatrix} 2 & 1 \\ 4 & 4 \end{pmatrix}$$

- Here we see that the operation is indeed multiplication when the `MatM` function is used.

In[176]:= **ToGroupoid[MatM[Z[4], 2]] // Operation**

Out[176]= Multiplication[-Mat$_2$(Z[4])-][#1, #2] &

- When `MatA` is used to create an extension, the inherent operation is addition.

In[177]:= **MatrixOperation[MatA[Z[3], {2, 5}]]**

Out[177]= Addition

■ 3.9.1 Individual matrices

The structure of a matrix in a matrix extension is identical to the usual *Mathematica* matrix structure and matrices are entered in exactly the same manner.

- Matrices are entered in the usual way.

In[178]:= **M = MatricesOver[Z[5], 2];**
A = {{4, 2}, {1, 4}}

Out[179]= {{4, 2}, {1, 4}}

ElementQ[*A, M*]	give `True` if *A* is an element of the matrix extension *M*, and `False` otherwise
RandomElement[*M*]	return a random nonzero matrix in the matrix extension *M*
RandomElement[*M, opts*]	return a random matrix in the matrix extension *M* according to the options specified in *opts*
RandomElements[*M, k, opts*]	return *k* random matrices in the matrix extension *M* according to the options specified in *opts*
RandomMatrix[*R, n*]	return a random *n*–by–*n* matrix over the ringoid *R* with no restrictions
RandomMatrix[*R, n, MatrixType → type*]	return a random *n*–by–*n* matrix over the ringoid *R* of the specified *type* (choices below)

Functions relating to individual matrices.

- Any matrix can be tested for membership in a matrix extension with `ElementQ`.

In[180]:= **ElementQ[A, M]**

Out[180]= True

- `ElementQ` returns `False` if the order of a matrix doesn't match that of the extension.

In[181]:= **ElementQ[{{1, 0, 0}, {1, 1, 1}, {1, 2, 4}}, M]**

Out[181]= False

- `RandomElement` acts on matrix extensions.

In[182]:= **RandomElement[M]**

Out[182]= {{4, 1}, {4, 0}}

option name	default value	
SelectFrom	NonZero	specifies restrictions on the random matrix that is selected; values are Any, NonZero, NonIdentity, and NonUnity
SelectBaseElements From	Any	specifies restrictions on the individual entries that appear in a random matrix; values are Any, NonZero, NonUnity, and NonIdentity

Options on `RandomElement` for matrices.

- A random nonzero matrix may have some zero entries. We can also get a matrix with all nonzero entries.

In[183]:= **Map[MatrixForm, {RandomElement[MatricesOver[Z[3], 5]],**
RandomElement[MatricesOver[Z[3], 5],
SelectBaseElementsFrom → NonZero]}]

Out[183]= $\left\{ \begin{pmatrix} 1 & 1 & 1 & 1 & 0 \\ 2 & 0 & 1 & 0 & 2 \\ 2 & 1 & 1 & 0 & 2 \\ 2 & 2 & 1 & 2 & 2 \\ 1 & 0 & 1 & 1 & 2 \end{pmatrix}, \begin{pmatrix} 2 & 1 & 2 & 1 & 1 \\ 1 & 2 & 2 & 2 & 2 \\ 1 & 1 & 2 & 1 & 1 \\ 1 & 1 & 1 & 1 & 1 \\ 1 & 1 & 1 & 2 & 1 \end{pmatrix} \right\}$

Whereas `RandomElement` requires its first argument to be a matrix extension, `RandomMatrix` uses the base ring for its first argument, and optionally one can specify the type of matrix being sought by giving a type with the `MatrixType` option. Here are some possibilities.

- Here some three-by-three matrices over $\mathbb{Z}_5$ illustrating the various types.

```
In[184]:= Map[TraditionalForm,
            examples = Map[RandomMatrix[Z[5], 3, MatrixType → #] &,
            { GL, SL, Diag, UT, LT, UTD, LTD , All}]]
```

$$
Out[184]= \left\{ \begin{pmatrix} 1 & 4 & 4 \\ 4 & 3 & 3 \\ 2 & 0 & 2 \end{pmatrix}, \begin{pmatrix} 3 & 3 & 4 \\ 4 & 0 & 4 \\ 0 & 4 & 0 \end{pmatrix}, \begin{pmatrix} 1 & 0 & 0 \\ 0 & 4 & 0 \\ 0 & 0 & 1 \end{pmatrix}, \begin{pmatrix} 0 & 1 & 1 \\ 0 & 0 & 4 \\ 0 & 0 & 0 \end{pmatrix}, \right.
$$
$$
\left. \begin{pmatrix} 0 & 0 & 0 \\ 4 & 0 & 0 \\ 0 & 1 & 0 \end{pmatrix}, \begin{pmatrix} 1 & 4 & 0 \\ 0 & 2 & 4 \\ 0 & 0 & 1 \end{pmatrix}, \begin{pmatrix} 2 & 0 & 0 \\ 0 & 3 & 0 \\ 4 & 3 & 3 \end{pmatrix}, \begin{pmatrix} 3 & 4 & 4 \\ 0 & 0 & 4 \\ 3 & 0 & 1 \end{pmatrix} \right\}
$$

■ Some of the determinants of these examples are predictable. (Which ones?)

```
In[185]:= Map[Det[Z[5], #] &, examples]
```

$$
Out[185]= \{4, 1, 4, 0, 0, 2, 3, 3\}
$$

`All`	any matrix (default)
`GL`	general linear (an invertible matrix)
`SL`	special linear (matrix with determinant equal to the unity of the base ring)
`Diag`	diagonal (an invertible diagonal matrix)
`UT`	strictly upper triangular (nonzero entries only above the diagonal)
`LT`	strictly lower triangular (nonzero entries only below the diagonal)
`UTD`	upper triangular (nonzero entries only above and on the diagonal)
`LTD`	lower triangular (nonzero entries only below and on the diagonal)

Possible values of `MatrixType`, an option of `RandomMatrix`.

`DiagQ[R, A]`	give `True` if the matrix A is a diagonal matrix over the ringoid R, and `False` otherwise
`GLQ[R, A]`	give `True` if the matrix A is an invertible matrix over the ringoid R, and `False` otherwise
`SLQ[R, A]`	give `True` if the matrix A is an invertible matrix over the ringoid R with determinant the unity of R, and `False` otherwise

Testing for special types of matrices.

■ Here we test the foregoing matrices with these three functions.

```
In[186]:= mats := {Map[MatrixForm, examples], Map[DiagQ[Z[5], #] &,
            examples], Map[GLQ[Z[5], #] &, examples],
            Map[SLQ[Z[5], #] &, examples]} // Transpose;
         {r1, r2} = Map[Transpose, {mats[[{1, 2, 3, 4}]],
            mats[[{5, 6, 7, 8}]]}];
```

```
TableForm[r1, TableSpacing → {0.5, 1}, TableHeadings →
  (heads = {{"matrix", "DiagQ", "GLQ", "SLQ"}, None})]
TableForm[r2, TableSpacing → {0.5, 1},
  TableHeadings → heads]
```

Out[188]//TableForm=

matrix	$\begin{pmatrix} 1 & 4 & 4 \\ 4 & 3 & 3 \\ 2 & 0 & 2 \end{pmatrix}$	$\begin{pmatrix} 3 & 3 & 4 \\ 4 & 0 & 4 \\ 0 & 4 & 0 \end{pmatrix}$	$\begin{pmatrix} 1 & 0 & 0 \\ 0 & 4 & 0 \\ 0 & 0 & 1 \end{pmatrix}$	$\begin{pmatrix} 0 & 1 & 1 \\ 0 & 0 & 4 \\ 0 & 0 & 0 \end{pmatrix}$
DiagQ	False	False	True	False
GLQ	True	True	True	False
SLQ	False	True	False	False

Out[189]//TableForm=

matrix	$\begin{pmatrix} 0 & 0 & 0 \\ 4 & 0 & 0 \\ 0 & 1 & 0 \end{pmatrix}$	$\begin{pmatrix} 1 & 4 & 0 \\ 0 & 2 & 4 \\ 0 & 0 & 1 \end{pmatrix}$	$\begin{pmatrix} 2 & 0 & 0 \\ 0 & 3 & 0 \\ 4 & 3 & 3 \end{pmatrix}$	$\begin{pmatrix} 3 & 4 & 4 \\ 0 & 0 & 4 \\ 3 & 0 & 1 \end{pmatrix}$
DiagQ	False	False	False	False
GLQ	False	True	True	True
SLQ	False	False	False	False

■ 3.9.2 Matrix arithmetic

For any matrix extension M, we can add and multiply matrices belonging to M, with products making sense only if we are in a square matrix extension. By specifying the base ring (and not a matrix extension), appropriately dimensioned nonsquare matrices can also be multiplied. The underlying ring operations determine the result of the matrix operations.

`Addition[M][A, B]`	return the sum of matrices A and B in the matrix extension M
`Addition[M, A, B]`	identical to `Addition[M][A, B]`
`Addition[R, A, B]`	if R is a ringoid or groupoid, identical to `Addition[MatricesOver[R]][A, B]`
`Multiplication[M][A, B]`	return the product of matrices A and B in the matrix extension M (and if the dimensions match appropriately when M is a ring)
`MatrixPower[M, A, k]`	return the kth power of the square matrix A in the matrix extension (or ring) M

Basic arithmetic functions over matrix extensions.

■ Here we form a matrix extension over $\mathbb{Z}_7$ to illustrate some of these functions.

In[190]:= **M3 = MatricesOver[Z[7], 3]**

Out[190]= $-Mat_3 (Z[7]) -$

■ Here is how a cyclic matrix can be created, as well as another matrix.

In[191]:= **Map[MatrixForm,**
          {A = Map[RotateRight[{1, 2, 6}, #] &, {0, 1, 2}],
          B = {{0, 0, 1}, {0, 1, 2}, {1, 0, 0}}}]

Out[191]= $\left\{ \begin{pmatrix} 1 & 2 & 6 \\ 6 & 1 & 2 \\ 2 & 6 & 1 \end{pmatrix}, \begin{pmatrix} 0 & 0 & 1 \\ 0 & 1 & 2 \\ 1 & 0 & 0 \end{pmatrix} \right\}$

Note that there are variations of the `Multiplication` function comparable to those for `Addition`.

■ Here are the three methods by which we can find the sum of matrices *A* and *B*.

In[192]:= **Map[MatrixForm, {Addition[M3][A, B],**
          Addition[M3, A, B], Addition[Z[7], A, B]}]

Out[192]= $\left\{ \begin{pmatrix} 1 & 2 & 0 \\ 6 & 2 & 4 \\ 3 & 6 & 1 \end{pmatrix}, \begin{pmatrix} 1 & 2 & 0 \\ 6 & 2 & 4 \\ 3 & 6 & 1 \end{pmatrix}, \begin{pmatrix} 1 & 2 & 0 \\ 6 & 2 & 4 \\ 3 & 6 & 1 \end{pmatrix} \right\}$

■ Here is their product.

In[193]:= **Map[MatrixForm, {Multiplication[M3][A, B],**
          Multiplication[M3, A, B], Multiplication[Z[7], A, B]}]

Out[193]= $\left\{ \begin{pmatrix} 6 & 2 & 5 \\ 2 & 1 & 1 \\ 1 & 6 & 0 \end{pmatrix}, \begin{pmatrix} 6 & 2 & 5 \\ 2 & 1 & 1 \\ 1 & 6 & 0 \end{pmatrix}, \begin{pmatrix} 6 & 2 & 5 \\ 2 & 1 & 1 \\ 1 & 6 & 0 \end{pmatrix} \right\}$

■ While M3 consisted of only three-by-three matrices, here we obtain a random three-by-two matrix over the same base ring, $\mathbb{Z}_7$, as well as a two-by-three matrix (also over the same base ring).

In[194]:= **{ (F = RandomElement[MatricesOver[Z[7], {3, 2}]]) //**
          MatrixForm,
          (G = RandomElement[MatricesOver[Z[7], {2, 3}]]) //
          MatrixForm}

Out[194]= $\left\{ \begin{pmatrix} 2 & 5 \\ 2 & 4 \\ 4 & 1 \end{pmatrix}, \begin{pmatrix} 4 & 2 & 4 \\ 2 & 1 & 1 \end{pmatrix} \right\}$

■ There is a variety of products that can now be computed with the matrices *A*, *B*, *F*, and *G*.

In[195]:= **Map[MatrixForm,**
    {Multiplication[Z[7], A, F], Multiplication[Z[7], G, A],
     Multiplication[Z[7], G, F], Multiplication[Z[7], F, G]}]

Out[195]= $\left\{ \begin{pmatrix} 2 & 5 \\ 1 & 1 \\ 6 & 0 \end{pmatrix}, \begin{pmatrix} 3 & 6 & 4 \\ 3 & 4 & 1 \end{pmatrix}, \begin{pmatrix} 0 & 4 \\ 3 & 1 \end{pmatrix}, \begin{pmatrix} 4 & 2 & 6 \\ 2 & 1 & 5 \\ 4 & 2 & 3 \end{pmatrix} \right\}$

- There are also some products that do not make sense.

In[196]:= **Multiplication[Z[7], F, A]**

  Multiplication::fail : A 3 by 2 matrix
    can not be multiplied by a 3 by 3 matrix.

Out[196]= $Failed

- Here are the powers -1 through 3 of the matrix $\begin{pmatrix} 2 & 3 \\ 4 & 5 \end{pmatrix}$ over $\mathbb{Z}_7$.

In[197]:= **Map[MatrixForm,**
      Table[MatrixPower[Z[7], {{2, 3}, {4, 5}}, k], {k, -1, 3}]]

Out[197]= $\left\{ \begin{pmatrix} 1 & 5 \\ 2 & 6 \end{pmatrix}, \begin{pmatrix} 1 & 0 \\ 0 & 1 \end{pmatrix}, \begin{pmatrix} 2 & 3 \\ 4 & 5 \end{pmatrix}, \begin{pmatrix} 2 & 0 \\ 0 & 2 \end{pmatrix}, \begin{pmatrix} 4 & 6 \\ 1 & 3 \end{pmatrix} \right\}$

As with polynomials, much of the basic ringoid functionality carries over into a matrix extension.

HasZeroQ[*M*]	give True if *M* has a zero, and False otherwise
Zero[*M*]	return the zero of *M*, if it exists, and $Failed otherwise
NegationOf[*M, A*]	return the negation of matrix *A* if all of its entries have a negation in the base ring, and $Failed otherwise
WithUnityQ[*M*]	give True if *M* has a unity, and False otherwise
Unity[*M*]	return the unity of *M* (actually the identity matrix) if it exists, and $Failed otherwise
UnitQ[*M, A*]	give True if *A* is a unit (has a multiplicative inverse in *M*), and False otherwise
ZeroDivisorQ[*M, A*]	give True if *A* is a zero divisor in *M*, and False otherwise

Using ringoid functions in a matrix extension.

- The negation of *A* is easily determined.

In[198]:= **Map[MatrixForm, {A, NegationOf[M3, A]}]**

$$Out[198]= \left\{ \begin{pmatrix} 1 & 2 & 6 \\ 6 & 1 & 2 \\ 2 & 6 & 1 \end{pmatrix}, \begin{pmatrix} 6 & 5 & 1 \\ 1 & 6 & 5 \\ 5 & 1 & 6 \end{pmatrix} \right\}$$

- The zero and one are well known.

$In[199]:=$ **Map[MatrixForm, {Zero[M3], Unity[M3]}]**

$$Out[199]= \left\{ \begin{pmatrix} 0 & 0 & 0 \\ 0 & 0 & 0 \\ 0 & 0 & 0 \end{pmatrix}, \begin{pmatrix} 1 & 0 & 0 \\ 0 & 1 & 0 \\ 0 & 0 & 1 \end{pmatrix} \right\}$$

- We can reuse *A* and *B* in a matrix extension over a different Ringoid in which the multiplication is always zero. (Observe that matrices do not carry with them the underlying ring, as do the polynomials discussed earlier.)

$In[200]:=$ **Map[MatrixForm, {A, B,**
　　　　　Multiplication[MatricesOver[TrivialZR[7], 3]][A, B]}]

$$Out[200]= \left\{ \begin{pmatrix} 1 & 2 & 6 \\ 6 & 1 & 2 \\ 2 & 6 & 1 \end{pmatrix}, \begin{pmatrix} 0 & 0 & 1 \\ 0 & 1 & 2 \\ 1 & 0 & 0 \end{pmatrix}, \begin{pmatrix} 0 & 0 & 0 \\ 0 & 0 & 0 \\ 0 & 0 & 0 \end{pmatrix} \right\}$$

■ 3.9.3 Determinants and inverses

The function Det is extended in AbstractAlgebra to compute the determinant of a square matrix over an arbitrary ringoid.

- Recall that for a two-by-two matrix, the determinant is the difference between two products: the product of the two diagonal elements and the product of the two off-diagonal elements.

$In[201]:=$ **Needs["Graphics`Arrow`","Arrow.m"];**
　　　　　{Line[{{-2,2},{-2,-2}}], Line[{{2,2},{2,-2}}],
　　　　　Text["a",{-1,1}], Text["c", {-1,-1}], Text["b",{1,1}],
　　　　　Text["d",{1,-1}], Arrow[{-1.5,1.9},{1.5,-1.1}],
　　　　　Arrow[{-1.5,-1.9},{1.5,1.1}], Text["a d - b c",{4.2,0}],
　　　　　Text["=",{2.45,0}], RGBColor[1,0,0],
　　　　　Arrow[{1.5,-1.5},{3,-0.2}], Arrow[{1.5,1.5},{4.2,0.2}]}//
　　　　　Show[Graphics[#],PlotRange→{{-2.3,5.5},{-2.3,2.3}}]&;

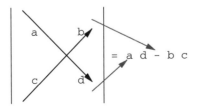

Det[*M*, *A*]	return the determinant of the square matrix *A* in the ring extension *M*
Det[*R*, *A*]	for a ringoid *R*, identical to Det[MatricesOver[*R*, Length[*A*]], *A*]
MultiplicativeInverse [*M*, *A*]	return the multiplicative inverse of *A* in the matrix extension (or ring) *M*, if it exists, and $Failed otherwise
Inverse[*R*, *A*]	identical to MultiplicativeInverse[*R*, *A*]

Extensions of Det and MultiplicativeInverse for matrix extensions.

- Here is the determinant of the matrix $A = \{\{4, 10\}, \{6, 10\}\}$. In this case, using mod 11 arithmetic, $4 * 10 - 10 * 6 = 7 - 5 = 2$.

In[203]:= **Det[MatricesOver[ZR[11], 2], A = {{4, 10}, {6, 10}}]**

Out[203]= 2

- A general formula for the determinant can be determined.

In[204]:= **Clear[a, b, c, d];**
           Det[M = MatricesOver[Z[11], 2], {{a, b}, {c, d}}]

Out[204]= Mod[-b c + a d, 11]

Caution: The evaluation of determinants of symbolic matrices does not work over all ringoids.

One of the fundamental theorems in linear algebra is that in the matrix ring over a field *F*, a matrix *A* has a multiplicative inverse if and only if the determinant of *A* is nonzero. The following theorem extends this to rings in general. The theorem can be used to determine whether a matrix is a unit.

Theorem. If *A* is a square matrix over a *commutative* ring with unity *R*, then *A* has a multiplicative inverse if and only if the determinant of *A* is a unit of *R*.

- The determinant of *A* (found above to be the value 2) indicates that it should have an inverse. Indeed, the product of *A* and A^{-1} should be the identity matrix.

In[205]:= **Map[MatrixForm, {Ai = MultiplicativeInverse[M, A],**
           Multiplication[M][A, Ai]}]

Out[205]= $\left\{ \begin{pmatrix} 5 & 6 \\ 8 & 2 \end{pmatrix}, \begin{pmatrix} 1 & 0 \\ 0 & 1 \end{pmatrix} \right\}$

- Here is what happens if a matrix is not invertible.

In[206]:= **MultiplicativeInverse[MatricesOver[Z[8]], {{1, 5}, {1, 7}}]**

```
Out[206]= MultiplicativeInverse[
             MatricesOver[Ringoid[{0, 1, 2, 3, 4, 5, 6, 7},
               Mod[#1 + #2, 8] &, Mod[#1 #2, 8] &]], {{1, 5}, {1, 7}}]
```

- Since the base ring of this matrix extension is not a field, it is apparent that this matrix has a determinant that is not a unit.

```
In[207]:= Det[Z[8], {{1, 5}, {1, 7}}]
```

```
Out[207]= 2
```

The determinant function uses an extended dot product to do its calculations.

Dot[R, u, v]	return the dot product of the vectors u and v whose coordinates are from the ringoid R

Dot product extension.

- Here are two examples using the generalized dot product function.

```
In[208]:= {Dot[Z[5], {2, 3, 4}, {1, 0, 2}],
           Dot[BooleanRing[2], {{}, {1, 2}}, {{1}, {2}}]}
```

```
Out[208]= {0, {2}}
```

■ 3.9.4 Matrix ringoids

Since the matrix extensions are finite, they can be represented as Ringoids. This is not practical for most situations, but some of the smaller ones can be converted using ToRingoid.

ToRingoid[MatricesOver[R, n]]	return the Ringoid consisting of the n–by–n matrices over the ringoid R
SizeLimit	option to restrict the size of a matrix ringoid, defaulting to 1000

Obtaining a Ringoid from a matrix extensions.

- This is an example of converting a matrix extension to a Ringoid.

```
In[209]:= M2 = ToRingoid[MatricesOver[Z[2], 2]]
```

```
Out[209]= Ringoid[{{{0, 0}, {0, 0}}, {{0, 0}, {0, 1}}, {{0, 0}, {1, 0}},
             {{0, 0}, {1, 1}}, {{0, 1}, {0, 0}}, {{0, 1}, {0, 1}},
             {{0, 1}, {1, 0}}, {{0, 1}, {1, 1}}, {{1, 0}, {0, 0}},
             {{1, 0}, {0, 1}}, {{1, 0}, {1, 0}}, {{1, 0}, {1, 1}},
             {{1, 1}, {0, 0}}, {{1, 1}, {0, 1}}, {{1, 1}, {1, 0}},
             {{1, 1}, {1, 1}}}, -Addition-, -Multiplication-]
```

■ Here are the invertible matrices in this ring.

In[210]:= **Map[MatrixForm, Units[M2]]**

$$Out[210]= \left\{ \begin{pmatrix} 0 & 1 \\ 1 & 0 \end{pmatrix}, \begin{pmatrix} 0 & 1 \\ 1 & 1 \end{pmatrix}, \begin{pmatrix} 1 & 0 \\ 0 & 1 \end{pmatrix}, \begin{pmatrix} 1 & 0 \\ 1 & 1 \end{pmatrix}, \begin{pmatrix} 1 & 1 \\ 0 & 1 \end{pmatrix}, \begin{pmatrix} 1 & 1 \\ 1 & 0 \end{pmatrix} \right\}$$

■ 3.9.5 Matrix groupoids

The multiplicative groupoid of many of the matrix extensions account for many interesting groups. These groupoids can be created using ToGroupoid.

■ Here is the multiplicative groupoid of the ring M2 that was generated in section 3.9.4.

In[211]:= **ToGroupoid[MatricesOver[Z[2], 2]] // Short**

Out[211]//Short=
    Groupoid[{{{0, 0}, {0, 0}}, {{0, 0}, {0, 1}},
        ≪12≫, {{1, 1}, {1, 0}}, {{1, 1}, {1, 1}}}, … -]

ToGroupoid[*M*]	form a Groupoid consisting of all matrices in the ring extension *M* using MatrixOperation[*M*] as the group operation (either Addition or Multiplication), allowing up to 1000 elements
ToGroupoid[*M*, SizeLimit →*n*]	form a Groupoid if Size[*M*] does not exceed *n*

Creation of a Groupoid by a matrix extension.

■ There are 81 two-by-two matrices over $\mathbb{Z}_3$. The SizeLimit option set to 80 prevents this groupoid from being formed.

In[212]:= **ToGroupoid[MatricesOver[Z[3], 2], SizeLimit → 80]**

    SizeLimit::toobig :
     With the present restriction on SizeLimit,
        there are too many elements to form a
        Groupoid that determines all of the elements.

Out[212]= $Failed

■ Here are the elements of the groupoid of strictly upper triangular three-by-three matrices over $\mathbb{Z}_2$.

In[213]:= **G = ToGroupoid[UT[Z[2], 3]]; Map[MatrixForm, Elements[G]]**

$$Out[213]= \left\{ \begin{pmatrix} 0 & 0 & 0 \\ 0 & 0 & 0 \\ 0 & 0 & 0 \end{pmatrix}, \begin{pmatrix} 0 & 0 & 0 \\ 0 & 0 & 1 \\ 0 & 0 & 0 \end{pmatrix}, \begin{pmatrix} 0 & 0 & 1 \\ 0 & 0 & 0 \\ 0 & 0 & 0 \end{pmatrix}, \begin{pmatrix} 0 & 0 & 1 \\ 0 & 0 & 1 \\ 0 & 0 & 0 \end{pmatrix}, \right.$$

$$\left. \begin{pmatrix} 0 & 1 & 0 \\ 0 & 0 & 0 \\ 0 & 0 & 0 \end{pmatrix}, \begin{pmatrix} 0 & 1 & 0 \\ 0 & 0 & 1 \\ 0 & 0 & 0 \end{pmatrix}, \begin{pmatrix} 0 & 1 & 1 \\ 0 & 0 & 0 \\ 0 & 0 & 0 \end{pmatrix}, \begin{pmatrix} 0 & 1 & 1 \\ 0 & 0 & 1 \\ 0 & 0 & 0 \end{pmatrix} \right\}$$

- Here is the groupoid of all two-by-two matrices over $\mathbb{Z}_3$ with determinant equal to one.

```
In[214]:= ToGroupoid[SL[Z[3], 2]] // Short
```

```
Out[214]//Short=
        Groupoid[{{{0, 1}, {2, 0}}, {{0, 1}, {2, 1}},
           ≪20≫, {{2, 2}, {1, 0}}, {{2, 2}, {2, 1}}}, … -]
```

The last two groupoids are examples of specialized collections of matrices. Following are methods of constructing various ring extensions that have some particular property. Each is a ring extension, not a ringoid or a groupoid. If the size is appropriate, however, they can be converted to a groupoid (or possibly a ringoid).

Diag$[R, n]$	return the extension of invertible diagonal n–by–n matrices over the ringoid R
GL$[R, n]$	return the extension of invertible n–by–n matrices over R
LT$[R, n]$	return the extension of n–by–n strictly lower triangular matrices (fully below the diagonal) over the ringoid R
LTD$[R, n]$	return the extension of n–by–n lower triangular matrices (including the diagonal) over the ringoid R
SL$[R, n]$	return the extension of n–by–n matrices over the ringoid R that are invertible and have determinant Unity$[R]$
UT$[R, n]$	return the extension of n–by–n strictly upper triangular matrices (fully above the diagonal) over the ringoid R
UTD$[R, n]$	return the extension of n–by–n upper triangular matrices (including the diagonal) over the ringoid R

Matrix subextensions

- Here we first form $GL_2(\mathbb{Z}_2)$ and then we convert it into a group.

```
In[215]:= {M = GL[Z[2], 2], ToGroupoid[M]}
```

```
Out[215]= {-GL₂(Z[2])-, Groupoid[{{{0, 1}, {1, 0}},
           {{0, 1}, {1, 1}}, {{1, 0}, {0, 1}}, {{1, 0}, {1, 1}},
           {{1, 1}, {0, 1}}, {{1, 1}, {1, 0}}}, -Operation-]}
```

There are a number of alternate approaches to forming ring extensions, as follows.

TypeName [*n*, *R*]	for any type of matrix called *TypeName*, this is identical to *TypeName*[*R*, *n*] (i.e., simply reverse the arguments for those in the previous list)
DiagonalMatrices [*args*]	identical to Diag [*args*]
GL [*n*, *k*]	identical to GL [ZR [*k*], *n*]
GeneralLinear [*args*]	identical to GL [*args*]
GeneralLinearGroup [*args*]	identical to GL [*args*]
SL [*n*, *k*]	identical to SL [ZR [*k*], *n*]
SpecialLinear [*args*]	identical to SL [*args*]
SpecialLinearGroup [*args*]	identical to SL [*args*]

Alternate names to the subextensions.

- Note that there are a number of alternate forms of generating $GL_2(\mathbb{Z}_2)$; here are two instances.

In[216]:= **{GL[2, Z[2]], GL[2, 2]}**

Out[216]= {-GL₂ (Z[2]) -, -GL₂ (Z[2]) -}

Another way to come up with some interesting groupoids is to evaluate Generate-Groupoid with random matrices, using matrix multiplication as the operation.

- The function unitMatrix generates a random invertible matrix. The second line forms the ring extension of all two-by-two matrices over $\mathbb{Z}_3$.

In[217]:= **unitMatrix := RandomMatrix[Z[3], 2, MatrixType → GL];**
      M = MatricesOver[Z[3], 2]

Out[218]= -Mat₂ (Z[3]) -

- Here are two random invertible matrices.

In[219]:= **Map[MatrixForm, {A1 = unitMatrix, A2 = unitMatrix}]**

Out[219]= $\left\{ \begin{pmatrix} 2 & 1 \\ 1 & 1 \end{pmatrix}, \begin{pmatrix} 1 & 0 \\ 0 & 2 \end{pmatrix} \right\}$

- Now we generate the subgroup generated by A1 and A2.

In[220]:= **G = GenerateGroupoid[{A1, A2},**
      Multiplication[M], "*", WideElements → True]

Out[220]= Groupoid[{{{0, 1}, {1, 0}}, {{0, 1}, {2, 0}},
      {{0, 2}, {1, 0}}, {{0, 2}, {2, 0}}, {{1, 0}, {0, 1}},

$$\{\{1, 0\}, \{0, 2\}\}, \{\{1, 1\}, \{1, 2\}\}, \{\{1, 1\}, \{2, 1\}\},$$
$$\{\{1, 2\}, \{1, 1\}\}, \{\{1, 2\}, \{2, 2\}\}, \{\{2, 0\}, \{0, 1\}\},$$
$$\{\{2, 0\}, \{0, 2\}\}, \{\{2, 1\}, \{1, 1\}\}, \{\{2, 1\}, \{2, 2\}\},$$
$$\{\{2, 2\}, \{1, 2\}\}, \{\{2, 2\}, \{2, 1\}\}\}, \text{-Operation-]}$$

■ 3.9.6 Miscellaneous functions

`IdentityMatrix[`M`]`	return the identity matrix, if it exists, for the matrix extension M
`IdentityMatrix[`R, n`]`	return the n–by–n identity matrix, if it exists, over the ringoid R
`MatrixOverQ[`R, A`]`	give `True` if the matrix A has its entries in the ringoid R, and `False` otherwise
`MatrixOperation[`M`]`	return the operation (s) inherent in the matrix extension M; values are `Addition`, `Multiplication`, or `Both`
`SizeOfMatrices[`M`]`	return the dimensions of the matrices in M

Miscellaneous functions.

■ The `IdentityMatrix` function has its natural generalization for any ringoid.

```
In[221]:= IdentityMatrix[MatricesOver[
              BooleanRing[{"cat", "dog"}], 3]] // MatrixForm
```

Out[221]//MatrixForm=

$$\begin{pmatrix} \{\text{cat, dog}\} & \{\} & \{\} \\ \{\} & \{\text{cat, dog}\} & \{\} \\ \{\} & \{\} & \{\text{cat, dog}\} \end{pmatrix}$$

■ Given a matrix, we can check to see what ring it may be based on by using the `MatrixOverQ` function.

```
In[222]:= Map[MatrixOverQ[#, {{2, 3}, {4, 5}}] &, {Z[5], Z[6], Z[7]}]
```

Out[222]= {False, True, True}

■ Here are three different matrix extensions, followed by the dimensions of each matrix in the extensions, as well as the operation(s).

```
In[223]:= {extensions =
              {Mat[Z[5], 2], MatA[Z[4], {3, 4}], MatM[Z[7], 3]},
           Map[SizeOfMatrices, extensions],
           Map[MatrixOperation, extensions]} // ColumnForm
```

Out[223]= {−Mat$_2$ (Z[5])−, −Mat$_{(3,4)}$ (Z[4])−, −Mat$_3$ (Z[7])−}

{{2, 2}, {3, 4}, {3, 3}}

{Both, Addition, Multiplication}

3.10 Functions on a ringoid

3.10.1 Function extensions and their elements

- The extension ring of functions over a ringoid is created with `Functions-Over`.

In[224]:= **T = FunctionsOver[ZR[5]]**

Out[224]= −Ring of Functions over Z[5]−

FunctionsOver [*R*]	return the extension of functions over *R*
Func [*images*]	represent a function where *images* represents the images of elements in a ringoid
ElementQ [Func [*images*], FunctionsOver [*R*]]	give True if Func [*images*] is a function in FunctionsOver [*R*], and False otherwise

Function rings and their elements.

The form of functions on a ringoid is a sequence of functional values consisting of the images of elements from the domain. The head of this sequence is Func.

- Here is one of the functions on $\mathbb{Z}_5$. We can verify that it is in the extension *T*.

In[225]:= **{f = Func[1, 2, 2, 3, 3], ElementQ[f, T]}**

Out[225]= {Func[1, 2, 2, 3, 3], True}

RandomElement [*F*]	return a random function in the function extension *F*
RandomElements [*F, n*]	return a list of *n* random functions in *F*
FuncToRules [*F, f*]	convert a function in the function extension *F* to a list of rules

Working with individual functions over ringoids.

- Here is a random function on $\mathbb{Z}_5$. Occasionally it is useful to convert a function to a list of rules.

In[226]:= **{g = RandomElement[T], FuncToRules[T, g]}**

Out[226]= {Func[4, 2, 0, 0, 3], {0 → 4, 1 → 2, 2 → 0, 3 → 0, 4 → 3}}

■ 3.10.2 Function arithmetic

Addition[F][f, g]	return the sum of functions f and g in the extension F
Addition[F, f, g]	identical to Addition[F][f, g]
Multiplication[F][f, g]	return the product of functions f and g
Multiplication[F, f, g]	identical to Multiplication[F][f, g]

Basic arithmetic functions.

HasZeroQ[F]	give True if F has a zero, and False otherwise
Zero[F]	return the zero of F if it exists, and \$Failed otherwise
NegationOf[F, f]	return the negation of f in F, if it exists, and \$Failed otherwise
WithUnityQ[F]	give True if F has a unity, and False otherwise
Unity[F]	return the unity of F, if it exists, and \$Failed otherwise
UnitQ[F, f]	give True if f is a unit of F, and False otherwise
ZeroDivisorQ[F, f]	give True if f is a zero divisor of F, and False otherwise

Other basic functions in function extensions.

■ Here is the ring of functions over $\mathbb{Z}_{11}$.

In[227]:= **V = FunctionsOver[Z[11]]**

Out[227]= -Ring of Functions over Z[11]-

■ Functions over larger rings can be generated in a variety of ways; here is one method.

In[228]:= **f = Apply[Func, Mod[Range[0, 10]^2, 11]]**

Out[228]= Func[0, 1, 4, 9, 5, 3, 3, 5, 9, 4, 1]

■ Every function on a field, such as $\mathbb{Z}_{11}$, is a polynomial function. Furthermore, a polynomial over any ringoid can be converted to a function with PolyToFunction.

In[229]:= **g = PolyToFunction[Z[11], Poly[Z[11], 1 + x + x^2]]**

Out[229]= Func[1, 3, 7, 2, 10, 9, 10, 2, 7, 3, 1]

Both addition and multiplication in function extensions are coordinate-wise operations. Over a ringoid with *n* elements, the operations each require *n* corresponding ringoid operations.

- Here is the sum of *f* and *g* .

In[230]:= `{f, g, Addition[V][f, g]} // ColumnForm`

Out[230]= Func[0, 1, 4, 9, 5, 3, 3, 5, 9, 4, 1]
Func[1, 3, 7, 2, 10, 9, 10, 2, 7, 3, 1]
Func[1, 4, 0, 0, 4, 1, 2, 7, 5, 7, 2]

- And here is their product.

In[231]:= `{f, g, Multiplication[V][f, g]} // ColumnForm`

Out[231]= Func[0, 1, 4, 9, 5, 3, 3, 5, 9, 4, 1]
Func[1, 3, 7, 2, 10, 9, 10, 2, 7, 3, 1]
Func[0, 3, 6, 7, 6, 5, 8, 10, 8, 1, 1]

- *V* has a zero function, as well as a unity, both being constant functions.

In[232]:= `{HasZeroQ[V], Zero[V], WithUnityQ[V], Unity[V]}`

Out[232]= {True, Func[0, 0, 0, 0, 0, 0, 0, 0, 0, 0, 0],
True, Func[1, 1, 1, 1, 1, 1, 1, 1, 1, 1, 1]}

- Adding *f* and the negation of *f* produces the zero function.

In[233]:= `h = NegationOf[V, f];`
`Addition[V, f, h]`

Out[234]= Func[0, 0, 0, 0, 0, 0, 0, 0, 0, 0, 0]

- A function is invertible if its images are invertible in the base ring.

In[235]:= `Map[UnitQ[V, #] &, {f, g}]`

Out[235]= {False, True}

- There is an inverse for *g*.

In[236]:= `MultiplicativeInverse[V, g]`

Out[236]= **Func[1, 4, 8, 6, 10, 5, 10, 6, 8, 4, 1]**

- But *f* does not have an inverse.

In[237]:= **MultiplicativeInverse[V, f]**

RingExtension::NoInverse :
 No mult. inverse in extension ring due
  to a lack of mult inverse in the base ring.

Out[237]= $Failed

- Units and zero divisors are disjoint.

In[238]:= **Map[ZeroDivisorQ[V, #] &, {f, g}]**

Out[238]= {True, False}

■ 3.10.3 Polynomial conversion and interpolation

PolyToFunction[R, p]	convert a polynomial p over R to a function over R
InterpolatingPolynomial[R, $\{\{x_1, y_1\}, \dots\}$]	return the polynomial of least degree passing through the list of points in $R \times R$ with distinct first coordinates, assuming that R is a field

Polynomial conversion and interpolation.

PolyToFunction can be used to check whether a polynomial has a linear factor. This is equivalent to checking whether it is irreducible for polynomials of degree two or three.

- Consider a typical polynomial p over $\mathbb{Z}_{11}$. Since the polynomial function corresponding to p has no zeros, p is irreducible over $\mathbb{Z}_{11}$.

In[239]:= **{p = Poly[Z[11], x³ + x + 6], f = PolyToFunction[Z[11], p]}**

Out[239]= {$6 + x + x^3$, Func[6, 8, 5, 3, 8, 4, 8, 4, 9, 7, 4]}

PolyToFunction and InterpolatingPolynomial are inverses of one another for a certain set of polynomials over a field.

- We start with a polynomial over $\mathbb{Z}_5$ and convert it to a function.

In[240]:= **h = PolyToFunction[Z[5], Poly[Z[5], x² + 1]]**

Out[240]= Func[1, 2, 0, 0, 2]

- Now generate a list of ordered pairs consisting of domain and range elements.

In[241]:= **hpairs = Transpose[{Elements[Z[5]], List@@h}]**

Out[241]= {{0, 1}, {1, 2}, {2, 0}, {3, 0}, {4, 2}}

■ A polynomial that "passes through" these points is the original one.

In[242]:= **InterpolatingPolynomial[Z[5], hpairs]**

Out[242]= $1 + x^2$

This use of InterpolatingPolynomial is an extension of the built-in function that works in the real and complex domain and is still available.

■ Using the foregoing points, the real interpolating polynomial is quite different.

In[243]:= **pr = InterpolatingPolynomial[hpairs, x] // Expand**

Out[243]= $1 + \dfrac{65\,x}{12} - \dfrac{151\,x^2}{24} + \dfrac{25\,x^3}{12} - \dfrac{5\,x^4}{24}$

■ But if the coefficients are reduced mod 5, the relationship is apparent.

In[244]:= **Map[Mod[#, 5] &, CoefficientList[pr, x] /.**
           {Rational[a_, b_] :> a PowerMod[b, -1, 5]}]

Out[244]= {1, 0, 1, 0, 0}

▦ 3.11 Finite fields

The finite fields, or Galois fields, are among the most important in abstract algebra. They can be most conveniently generated with the specialized functions in the Abstract-Algebra`FiniteFields package. A portion of the code in this package comes from the standard package Algebra`FiniteFields (mainly the code to implement IrreduciblePolynomial), but the output is fundamentally different. Our package returns a Ringoid, and the polynomials used for inputs can be standard polynomials or polynomials created with the Poly function.

GF $[p^d]$	return the Galois field of order p^d for prime p
GF $[p, d]$	identical to GF $[p^d]$
GF $[p^d, poly]$	return the finite field using the specified irreducible polynomial of degree d
GF $[p, d, poly]$	identical to GF $[p^d, poly]$

Creation of Galois fields using GF.

■ All fields of order 25 are isomorphic, so F is unique.

In[245]:= **F = GF[25]**

Out[245]= Ringoid[{0, x, 2 x, 3 x, 4 x, 1, 1 + x, 1 + 2 x, 1 + 3 x,
        1 + 4 x, 2, 2 + x, 2 + 2 x, 2 + 3 x, 2 + 4 x, 3, 3 + x,
        3 + 2 x, 3 + 3 x, 3 + 4 x, 4, 4 + x, 4 + 2 x, 4 + 3 x, 4 + 4 x},
        -Addition-, -Multiplication-]

- The `Indeterminate` option can specify the symbol to be used for the indeterminate.

In[246]:= **F2 = GF[3, 2, Indeterminate → y]**

Out[246]= Ringoid[{0, y, 2 y, 1, 1 + y, 1 + 2 y, 2, 2 + y, 2 + 2 y},
        -Addition-, -Multiplication-]

- The index provided to GF must be a power of a prime.

In[247]:= **GF[12]**

    GF::badindex : The index for GF needs
      to be a power of a prime, which 12 is not.

Out[247]= $Failed

Here are some functions for working with Galois fields.

`IrreduciblePolyOverZpQ[` *poly, p*]	give `True` if the polynomial *poly* (formed as an ordinary polynomial or with `Poly`) is irreducible over the ring $\mathbb{Z}_p$, and `False` otherwise
`IrreduciblePolynomial[` *ind, p, d*]	return an irreducible polynomial in the indeterminate *ind* of degree *d* over $\mathbb{Z}_p$
`FieldIrreducible[GF[`*n*`]]`	return the irreducible polynomial used in establishing the Galois field GF[*n*]
`ExtensionDegree[GF[`*n*`]]`	give the degree of the extension of GF[*n*]
`GaloisFieldQ[`*R*`]`	give `True` if the ring *R* is a Galois field and was created using the GF function, and `False` otherwise

Functions used in the creation and identification of Galois fields.

- If a polynomial is provided, it must be irreducible over the integers mod *p*.

In[248]:= **GF[5, Poly[Z[5], x^2 + 4]]**

    GF::irr :
      The polynomial $4 + x^2$ needs to be irreducible over Z[5].

Out[248]= $Failed

- The error in the preceding example can be anticipated.

In[249]:= **IrreduciblePolyOverZpQ[x^2 + 4, 5]**

Out[249]= False

- Here is a quadratic we could use for GF[25].

In[250]:= **IrreduciblePolynomial[x, 5, 2]**

Out[250]= $3 + 2 x + x^2$

- IrreduciblePolynomial returns the default polynomial for GF[25]. Note that this is the same as the earlier result. Also observe that F is a quadratic extension.

In[251]:= **{FieldIrreducible[F = GF[25]], ExtensionDegree[F]}**

Out[251]= $\{3 + 2 x + x^2, 2\}$

- Galois fields created outside the package are not recognized.

In[252]:= **{GaloisFieldQ[Z[7]], GaloisFieldQ[GF[7]]}**

Out[252]= {False, True}

PrimitivePolynomials [GF[n]]	return a list of primitive polynomials in GF[n] (generators of the multiplicative group of GF[n])
TableOfPowers[GF[n]]	return a table, starting first with {0, 0}, followed by pairs of the form $\{q^j, r\}$ where r is an element in GF[n], q is the "simplest" primitive polynomial for this ring, and j is the power to which q needs to be raised to be equal to r
PowerList[GF[n]]	equivalent to TableOfPowers, added for compatibility with the Algebra`FiniteFields` package

Powers in a Galois field.

- The primitive polynomial chosen for GF[16] was x, but there are seven other choices for a primitive polynomial. (Note, however, that x is indeed the simplest in this case.) In general, there are $\phi(p^n - 1)$ primitive polynomials in GF[p^n], where ϕ is Euler's totient function.

In[253]:= **{EulerPhi[15], pp = PrimitivePolynomials[GF[16]]}**

Out[253]= $\{8,$
$\{x, x^2, x + x^2, 1 + x + x^2, 1 + x^3, x^2 + x^3, 1 + x^2 + x^3, x + x^2 + x^3\}\}$

- Here are the powers of x in GF[16].

In[254]:= **TableOfPowers[GF[16]] // MatrixForm**

Out[254]//MatrixForm=

$$\begin{pmatrix} 0 & 0 \\ x & x \\ x^2 & x^2 \\ x^3 & x^3 \\ x^4 & 1 + x^3 \\ x^5 & 1 + x + x^3 \\ x^6 & 1 + x + x^2 + x^3 \\ x^7 & 1 + x + x^2 \\ x^8 & x + x^2 + x^3 \\ x^9 & 1 + x^2 \\ x^{10} & x + x^3 \\ x^{11} & 1 + x^2 + x^3 \\ x^{12} & 1 + x \\ x^{13} & x + x^2 \\ x^{14} & x^2 + x^3 \\ 1 & 1 \end{pmatrix}$$

- We can anticipate that the order of any of the primitive polynomials in GF[16] is 15.

In[255]:= **OrderOfElement[**
   MultiplicativeGroupoid[GF[16]], RandomElement[pp]]

Out[255]= 15

- This function can also be used to find the highest-order elements in the multiplicative groupoid of an ordinary ringoid.

In[256]:= **{Orders[NonZeroMGroupoid[Z[5]]],**
   PrimitivePolynomials[Z[5]]} // ColumnForm

Out[256]= {{1, 1}, {2, 4}, {3, 4}, {4, 2}}
   {2, 3}

AdditiveToMultiplicative[R, *add*]	convert the element *add* (in additive form) in the field *R* to the multiplicative form
MultiplicativeToAdditive[R, *mult*]	convert the element *mult* (in multiplicative form) in the field *R* to the additive form

Changing representations of a field element.

- In the field, GF[16], the element expressed additively as $1 + x^2 + x^3$ can be expressed multiplicatively as x^{11}.

In[257]:= **AdditiveToMultiplicative[GF[16], 1 + x^2 + x^3]**

Out[257]= x^{11}

- Similarly, we can convert the multiplicative form x^{11} to the additive form $1 + x^2 + x^3$.

In[258]:= **MultiplicativeToAdditive[GF[16], x^{11}]**

Out[258]= $1 + x^2 + x^3$

Chapter 4

Morphoids

Given two groups (or rings), we may want to establish some function between them. Usually, the interest is not just how the function interacts with the elements of the two structures, but also how it relates to the operations. Given groups $(G_1, *)$ and $(G_2, \#)$, we are often interested in a function $f : G_1 \to G_2$ where $f(x * y) = f(x) \# f(y)$ for all x and y in G_1. In this case, we say f is a *homomorphism* between the two groups. Clearly not all functions between groups are homomorphisms. In `AbstractAlgebra`, we define a data structure called a `Morphoid` that reflects a mathematical function between two groupoids or ringoids. In this chapter, we consider how to form and explore `Morphoids`.

▣ 4.1 Forming Morphoids

`FormMorphoid[f, S₁, S₂]`	give the `Morphoid` determined by the function f between structures S_1 and S_2
`FormMorphoid[rules, S₁, S₂]`	give the `Morphoid` determined by the list of rules given in *rules* that gives a mapping between structures S_1 and S_2
`FormMorphoid[positionList, S₁, S₂]`	give the `Morphoid` determined by the indices found in *positionList* that gives a mapping between structures S_1 and S_2
`FormMorphoid[f, S₁, S₂,` `FormatFunction → val]`	after forming the `Morphoid`, format it according to *val*, with `False` being the default value
`FormMorphoidSetup[S₁, S₂]`	provide a graphic to assist in using the position-list approach to forming a `Morphoid`

Methods of forming a `Morphoid`.

- The function between two groupoids (or ringoids) used to determine a `Morphoid` can be a built-in function or a pure function.

```
In[1]:= {f1 = FormMorphoid[Identity, Z[5], Z[5]],
        f2 = FormMorphoid[Mod[# + 2, 5] &, Z[5], Z[5]]}

Out[1]= {Morphoid[Identity[#1] &, -Z[5]-, -Z[5]-],
         Morphoid[Mod[#1 + 2, 5] &, -Z[5]-, -Z[5]-]}
```

- One can also define a function on each element of the domain and then construct a Morphoid. Here we also illustrate the Visual mode of FormMorphoid.

```
In[2]:= Clear[g];
        Do[g[i] = Mod[i + 2, 5], {i, 0, 4}]
        f3 = FormMorphoid[g, Z[5], Z[5], Mode → Visual]
```

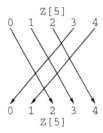

```
Out[3]= Morphoid[g[#1] &, -Z[5]-, -Z[5]-]
```

- Although f2 and f3 were produced by different means, they represent the same mathematical function and are equal as Morphoids.

```
In[4]:= EqualMorphoidQ[f2, f3]

Out[4]= True
```

- Using a list of rules (showing how domain elements are to be mapped to codomain elements) is another approach to forming a Morphoid.

```
In[5]:= f4 = FormMorphoid[
          {0 → 2, 1 → 3, 2 → 4, 3 → 0, 4 → 1}, Z[5], Z[5]]

Out[5]= Morphoid[{0 → 2, 1 → 3, 2 → 4, 3 → 0, 4 → 1}, -Z[5]-, -Z[5]-]
```

- The listing of the rules is automatically suppressed if the list is too long; we can also manually suppress the list from being displayed.

```
In[6]:= FormMorphoid[{0 → 2, 1 → 3, 2 → 4, 3 → 0, 4 → 1},
          Z[5], Z[5], FormatFunction → True]

Out[6]= Morphoid[-Rules-, -Z[5]-, -Z[5]-]
```

- Applying EqualMorphoidQ over the Cartesian product of two lists both containing f2, f3, and f4 shows that these three Morphoids are identical.

```
In[7]:= CloseSets[{f2, f3, f4}, {f2, f3, f4}, EqualMorphoidQ]
```

```
Out[7]= {True}
```

- Since S_3 and D_3 are isomorphic, we should be able to find an isomorphism between them. Since there is no natural formula to define this function, and setting up a list of rules is a bit tedious for these two groups, it is useful to formulate the `Morphoid` by using a list of positions in the codomain to give the pairing. The `FormMorphoidSetup` function is used to begin this process.

```
In[8]:= FormMorphoidSetup[Symmetric[3], Dihedral[3]];
```

Domain			Codomain
{1, 2, 3}	1	1	1
{1, 3, 2}	2	2	Rot
{2, 1, 3}	3	3	Rot^2
{2, 3, 1}	4	4	Ref
{3, 1, 2}	5	5	Rot**Ref
{3, 2, 1}	6	6	Rot^2**Ref

- Using the preceding graphic, suppose that we want to map $\{1, 2, 3\}$ to 1, $\{1, 3, 2\}$ to `Ref`, $\{2, 1, 3\}$ to `Rot ** Ref`, $\{2, 3, 1\}$ to `Rot`, $\{3, 1, 2\}$ to Rot^2, and finally $\{3, 2, 1\}$ to Rot^2 `** Ref`. Note that the (ordered) images of this map occur in positions $\{1, 4, 6, 2, 3, 5\}$ in the list of elements in D_3.

```
In[9]:= f5 = FormMorphoid[
            {1, 4, 5, 2, 3, 6}, Symmetric[3], Dihedral[3]]
```

```
Out[9]= Morphoid[{{1, 2, 3} → 1, {1, 3, 2} → Ref,
            {2, 1, 3} → Rot ** Ref, {2, 3, 1} → Rot, {3, 1, 2} → Rot²,
            {3, 2, 1} → Rot² ** Ref}, -S[3]-, -D[3]-]
```

- The following shows that this is an isomorphism.

```
In[10]:= IsomorphismQ[f5, Cautious → True]
```

```
Out[10]= True
```

- Recall how the `Parity` function works.

```
In[11]:= parityList = Map[{#, Parity[#]} &, Elements[Symmetric[3]]]
```

```
Out[11]= {{{1, 2, 3}, 1}, {{1, 3, 2}, -1}, {{2, 1, 3}, -1},
            {{2, 3, 1}, 1}, {{3, 1, 2}, 1}, {{3, 2, 1}, -1}}
```

- Now consider a "natural" morphism with S_3 as its domain and $\{\pm 1\}$ as the codomain.

```
In[12]:= FormMorphoidSetup[Symmetric[3], IntegerUnits];
```

```
Domain
{1, 2, 3}    1
{1, 3, 2}    2              Codomain
{2, 1, 3}    3  | 1           1
{2, 3, 1}    4  | 2          -1
{3, 1, 2}    5
{3, 2, 1}    6
```

- We want the elements with parity -1 to go to position 2 in the list of the codomain elements.

In[13]:= **positions = Last[Transpose[parityList]] /. -1 → 2**

Out[13]= {1, 2, 2, 1, 1, 2}

- Now we can easily form the Morphoid using this list of positions.

In[14]:= **f6 = FormMorphoid[positions, Symmetric[3], IntegerUnits]**

Out[14]= Morphoid[
         {{1, 2, 3} → 1, {1, 3, 2} → -1, {2, 1, 3} → -1, {2, 3, 1} → 1,
          {3, 1, 2} → 1, {3, 2, 1} → -1}, -S[3]-, -IntegerUnits-]

- This is indeed a homomorphism.

In[15]:= **MorphismQ[f6]**

Out[15]= True

- In fact, the kernel of this homomorphism is the alternating group on three letters.

In[16]:= **Elements[Kernel[f6]] === Elements[Alternating[3]]**

Out[16]= True

- If the domain is a cyclic groupoid, we can get by with just specifying a single defining rule. (In this case, this will define the other rules since "g" is a generator of CyclicGroup[5]. Note also that ZG[5] is the group $\mathbb{Z}_5$.)

In[17]:= **f7 = FormMorphoid["g" → 3, CyclicGroup[5], ZG[5]]**

Out[17]= Morphoid[g → 3, -Cyclic[5]-, -Z[5]-]

- As expected, these two groups are isomorphic.

In[18]:= **IsomorphismQ[f7]**

Out[18]= True

▓ 4.2 Structure of Morphoids

There are three fundamental parts of a Morphoid: the function, domain, and codomain.

MorphoidFunction[*f*]	give the function (or list of rules) of the Morphoid *f*
MorphoidRules[*f*]	give the list of rules (or function) of the Morphoid *f*
Domain[*f*]	give the domain of the Morphoid *f*
Codomain[*f*]	give the codomain (target space) of the Morphoid *f*

Extracting parts of a Morphoid.

- Define the Morphoid that takes any element x in $\mathbb{Z}_{10}$ and maps it to $3\,x$ in $\mathbb{Z}_{30}$.

In[19]:= **g1 = FormMorphoid[Mod[3 #, 30] &, Z[10], Z[30]]**

Out[19]= Morphoid[Mod[3 #1, 30] &, -Z[10]-, -Z[30]-]

- Here is how we extract the function, domain, and codomain of g1.

In[20]:= **{MorphoidFunction[g1], Domain[g1], Codomain[g1]}**

Out[20]= {Mod[3 #1, 30] &,
    Groupoid[{0, 1, 2, 3, 4, 5, 6, 7, 8, 9}, Mod[#1 + #2, 10] &],
    Groupoid[{0, 1, 2, 3, 4, 5, 6, 7, 8, 9, 10,
      11, 12, 13, 14, 15, 16, 17, 18, 19, 20, 21, 22,
      23, 24, 25, 26, 27, 28, 29}, Mod[#1 + #2, 30] &]}

Sometimes we may wish to change the way we view a function or Morphoid; here are some methods of doing so.

- This is the result of converting the Morphoid we previously defined into a rules-based Morphoid.

In[21]:= **g2 = ToRules[g1]**

Out[21]= Morphoid[{0 → 0, 1 → 3, 2 → 6, 3 → 9, 4 → 12, 5 → 15,
    6 → 18, 7 → 21, 8 → 24, 9 → 27}, -Z[10]-, -Z[30]-]

- MorphoidRules extracts just the list of rules.

In[22]:= **someRules = MorphoidRules[g2]**

Out[22]= {0 → 0, 1 → 3, 2 → 6, 3 → 9, 4 → 12,
    5 → 15, 6 → 18, 7 → 21, 8 → 24, 9 → 27}

ToRules[*f*]	convert the function-based Morphoid *f* to a rules-based Morphoid
ToRules[*f, D*]	convert the function *f* with domain *D* (where *D* can be a set, groupoid, or ringoid) to a list of rules of the form x → f[x]
ToRules[*f, D, C*]	convert the function *f* with domain *D* to a list of rules of the form x → f[x], guaranteeing that the images fall in the codomain *C* (returning $Failed if the codomain *C* does not contain all of the images)
ToFunction[*f*]	convert the rules-based Morphoid *f* to a function-based Morphoid
ToFunction[*f, g*]	convert the rules-based Morphoid *f* to a function-based Morphoid named *g*
ToFunction[*rules*]	convert a list of rules into a function named ffx (where x is an integer)
ToFunction[*rules, g*]	convert a list of rules into a function named *g*

Converting between rules and functions.

■ ToFunction works in different ways with different types of input.

In[23]:= **{g5 = ToFunction[g2], ToFunction[g2, g3],**
         ToFunction[someRules], ToFunction[someRules, g4]}

Out[23]= {Morphoid[ff3[#1] &, -Z[10]-, -Z[30]-],
           Morphoid[g3[#1] &, -Z[10]-, -Z[30]-], ff4, g4}

Note that the function listed in the Morphoid g5 (starting with ff) can be inspected with Information or ?.

EqualMorphoidQ[*f, g*]	give True if Morphoids *f* and *g* are equal as mathematical functions, and False otherwise
MorphoidComposition[*g, f*]	give the Morphoid resulting from composing *f* followed by *g*

Functions on pairs of Morphoids.

■ Given two Morphoids, we may want to compose these two functions.

In[24]:= **f1 = FormMorphoid[Mod[# + 3, 8] &, Z[4], Z[7]];**
         f2 = FormMorphoid[Mod[#^2 + 1, 12] &, Z[7], Z[12]];
         MorphoidComposition[f2, f1]

Out[24]= Morphoid[{0 → 10, 1 → 5, 2 → 2, 3 → 1}, -Z[4]-, -Z[12]-]

■ Note that the order of the operands is important; the composition does not make sense in the other direction. `MorphoidComposition[g, f]` returns a function *h* where we have $h(x) = g(f(x))$.

In[25]:= **MorphoidComposition[f1, f2]**

> *MorphoidComposition:notdef :*
> *Composition is not defined because the image*
> *of the first map is Z[12] while the domain of*
> *the second map is Z[4], which are not the same.*

Out[25]= $Failed

■ 4.3 Built-in Morphoids

There are several built-in `Morphoids` that reflect commonly used functions.

`ZMap[m, n]`	form the `Morphoid` from $\mathbb{Z}_m$ to $\mathbb{Z}_n$ with the function `Mod[#, n]&` (i.e., reduce mod *n*), where the current value of `DefaultStructure` specifies whether these structures are to be considered groups or rings
`ZMap[m, n, g → h]`	for *g* a generator of $\mathbb{Z}_m$, form the `Morphoid` from $\mathbb{Z}_m$ to $\mathbb{Z}_n$ defined by the rule $g \to h$
`ZMap[m, n,` `Structure → StType]`	form the `Morphoid` `ZMap[m, n]` using the structure *StType* (with value `Group` or `Ring`)

Variations of the `ZMap` function.

■ There is a natural morphism from $\mathbb{Z}_{18}$ into $\mathbb{Z}_3$.

In[26]:= **f = ZMap[18, 3]**

Out[26]= Morphoid[1 → 1, -Z[18]-, -Z[3]-]

■ Here is how the function actually works on each domain element.

In[27]:= **ToRules[f]**

Out[27]= Morphoid[{0 → 0, 1 → 1, 2 → 2, 3 → 0, 4 → 1, 5 → 2,
    6 → 0, 7 → 1, 8 → 2, 9 → 0, 10 → 1, 11 → 2, 12 → 0, 13 → 1,
    14 → 2, 15 → 0, 16 → 1, 17 → 2}, -Z[18]-, -Z[3]-]

■ The following visualization also makes it clear what is happening with the function *f*.

In[28]:= **VisualizeMorphoid[f, ColorCodomain → Automatic];**

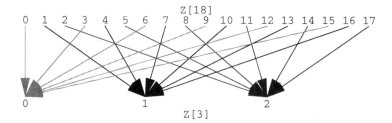

- With the same domain and codomain, we can also set up a map defined by
  sending the generator 5 in $\mathbb{Z}_{18}$ to the generator 1 in $\mathbb{Z}_3$.

In[29]:= **VisualizeMorphoid[**
       ZMap[18, 3, 5 → 1], ColorCodomain → Automatic];

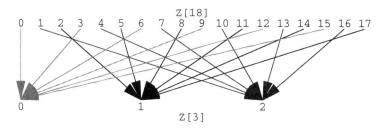

Given any morphism $f : G \to H$ between groups (or rings), there are two natural mor-
phisms induced from f. The first is $g : G \to G/\mathrm{Ker}\,(f)$ defined by $x \mapsto x\,\mathrm{Ker}(f)$ and the
second is $h : G/\mathrm{Ker}\,(f) \to \mathrm{Im}\,(f)$ defined by $x\,\mathrm{Ker}(f) \mapsto f(x)$. What follows is a typical
diagram to illustrate the relations between these morphisms.

In[30]:= **Show[Graphics[{Text["G", {0, 0}], Text["H", {5, 0}],**
       Text["G/Ker(f)", {2.5, -3.5}], Text["f", {2.5, .15}],
       Text["g", {1.0, -1.75}], Text["h", {4, -1.75}],
       RGBColor[0, 0, 1], Arrow[{.25, 0}, {4.75, 0}],
       Arrow[{.15, -.15}, {2.35, -3.35}],
       Arrow[{2.65, -3.35}, {4.85, -.15}]}]];

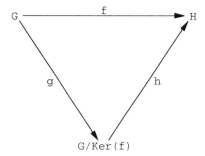

- Here is the induced canonical homomorphism based on `ZMap[18, 6]`.

In[31]:= **f3 = InducedCanonical[ZMap[18, 6]]**

QuotientGroup:NS :
This quotient group uses NS to represent the normal subgroup
{0, 6, 12} that you specified Use CosetToListto
convert this coset representationto a list of elements.

Out[31]= Morphoid[{0 → NS, 1 → 1 + NS, 2 → 2 + NS, 3 → 3 + NS, 4 → 4 + NS,
5 → 5 + NS, 6 → NS, 7 → 1 + NS, 8 → 2 + NS, 9 → 3 + NS,
10 → 4 + NS, 11 → 5 + NS, 12 → NS, 13 → 1 + NS, 14 → 2 + NS,
15 → 3 + NS, 16 → 4 + NS, 17 → 5 + NS}, -Z[18]-, -Z[18]/NS-]

InducedCanonical[*f*]	given a morphism $f : G \to H$, return the induced Morphoid $g : G \to G/\text{Kernel}[f]$
InducedIsomorphism[*f*]	given a morphism $f : G \to H$, return the induced Morphoid $h : G/\text{Kernel}[f] \to \text{Image}[f]$

Induced Morphoids.

- This morphism sends an element x in $\mathbb{Z}_{18}$ to the coset $x + \{0, 6, 12\}$.

In[32]:= **VisualizeMorphoid[f3, ColorCodomain → Automatic];**

KEY for Z[18]/NS: label used →
  element: {h1 → NS, h2 → 1 + NS, h3 → 2 +
  NS, h4 → 3 + NS, h5 → 4 + NS, h6 → 5 + NS}

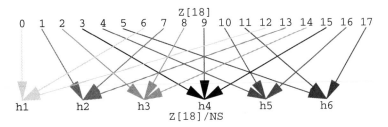

- In contrast, the induced isomorphism maps from $\mathbb{Z}_{18} / \{0, 6, 12\}$ to $\mathbb{Z}_6$ (the image, in this case).

In[33]:= **f4 = InducedIsomorphism[ZMap[18, 6]]**

Out[33]= Morphoid[{NS → 0, 1 + NS → 1, 2 + NS → 2,
3 + NS → 3, 4 + NS → 4, 5 + NS → 5}, -Z[18]/NS-, -Z[6]-]

- Now let's compose f3 and f4.

In[34]:= **f5 = MorphoidComposition[f4, f3]**

Out[34]= Morphoid[{0 → 0, 1 → 1, 2 → 2, 3 → 3, 4 → 4, 5 → 5,
        6 → 0, 7 → 1, 8 → 2, 9 → 3, 10 → 4, 11 → 5, 12 → 0, 13 → 1,
        14 → 2, 15 → 3, 16 → 4, 17 → 5}, -Z[18]-, -Z[6]-]

- Note that this composition is the same as ZMap[18, 6].

In[35]:= **EqualMorphoidQ[f5, ZMap[18, 6]]**

Out[35]= True

Sgn[*G*]	given a permutation group *G*, return the Morphoid determined by the Parity function

Other built-in Morphoids.

- Under the Sgn function, odd permutations go to −1, while even permutations go to 1.

In[36]:= **VisualizeMorphoid[Sgn[SymmetricGroup[3]]];**

        KEY for S[3]: label used → element: {g1 →
        {1, 2, 3}, g2 → {1, 3, 2}, g3 → {2, 1, 3},
        g4 → {2, 3, 1}, g5 → {3, 1, 2}, g6 → {3, 2, 1}}

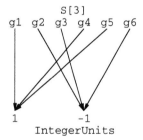

4.4 Properties

4.4.1 Surjectivity and injectivity

There are two standard properties of functions that are often of interest. A function $f : A \to B$ is said to be *injective* if $f(x) = f(y)$ implies $x = y$ and *surjective* if for all $b \in B$ there exists an $a \in A$ such that $f(a) = b$.

- As is the case over the integers, the squaring function is also not injective over $\mathbb{Z}_{10}$.

In[37]:= **InjectiveQ[f = FormMorphoid[Mod[#^2, 10] &, Z[10], Z[10]]]**

Out[37]= False

- ZMap[*m*, *n*] is always surjective as long as $m \geq n$.

In[38]:= **SurjectiveQ[g = ZMap[12, 5]]**

Out[38]= True

InjectiveQ[*f*]	give True if the Morphoid *f* is injective (one-to-one), and False otherwise
SurjectiveQ[*f*]	give True if the Morphoid *f* is surjective (onto), and False otherwise
OneToOneQ[*f*]	identical to InjectiveQ[*f*]
OntoQ[*f*]	identical to SurjectiveQ[*f*]

Basic properties of functions.

■ 4.4.2 Preserving operations

While injectivity and surjectivity are general properties for any functions, the possibility of operation preserving occurs when the function is between two structured sets, such as groups or rings. Given a function $f : (G_1, *) \to (G_2, \#)$ between two groups, we say f preserves the operation of group G_1 for the pair (x, y) (for x and y in G_1) if $f(x * y) = f(x) \# f(y)$. If the operation is preserved for all pairs coming from the domain, then the function is a homomorphism (or morphism). For rings, we require both the addition and the multiplication to be preserved.

PreservesQ[*f*, {*a*, *b*}]	give True if the Morphoid *f* preserves the binary operation (s) for the pair (a, b) from the domain of *f*, and False otherwise
PreservesQ[*f*, {*a*, *b*}, Mode → Visual]	in addition to giving True or False, give a visualization illustrating the process
MorphismQ[*f*]	give True if the Morphoid *f* preserves the binary operation (s) for all pairs (a, b) from the domain of *f*, and False otherwise
MorphismQ[*f*, Mode → Visual]	in addition to giving True or False, give a visualization illustrating which pairs preserve the operations and which do not
HomomorphismQ[*f*]	identical to MorphismQ[*f*]

Functions to test for operation preservation.

- Consider the following two Morphoids, whose only difference is the structure used (group versus ring).

In[39]:= **{f$_g$ = ZMap[10, 10, 1 → 3, Structure → Group],
         f$_r$ = ZMap[10, 10, 1 → 3, Structure → Ring]}**

Out[39]= {Morphoid[1 → 3, -Z[10]-, -Z[10]-],
         Morphoid[1 → 3, -Z[10]-, -Z[10]-]}

- As we track what happens to the elements 2 and 5 in the following diagram, we see that with the group Morphoid f_g indeed we have $f_g(2 + 5) = f_g(2) + f_g(5)$. Therefore, the operation is preserved for this pair.

In[40]:= **PreservesQ[f$_g$, {2, 5}, Mode → Visual]**

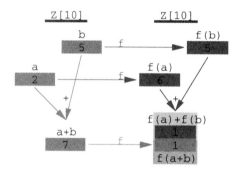

Out[40]= True

- We can test every pair of elements in the domain by using the MorphismQ function.

In[41]:= **MorphismQ[f$_g$]**

Out[41]= True

- When a Morphoid is based on a ring, one needs to check both the addition and the multiplication. Here is the Visual mode of MorphismQ for the ring Morphoid f_r.

In[42]:= **MorphismQ[f$_r$, Mode → Visual]**

The table entry corresponding to the sum a+b
(resp. product a*b) in the domain of the morphoid
is colored if and only if addition (resp.
multiplication) of the pair {a,b} is preserved by
the morphoid; i.e., f(a+b) = f(a)+f(b) (resp.
f(a*b) = f(a)*f(b))

Add(Z[10])   x * y

+	0	1	2	3	4	5	6	7	8	9
0	0	1	2	3	4	5	6	7	8	9
1	1	2	3	4	5	6	7	8	9	0
2	2	3	4	5	6	7	8	9	0	1
3	3	4	5	6	7	8	9	0	1	2
4	4	5	6	7	8	9	0	1	2	3
5	5	6	7	8	9	0	1	2	3	4
6	6	7	8	9	0	1	2	3	4	5
7	7	8	9	0	1	2	3	4	5	6
8	8	9	0	1	2	3	4	5	6	7
9	9	0	1	2	3	4	5	6	7	8

Mult(Z[10])   x * y

*	0	1	2	3	4	5	6	7	8	9
0	0	0	0	0	0	0	0	0	0	0
1	0	1	2	3	4	5	6	7	8	9
2	0	2	4	6	8	0	2	4	6	8
3	0	3	6	9	2	5	8	1	4	7
4	0	4	8	2	6	0	4	8	2	6
5	0	5	0	5	0	5	0	5	0	5
6	0	6	2	8	4	0	6	2	8	4
7	0	7	4	1	8	5	2	9	6	3
8	0	8	6	4	2	0	8	6	4	2
9	0	9	8	7	6	5	4	3	2	1

Out[42]= False

- From the preceding tables, we see that the pair (2, 3) does not preserve the multiplication operation. Here is an illustration of an explanation.

In[43]:= **PreservesQ[f$_r$, {2, 3}, Mode → Visual]**

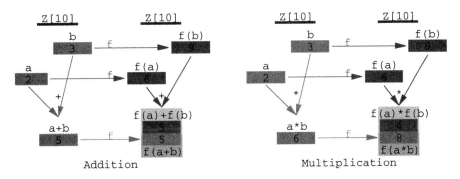

Out[43]= False

Whereas MorphismQ checks every pair of elements to see if the operations are preserved, ProbableMorphismQ pursues the same question randomly by testing a shorter list of pairs.

ProbableMorphismQ[*f*]	give True if the Morphoid *f* is probably a morphism, and False otherwise
ProbableMorphismQ[*f, opts*]	use the options *opts* to determine if *f* is probably a morphism

Probabilistic approach to checking if a morphism.

- We can use a probabilistic approach to check for morphisms with the ProbableMorphismQ function.

In[44]:= **ProbableMorphismQ[f$_r$]**

```
ProbableMorphismQ:warning :
  The ProbableMorphismQfunction is being used; '
    True' results are only probable, not certain.
```

Out[44]= False

There are several options to consider for this function.

option name	default value	
PrintMessage	True	specifies whether a warning should be given about using a probabilistic approach
SampleSize	5	specifies the size of sample to choose when randomly checking
SamplePairs	Random	specifies the method by which pairs should be checked (alternatively, Default uses a built-in list, or a list of pairs of *positions* can be provided)

Options for `ProbableMorphismQ`.

- If we choose pairs carefully, we can obtain erroneous results with this function. Here we use pairs of positions in the domain, where {3, 6} indicates that we are testing the third element (2) and the sixth element (5).

In[45]:= **ProbableMorphismQ[f$_r$, PrintMessage → False,**
        SamplePairs → {{1, 1}, {1, 6}, {3, 6}, {6, 8}}]

Out[45]= True

- Setting the SampleSize is immaterial when we use the built-in list of pairs specified by Default.

In[46]:= **ProbableMorphismQ[f$_r$, SampleSize → 2000,**
        SamplePairs → Default, PrintMessage → False]

Out[46]= False

A function *f* is an *isomorphism* if it is an injective and surjective (i.e., bijective) homomorphism.

IsomorphismQ[*f*]	give True if the Morphoid *f* is an isomorphism (using ProbableMorphismQ), and False otherwise
IsomorphismQ[*f*, Cautious → True]	give True if the Morphoid *f* is an isomorphism (using MorphismQ), and False otherwise

`IsomorphismQ` variations.

- The function on $\mathbb{Z}_{20}$ defined by the rule $1 \to 5$ is not surjective, thus failing one of the requirements to be an isomorphism.

```
In[47]:= IsomorphismQ[ZMap[20, 20, 1 → 5]]

        Morphoid::notonto :
          Since the Morphoid is not onto, it can not be an isomorphism

Out[47]= False
```

- If we send 1 to a generator such as 7, we do obtain an isomorphism. (At least we are informed that it is *probably* an isomorphism.)

```
In[48]:= IsomorphismQ[ZMap[20, 20, 1 → 7]]

        ProbableMorphismQ::"warning" :
          "The ProbableMorphismQ function is being used; 'True'
          results are only probable, not certain.

Out[48]= True
```

- If we want to be certain, we can use the Cautious option. (Using SetOptions, we can change it so that True is the default value for this option).

```
In[49]:= IsomorphismQ[ZMap[20, 20, 1 → 7], Cautious → True]

Out[49]= True
```

■ 4.5 Kernel, Image, and InverseImage

Kernel[f]	return the kernel (as a groupoid or ringoid) of the Morphoid f (assuming the codomain has an identity element)
Image[f]	return the image (as a groupoid or ringoid) of the Morphoid f
Image[f, S]	return the image (as a groupoid or ringoid) of the subset S of the domain under the Morphoid f

Kernel and Image functions.

- Define the following Morphoid that is not a homomorphism.

```
In[50]:= f = FormMorphoid[Mod[# - 2, 13] &, Z[5], Z[13]]

Out[50]= Morphoid[Mod[#1 - 2, 13] &, -Z[5]-, -Z[13]-]
```

- It still makes sense to ask about the kernel, even if f does not satisfy MorphismQ.

```
In[51]:= {MorphismQ[f], Kernel[f]}
```

Out[51]= {False, Groupoid[{2}, Mod[#1 + #2, 5] &]}

- The Kernel function has a Visual mode.

In[52]:= **K = Kernel[ZMap[12, 4], Mode → Visual]**

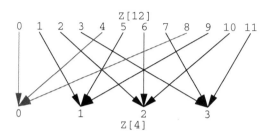

Out[52]= Groupoid[{0, 4, 8}, Mod[#1 + #2, 12] &]

- Using the Parity function, the image of all elements in A_3 is 1.

In[53]:= **Image[FormMorphoid[Parity, Alternating[3], IntegerUnits]]**

Out[53]= Groupoid[{1}, Times]

- We make a quotient group with the kernel K as a normal subgroup of $\mathbb{Z}_{12}$ and form a morphism between it and $\mathbb{Z}_4$.

In[54]:= **FormMorphoidSetup[QuotientGroup[Z[12], K], Z[4]];**

QuotientGroup :NS :
  This quotient group uses NS to represent the normal subgroup
  {0, 4, 8} that you specified Use CosetToListto
  convert this coset representationto a list of elements.

Domain			Codomain
NS	1	1	0
1 + NS	2	2	1
2 + NS	3	3	2
3 + NS	4	4	3

- The First Isomorphism Theorem guarantees that this is an isomorphism; *Mathematica* agrees.

In[55]:= **IsomorphismQ[**
    FormMorphoid[{1, 2, 3, 4}, QuotientGroup[Z[12], K], Z[4]]]

Out[55]= True

- Note that the codomain is the target set, while the image is the set that is actually hit.

In[56]:= **f = ZMap[12, 24, 1 → 2];**
Map[Elements, {Codomain[f], Image[f]}]

Out[56]= {{0, 1, 2, 3, 4, 5, 6, 7, 8, 9, 10, 11,
        12, 13, 14, 15, 16, 17, 18, 19, 20, 21, 22, 23},
       {0, 2, 4, 6, 8, 10, 12, 14, 16, 18, 20, 22}}

InverseImage[f, y]	give the list of elements in the domain of the Morphoid f whose image is y
InverseImage[f, {y_1, y_2, ...}]	give the list of elements in the domain of the Morphoid f whose image is in {y_1, y_2, ...}
InverseImages[f]	equivalent to InverseImage[f, Elements[Codomain[f]]] but partitioned by preimages of elements in codomain
InverseImages[f, WithImages → True]	list the image elements with the list of preimages for each image element
Fiber[f, S]	equivalent to InverseImage[f, S]

Functions related to inverse images.

- The Visual mode of InverseImage illustrates the elements that are mapped onto 2 for the Morphoid ZMap[12,4].

In[57]:= **InverseImage[ZMap[12, 4], 2, Mode → Visual]**

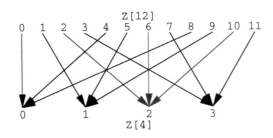

Out[57]= {2, 6, 10}

- Two observations should be made from the following output: (1) elements not in the codomain are ignored (but a warning message is given); (2) {0, 2} is a subgroup of $\mathbb{Z}_4$ and the set of inverse images for it is also a subgroup (in $\mathbb{Z}_{12}$, in this case).

In[58]:= **{SubgroupQ[**
inv = InverseImage[ZMap[12, 4], {0, 2, 5}], Z[12]], inv}

> *MemberQ::elmnt : 5 is not an element of Z[4].*

Out[58]= {True, {0, 4, 8, 2, 6, 10}}

- The `WithImages` option of `InverseImages` is intended to make it explicit which codomain elements are associated with which domain elements.

In[59]:= **InverseImages[FormMorphoid[Mod[#^2, 5] &, Z[7], Z[7]],**
     WithImages → True]

Out[59]= {{{0, 5}, 0}, {{1, 4, 6}, 1}, {{}, 2},
     {{}, 3}, {{2, 3}, 4}, {{}, 5}, {{}, 6}}

▦ 4.6 Automorphisms

Whereas a morphism describes a function between two groups, an automorphism describes an isomorphism between a group and itself.

`Automorphism[G, x → y]`	for a cyclic group *G*, form the automorphism determined by the rule $x \to y$
`AutomorphismGroup[G]`	return the group of automorphisms of the cyclic group *G*
`Aut[G]`	identical to `AutomorphismGroup[G]`
`InnerAutomorphism[G, g]`	given an element *g* in the group *G*, return the inner automorphism induced by *g* (via conjugation)
`InnerAutomorphism[G, g, FunctionForm → type]`	express the inner automorphism using rules if *type* is `Rules` and as a function if *type* is `Function`
`InnerAutomorphismGroup[G]`	return the group of inner automorphisms of the group *G*
`Inn[G]`	identical to `InnerAutomorphismGroup[G]`

Automorphism functions.

- The `Textual` mode of the `Automorphism` function provides some extra details regarding the map.

In[60]:= **g1 = Automorphism[Z[10], 1 → 3, Mode → Textual]**

     2 = 1+1 is mapped to 3+3 = 6
     3 = 1+1+1 is mapped to 3+3+3 = 9
     4 = 1+1+<<1>>+1 is mapped to 3+3+<<1>>+3 = 2
     5 = 1+1+<<2>>+1 is mapped to 3+3+<<2>>+3 = 5
     6 = 1+1+<<3>>+1 is mapped to 3+3+<<3>>+3 = 8
     7 = 1+1+<<4>>+1 is mapped to 3+3+<<4>>+3 = 1
     8 = 1+1+<<5>>+1 is mapped to 3+3+<<5>>+3 = 4

```
        9 = 1+1+<<6>>+1 is mapped to 3+3+<<6>>+3 = 7
        0 = 1+1+<<7>>+1 is mapped to 3+3+<<7>>+3 = 0
Out[60]= Morphoid[1 → 3, -Z[10]-, -Z[10]-]
```

- We can see that the rules governing this function are exactly as described.

In[61]:= **MorphoidRules[g1]**

Out[61]= {0 → 0, 1 → 3, 2 → 6, 3 → 9, 4 → 2, 5 → 5, 6 → 8, 7 → 1, 8 → 4, 9 → 7}

- The defining rule must involve two generators.

In[62]:= **Automorphism[Z[10], 1 → 4]**

```
Automorphism :badrule :
    The rule provided does not uniquely define an automorphism
      on Z[10]. The rule must map a generator to a generator.
```

Out[62]= $Failed

- Since there are four generators in $\mathbb{Z}_{12}$, the group of automorphisms has four elements. The following should make it clear which group of order four this is.

In[63]:= **Orders[AutomorphismGroup[Z[12]]]**

```
Out[63]= {{Morphoid[1 → 1, -Z[12]-, -Z[12]-], 1},
          {Morphoid[1 → 5, -Z[12]-, -Z[12]-], 2},
          {Morphoid[1 → 7, -Z[12]-, -Z[12]-], 2},
          {Morphoid[1 → 11, -Z[12]-, -Z[12]-], 2}}
```

- Since the inner automorphism on $\mathbb{Z}_7$ induced by 4 is simply conjugation by 4, for this Abelian group this amounts to the Identity map.

In[64]:= **EqualMorphoidQ[InnerAutomorphism[Z[7], 4],**
       FormMorphoid[Identity, Z[7], Z[7]]]

Out[64]= True

- By default, Morphoids based on inner automorphisms are formatted.

In[65]:= **InnerAutomorphism[Symmetric[3], {3, 1, 2}]**

Out[65]= Morphoid[Conjugation by {3, 1, 2}, -S[3]-, -S[3]-]

- This formatting can be overridden, though, giving the list of rules instead.

In[66]:= **g2 = InnerAutomorphism[Symmetric[3],**
         {3, 1, 2}, FormatFunction → False]

```
Out[66]= Morphoid[{{1, 2, 3} → {1, 2, 3},
            {1, 3, 2} → {3, 2, 1}, {2, 1, 3} → {1, 3, 2},
            {2, 3, 1} → {2, 3, 1}, {3, 1, 2} → {3, 1, 2},
            {3, 2, 1} → {2, 1, 3}}, -S[3]-, -S[3]-]
```

- If the actual function definition is of interest, you can see it by changing the
  FunctionForm (which defaults to Rules).

```
In[67]:= InnerAutomorphism[Symmetric[3], {3, 1, 2},
            FormatFunction → False, FunctionForm → Function]
```

```
Out[67]= Morphoid[
            Operation[Groupoid[{{1, 2, 3}, {1, 3, 2}, {2, 1, 3}, {2,
                3, 1}, {3, 1, 2}, {3, 2, 1}}, -Operation-]][
              Operation[Groupoid[{{1, 2, 3}, {1, 3, 2}, {2, 1, 3},
                  {2, 3, 1}, {3, 1, 2}, {3, 2, 1}}, -Operation-]][
                GroupInverse[Groupoid[{{1, 2, 3}, {1, 3, 2}, {2, 1, 3},
                    {2, 3, 1}, {3, 1, 2}, {3, 2, 1}}, -Operation-],
                  {3, 1, 2}], #1], {3, 1, 2}] &, -S[3]-, -S[3]-]
```

- Here is the group of inner automorphisms on D_3.

```
In[68]:= innd3 = InnerAutomorphismGroup[Dihedral[3]]
```

```
Out[68]= Groupoid[{-Elements-}, -Operation-]
```

- Which group of order six is this?

```
In[69]:= CayleyTable[innd3, Mode → Visual];
```

```
            KEY for Inn[D[3]]: label used → element: {g1 → Morphoid[
              Conjugation by 1, -D[3]-, -D[3]-], g2 → Morphoid[
              Conjugation by Rot, -D[3]-, -D[3]-], g3 → Morphoid[
              Conjugation by Rot^2, -D[3]-, -D[3]-], g4 → Morphoid[
              Conjugation by Ref, -D[3]-, -D[3]-], g5 → Morphoid[
              Conjugation by Rot**Ref, -D[3]-, -D[3]-], g6 →
              Morphoid[Conjugation by Rot^2**Ref, -D[3]-, -D[3]-]}
```

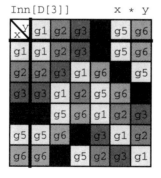

▣ 4.7 Visualizing Morphoids

`VisualizeMorphoid[f]`	give a graphic illustrating the Morphoid *f*
`VisualizeMorphoid[f,` `ColorDomain → {{color₁, dom₁}, ...}]`	in the graphic, color the arrow from the domain element *dom$_k$* using *color$_k$*

`VisualizeMorphoid[`
`f, ColorCodomain →`
`{{color₁, cod₁}, ...}]`
	in the graphic, color the arrow (s) to the codomain element *cod$_k$* using *color$_k$*

`VisualizeMorphoid[f,` `ColorDomain → Automatic]`	color the arrows from the domain elements using Hue across the length of the domain
`VisualizeMorphoid[f,` `ColorCodomain → Automatic]`	color the arrows to the codomain elements using Hue across the length of the codomain

Variations of `VisualizeMorphoid`.

- Individual elements can receive special coloring as follows. Note that Codomain requests take priority over Domain requests and any nonsense is ignored.

In[70]:= `VisualizeMorphoid[ZMap[12, 6], ColorDomain →`
  `{{Green, 2}, {junk}, {Red, 15}, {Green, 9}},`
  `ColorCodomain → {{Black, 2}, {Magenta, 5}}];`

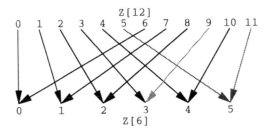

- Coloring the Codomain using the value Automatic readily shows inverse images of values in the codomain.

In[71]:= `VisualizeMorphoid[ZMap[12, 6], ColorCodomain → Automatic];`

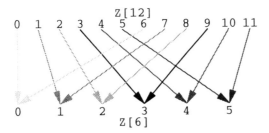

Chapter 5

Additional Functionality

In the final chapter, we look at some additional functionality that does not neatly fit into the chapters on groupoids, ringoids, or morphisms.

▥ 5.1 Global variables and options

There are several global variables and options that are worth noting.

name	initial value	
DefaultStru cture	Group	context (groups versus rings) that is assumed for the functions in use
VisualTextS hown	2	specifies how often accompanying text for visualizations is shown
BackgroundC olors	(see below)	(ordered) list of colors that are used for Cayley tables
Output	Computational	available for some functions when the Visual mode is being used in order to change the output from the computation to the visual graphic (using the value Graphics, or GraphicsArray for a few functions)
DefaultOrder	RightToLeft	default direction for multiplying permutations

Some global variables.

■ We can always see the default structure that is being assumed.

In[1]:= **DefaultStructure**

Out[1]= Group

- If we wish to change the value of this variable, it is best to use the Switch-StructureTo function because it redefines some options for various functions.

In[2]:= **SwitchStructureTo[Group]**

Out[2]= Group

{Yellow, Orange, Violet, Blue, Mint, Turquoise, EmeraldGreen, GreenDark, Pink, BlueLight, Banana, Green, Brown, Gray, Red, Purple, CadmiumYellow, Maroon, Navy, Salmon, Aquamarine, Indigo, Lavender, Antique, Bisque, Burlywood, Eggshell, Khaki, BlueViolet, CadmiumOrange, CadmiumRedDeep, Cerulean, Chartreuse, Cyan, DeepPink, Magenta, OrangeRed, Peacock, SkyBlueDeep, TurquoiseDark, Ultramarine}

BackgroundColors.

- By adding Output → Graphics to a function when using the Visual mode, the computational output is suppressed and the graphic becomes the output.

In[3]:= **gr1 = CayleyTable[Z[4], Mode → Visual, Output → Graphics]**

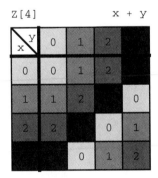

Out[3]= - Graphics -

- Now this graphic image can be used along side another graphic using GraphicsArray.

In[4]:= **gr2 = LeftCosets[Z[8], {0, 4}, Mode → Visual,**
**   Output → Graphics, DisplayFunction → Identity];**
**   Show[GraphicsArray[{gr1, gr2}],**
**   DisplayFunction → $DisplayFunction]**

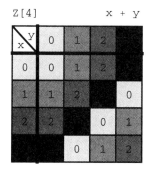

Out[5]=   **- GraphicsArray -**

▨ 5.2 Working with permutations and cycles

▪ 5.2.1 Introduction

A permutation of a set S is a bijection (one-to-one and onto function) from S to S. Frequently, either the set S is the first n positive integers or it can be viewed as such. For the most part, in these packages we view a permutation σ as a function on Range[n]. Therefore, we describe it by the ordered list of the functional *images* of the domain $\{1, 2, ..., n\}$. In other words, the permutation σ is given as a rearrangement of this set. For example, let σ be the permutation that takes 1 to 4, 2 to 2, 3 to 1, and 4 to 3.

- We define a permutation by using the ordered set of images of the domain $\{1, 2, ..., n\}$. Here, $\sigma(1) = 4$, $\sigma(2) = 2$, $\sigma(3) = 1$, and $\sigma(4) = 3$.

In[6]:=   σ = **{4, 2, 1, 3}**

Out[6]= $\{4, 2, 1, 3\}$

- The function `PermutationMatrix` allows us to view both the domain (first row) and image (second row) of a permutation, as it is sometimes presented.

In[7]:=   **PermutationMatrix[σ]**

Out[7]//MatrixForm=

$$\begin{pmatrix} 1 & 2 & 3 & 4 \\ 4 & 2 & 1 & 3 \end{pmatrix}$$

- Note that a permutation must be a bijection. Therefore, a list of length n is considered a permutation only if it is the list $\{1, 2, ..., n\}$ after the elements are sorted.

In[8]:=   **PermutationQ[{5, 2, 1, 3}]**

Out[8]=   False

A number of functions deal with permutations (and cycles). Note that while the functions listed here are compatible with those found in the standard package `DiscreteMath`-`Permutations``, the functionality is greatly expanded here.

■ Below is a list of functions that have either the word `Permutation` or `Cycle` in their name (some of which are not related to these packages).

In[9]:= **Union[Names["*Permutation*"], Names["*Cycle*"]]**

Out[9]= {AnimationCycleOffset, AnimationCycleRepetitions, Cycle, CycleAs, DisjointCyclesQ, EvenPermutationQ, ExtendPermutation, FormGroupoidFromCycles, FromCycles, MultiplyCycles, MultiplyPermutations, OddPermutationQ, PermutationComposition, PermutationGroup, PermutationImage, PermutationInverse, PermutationMatrix, PermutationQ, Permutations, PermutationToPower, RandomPermutation, SamePermutationQ, ShowColoredPermutation, ShowPermutation, ShowPossiblePermutations, TestPermutationQ, ToCycles, ToPermutation}

PermutationQ [*list*]	give True if *list* represents a permutation, and False otherwise
RandomPermutation [*n*]	give a random permutation on {1, 2, ..., *n*}
Ordering [*list*]	give the permutation that puts the elements in *list* in order

Basic functions for working with permutations; these are almost equivalent to those in `DiscreteMath`Permutations``.

■ The `Ordering` function does not require the list to be a permutation on the integers 1, 2, ..., *n*. The following list is considered ordered when the elements appear in the order {1, 3, 2}.

In[10]:= **Ordering[{"aleph", "gimel", "beth"}]**

Out[10]= {1, 3, 2}

■ When the list is a permutation, `Ordering` effectively finds the inverse of the permutation.

In[11]:= **q = Ordering[p = {1, 3, 2}]**

Out[11]= {1, 3, 2}

■ The following verifies that *p* and *q* are inverses.

In[12]:= **PermutationComposition[q, p]**

Out[12]= {1, 2, 3}

- And here is another way of verifying it.

In[13]:= **q == PermutationInverse[p]**

Out[13]= True

■ 5.2.2 Permutation operations

Since we are viewing a permutation of length *n* as a *function* on Range[*n*], we multiply two permutations by composing the two functions. Since there are two different traditions as to the order in which one composes two functions, the order for multiplying permutations is likewise in dispute. To remedy this, the function MultiplyPermutations has an option called ProductOrder that permits the values RightToLeft and LeftToRight.

MultiplyPermutations[*perm2*, *perm1*]	give the product of the permutation *perm1* followed by the permutation *perm2*, working from right to left
MultiplyPermutations[*perm2*, *perm1*, ProductOrder → LeftToRight]	give the product of the permutation *perm2* followed by the permutation *perm1*, working from left to right
PermutationComposition[*perm2*, *perm1*]	identical to MultiplyPermutations[*perm2*, *perm1*, ProductOrder → RightToLeft]

Different ways of multiplying permutations.

- Here are two randomly chosen permutations of length eight and four respectively.

In[14]:= **{p = RandomPermutation[8], q = RandomPermutation[4]}**

Out[14]= {{8, 7, 5, 6, 3, 1, 4, 2}, {3, 2, 1, 4}}

- A permutation in S_m can be extended to one in S_n, $n > m$, with ExtendPermutation. For example, we can extend *q* to a permutation in S_8, extending the function as the identity on the last four entries.

In[15]:= **{q, ExtendPermutation[q, 8]} // ColumnForm**

Out[15]= {3, 2, 1, 4}
{3, 2, 1, 4, 5, 6, 7, 8}

- ExtendPermutation is used automatically when we multiply *p* and *q*. This is the result when performing the product from right to left, the default approach.

In[16]:= **MultiplyPermutations[p, q]**

Out[16]= {5, 7, 8, 6, 3, 1, 4, 2}

- Using the same permutations, we (generally) get a different result when we perform the product working from left to right.

In[17]:= **MultiplyPermutations[p, q, ProductOrder → LeftToRight]**

Out[17]= {8, 7, 5, 6, 1, 3, 4, 2}

- Note that permutations can be entered into `MultiplyPermutations` either as a list of rules or in the standard permutation form. Note also that the `Textual` mode indicates the rule form of the product but returns the standard form.

In[18]:= **MultiplyPermutations[{1 → 4, 2 → 5, 3 → 3, 4 → 2, 5 → 1},**
{3, 1, 4, 2}, Mode → Textual]

```
The permutation resulting from the product of
    {3, 1, 4, 2}
followed by
    {1 → 4, 2 → 5, 3 → 3, 4 → 2, 5 → 1}
can be given as
    {1 → 3, 2 → 4, 3 → 2, 4 → 5, 5 → 1}
or as
```

Out[18]= {3, 4, 2, 5, 1}

Here is a summary of some miscellaneous functions dealing with permutations, several of which we have already seen in use.

`ExtendPermutation[p, n]`	for a permutation p of length m, return the equivalent, extended permutation of length n ($n > m$) that is the identity on positions $m+1, ..., n$
`PermutationInverse[p]`	give the permutation that is the inverse of permutation p
`PermutationMatrix[p]`	show a permutation p in matrix form where the bottom row is p and the top row consists of 1, 2, ..., `Length[p]`
`PermutationToPower[p, n]`	give the nth power of the permutation p (where is n is any integer)

Miscellaneous functions.

- `PermutationToPower` is analogous to `ElementToPower`; both can be used to raise an element to any positive or negative power.

In[19]:= **TableForm[**
Table[{k, PermutationToPower[p, k]}, {k, -1, 4}],
TableDepth → 2,
TableHeadings → {None, {"k", "\!\(p\^k\)"}}]

```
Out[19]//TableForm=
         k          pᵏ
         -1         {6, 8, 5, 7, 3, 4, 2, 1}
         0          {1, 2, 3, 4, 5, 6, 7, 8}
         1          {8, 7, 5, 6, 3, 1, 4, 2}
         2          {2, 4, 3, 1, 5, 8, 6, 7}
         3          {7, 6, 5, 8, 3, 2, 1, 4}
         4          {4, 1, 3, 2, 5, 7, 8, 6}
```

■ 5.2.3 Representing permutations

Since permutations are functions from `Range[n]` to itself, it is natural to represent a permutation by a list of rules.

ToRules [*perm*]	for a permutation *perm* of length *n*, give a list of rules of the form $x \to y$ where x is in Range[n] and y is *perm*(x), the image of x under the permutation
ToPermutation [*list*]	when *list* is a list of rules in the form $x \to y$ or $\{x \to y\}$, give the permutation, if possible, corresponding to the rules

Rules and permutations can be interchanged.

- The following converts a permutation into a list of rules corresponding to the permutation.

In[20]:= **{q = RandomPermutation[4], qrules = ToRules[q]}**

Out[20]= {{1, 4, 2, 3}, {1 → 1, 2 → 4, 3 → 2, 4 → 3}}

- `ToPermutation` converts a list of rules into a permutation. Here we see that this function is the inverse to `ToRules`.

In[21]:= **ToPermutation[qrules] == q**

Out[21]= True

■ 5.2.4 Cycles

Consider the permutation $\sigma = \{3, 1, 4, 2\}$. This represents the function that takes 1 to 3, 2 to 1, 3 to 4 and 4 to 2. Note that σ is a *cyclic permutation* in the sense that $\sigma(1) = 3$, $\sigma(3) = \sigma(\sigma(1)) = 4$, $\sigma(4) = \sigma(\sigma(\sigma(1))) = 2$, and then $\sigma(2)$ takes us back to 1, cycling through all of the elements. Often such a cycle is represented in the mathematical literature as (1 3 4 2). Since juxtaposition implies multiplication in *Mathematica*, and since parentheses cannot be used as delimiters, we need another structure to represent a *cycle*. In these packages, this cycle is represented by Cycle[1, 3, 4, 2], with the head Cycle and the elements separated by commas, as in a list. Cycles are important in abstract algebra (for example, any permutation can be written as a product of cycles) and consequently receive some attention in these packages.

The `DiscreteMath`Permutations`` package contains two functions for converting between permutations and cycles: `ToCycles` and `FromCycles`. For compatibility reasons, we can still operate using the approach presented there, but because the approach here is clearer and more robust, you are encouraged to adopt the defaults used here.

`ToCycles[perm]`	give the decomposition of *perm* into a list of cycles of the form `Cycle[a, b, ...]`
`ToCycles[perm, CycleAs → List]`	give the decomposition of *perm* into a list of cycles of the form found in the `DiscreteMath` package
`FromCycles[{cyc₁, cyc₂, ...}]`	give the permutation corresponding to the given cyclic decomposition

Converting between cycles and permutations.

option name	default value	
`CycleAs`	`Cycle`	specifies whether a cycle should be represented using the `Cycle[a, b, ...]` structure or as a list of lists (using `List`)
`Normalize`	`True`	when set to `True`, specifies that all cycles of length one are dropped (unless needed to show the length of the original permutation) and a normalization is applied to all remaining cycles (only valid if we have `CycleAs → Cycle`)

Options for `ToCycles`.

- Consider the following permutation as an example.

```
In[22]:= r = {8, 2, 5, 6, 4, 3, 1, 9, 7, 10, 11}

Out[22]= {8, 2, 5, 6, 4, 3, 1, 9, 7, 10, 11}
```

- Here is how `ToCycles` works in `DiscreteMath`Permutations``.

```
In[23]:= ToCycles[r, CycleAs → List]

Out[23]= {{8, 9, 7, 1}, {2}, {5, 4, 6, 3}, {10}, {11}}
```

- Here is how the new version of `ToCycles` works. Note the omission of the two singleton cycles {2} and {10} but retention of {11}. Note also that the other cycles are "rotated" until the smallest element in the cycle appears first.

```
In[24]:= rcycles = ToCycles[r]
```

Out[24]= {Cycle[1, 8, 9, 7], Cycle[3, 5, 4, 6], Cycle[11]}

- Setting Normalize to False retains the structure of the DiscreteMath approach but changes the lists to having the head Cycle.

In[25]:= **ToCycles[r, Normalize → False]**

Out[25]= {Cycle[8, 9, 7, 1], Cycle[2],
          Cycle[5, 4, 6, 3], Cycle[10], Cycle[11]}

- FromCycles takes any output of ToCycles and acts as an inverse.

In[26]:= **{FromCycles[ToCycles[r, CycleAs → List]] == r,**
          FromCycles[ToCycles[r]] == r,
          FromCycles[ToCycles[r, Normalize → False]] == r}

Out[26]= {True, True, True}

- The list of cycles needs to be disjoint before one can hope for a permutation to represent the cycle list. (This is a failure in the DiscreteMath package; this is not checked.)

In[27]:= **FromCycles[{Cycle[2, 3], Cycle[1, 2]}]**

          Cycle::disjoint :
           The cycles in the list {Cycle[2, 3], Cycle[1, 2]}
             need to be disjoint to use this function.

Out[27]= $Failed

- ToCycles can also work on a *list* of permutations, returning a list of cycle decompositions.

In[28]:= **ToCycles[{RandomPermutation[8], RandomPermutation[8]}] //**
          ColumnForm

Out[28]= {Cycle[1, 7, 3, 8, 6]}
          {Cycle[1, 8, 2, 6, 7], Cycle[3, 5, 4]}

- The permutation given to ToCycles can be given as a list of rules (in various forms).

In[29]:= **{ToCycles[{1 → 4, 2 → 3, 3 → 1, 4 → 2}],**
          ToCycles[1 → 4, 2 → 3, 3 → 1, 4 → 2],
          ToCycles[{{1 → 4}, {2 → 3}, {3 → 1}, {4 → 2}}]}

Out[29]= {{Cycle[1, 4, 2, 3]}, {Cycle[1, 4, 2, 3]}, {Cycle[1, 4, 2, 3]}}

- ToPermutation can act on a cycle (as well as a list of rules representing a permutation, as we saw earlier).

In[30]:= **ToPermutation[Cycle[2, 3, 5]]**

Out[30]= {1, 3, 5, 4, 2}

■ 5.2.5 Cycle operations

Cycles can be multiplied just as permutations can. In the following functions, the cycles can be either in list form or `Cycle` form. These cycles are multiplied in the same order used by `MultiplyPermutations`, which is `RightToLeft` by default.

`MultiplyCycles[` cyc_2, cyc_1, n`]`	give the product of cycle cyc_1 followed by cyc_2 as a permutation of length n (in S_n), multiplying from right to left
`MultiplyCycles[`cyc_2, cyc_1`]`	give the product of cycle cyc_1 followed by cyc_2 as a permutation of length n, where n is the maximum value appearing in either cycle, multiplying from right to left
`MultiplyCycles[` cyc_k, ..., cyc_2, cyc_1`]`	give the product of the cycles cyc_1 followed by cyc_2 followed by ... cyc_k, as a permutation of length n, where n is the maximum value appearing in any cycle
`MultiplyCycles[` {cyc_k, ..., cyc_2, cyc_1}`]`	equivalent to `MultiplyCycles[`cyc_k, ..., cyc_2, cyc_1`]`

`MultiplyCycles[...,`
`ProductOrder →`
`LeftToRight]`
	multiply the cycles by multiplying from left to right

cyc_2 @ cyc_1	equivalent to `MultiplyCycles[`cyc_2, cyc_1`]`

Variations in the `MultiplyCycles` function.

■ Here we obtain two different cycles by choosing the first cycle in the decomposition of two random permutations.

In[31]:= {**c1 = First[ToCycles[RandomPermutation[12]]],**
        c2 = First[ToCycles[RandomPermutation[12]]]}

Out[31]= {Cycle[1, 5], Cycle[1, 4, 12]}

■ The product of the two cycles is performed by multiplying the cycles from right to left (by default). One can use the function name (`MultiplyCycles`) or the shortcut.

In[32]:= {**MultiplyCycles[c2, c1], c2 @ c1**}

Out[32]= {{5, 2, 3, 12, 4, 6, 7, 8, 9, 10, 11, 1},
        {5, 2, 3, 12, 4, 6, 7, 8, 9, 10, 11, 1}}

- The same result can be obtained by converting the cycles to permutations and then multiplying them.

```
In[33]:= p1 = FromCycles[{c1}];
         p2 = FromCycles[{c2}];
         MultiplyPermutations[p2, p1]
```

```
Out[33]= {5, 2, 3, 12, 4, 6, 7, 8, 9, 10, 11, 1}
```

- Actually, `MultiplyPermutations` multiplies the cycles without first doing the conversion.

```
In[34]:= MultiplyPermutations[c2, c1]
```

```
Out[34]= {5, 2, 3, 12, 4, 6, 7, 8, 9, 10, 11, 1}
```

- To convey that these cycles are to be multiplied as elements in S_9, add 9 as the final argument.

```
In[35]:= MultiplyCycles[Cycle[2, 3], Cycle[4, 6], 9]
```

```
Out[35]= {1, 3, 2, 6, 5, 4, 7, 8, 9}
```

- When the cycles are not disjoint, the order in which these are multiplied matters.

```
In[36]:= {MultiplyCycles[
           {Cycle[2, 6, 4], Cycle[4, 3, 1], Cycle[2, 5, 1]}],
         MultiplyCycles[{Cycle[2, 5, 1],
           Cycle[2, 6, 4], Cycle[4, 3, 1]}]}
```

```
Out[36]= {{6, 5, 1, 3, 2, 4}, {5, 6, 2, 3, 1, 4}}
```

- However, when the cycles are disjoint, the order of the factors is immaterial.

```
In[37]:= MultiplyCycles[{Cycle[2, 6, 4], Cycle[3, 5, 1]}] ==
         MultiplyCycles[{Cycle[3, 5, 1], Cycle[2, 6, 4]}]
```

```
Out[37]= True
```

■ 5.2.6 Other cycle-related functions

- The `DisjointCyclesQ` function can ascertain whether the cycles in a list (or sequence) are disjoint.

```
In[38]:= DisjointCyclesQ[{Cycle[2, 6, 4], Cycle[3, 5, 1]}]
```

```
Out[38]= True
```

- Representing a permutation with cycles is not unique. The function `SamePermutationQ` determines whether two lists of cycles represent the same permutation.

```
In[39]:= {SamePermutationQ[{{1, 8}, {3, 2, 5}}, {{5, 3, 2}, {8, 1}}],
        SamePermutationQ[{Cycle[1, 8], Cycle[3, 2, 5]},
        {Cycle[5, 3, 2], Cycle[8, 1]}]}

Out[39]= {True, True}
```

DisjointCyclesQ[cyc_1, cyc_2, ...]	give True if the cycles cyc_1, cyc_2,... are disjoint, and False otherwise
DisjointCyclesQ[{ cyc_1, cyc_2, ... }]	give True if the cycles cyc_1, cyc_2,... are disjoint, and False otherwise
SamePermutationQ[$cyclist_1$, $cyclist_2$]	give True if the cycles in $cyclist_1$ represent the same permutation as the cycles in $cyclist_2$, and False otherwise

Other functions related to cycles.

Cycles of length two are called transpositions. Any cycle can be written as a product of transpositions, and consequently so can any permutation. While the number of transpositions in a given permutation is not unique, whether this number is odd or even is fixed.

ToTranspositions[cyc]	give the cycle cyc as a product of transpositions
ToTranspositions[$perm$]	give the permutation $perm$ as a product of transpositions

Converting to transpositions.

- Here we convert a random permutation of length six into transpositions.

```
In[40]:= ToTranspositions[p = RandomPermutation[6]]

Out[40]= {Cycle[1, 6], Cycle[1, 3],
         Cycle[1, 5], Cycle[1, 4], Cycle[1, 2]}
```

- In this case, first we convert the same permutation to cycles and then convert each cycle into transpositions.

```
In[41]:= Map[ToTranspositions, ToCycles[p]]

Out[41]= {{Cycle[1, 6], Cycle[1, 3],
          Cycle[1, 5], Cycle[1, 4], Cycle[1, 2]}}
```

The number of transpositions in a permutation gives some information about the permutation. Actually, whether the number of transpositions is even or odd is what is important. The following functions can be used in ascertaining this.

EvenPermutationQ[*perm*]	give True if *perm* is an even permutation, and False otherwise
OddPermutationQ[*perm*]	give True if *perm* is an odd permutation, and False otherwise
Parity[*perm*]	give 1 if *perm* is an even permutation, and −1 if an odd permutation

Functions dealing with the parity of a permutation.

- First, let's create five random permutations of length six.

```
In[42]:= Clear[p]
         Table[p[k] = RandomPermutation[6], {k, 5}]
```

```
Out[43]= {{1, 2, 5, 4, 6, 3}, {3, 2, 4, 5, 6, 1},
          {1, 4, 2, 5, 3, 6}, {2, 1, 4, 5, 3, 6}, {5, 4, 2, 1, 6, 3}}
```

- Now consider the following table that shows how the oddness/evenness of a permutation is related to the number of transpositions.

```
In[44]:= TableForm[
           Table[{p[k], Parity[p[k]], OddPermutationQ[p[k]],
             Length[ToTranspositions[p[k]]]}, {k, 5},
           TableDepth → 2, TableHeadings →
             {None, {"p", "parity", "odd?", "# of transp.\n"}},
           TableSpacing → {0.5, 2}]
```

```
Out[44]//TableForm=
          p                  parity   odd?    # of transp.

          {1, 2, 5, 4, 6, 3}   1       False   2
          {3, 2, 4, 5, 6, 1}   1       False   4
          {1, 4, 2, 5, 3, 6}  -1       True    5
          {2, 1, 4, 5, 3, 6}  -1       True    5
          {5, 4, 2, 1, 6, 3}  -1       True    5
```

■ 5.2.7 Stabilizers and orbits

Since a permutation can be regarded as a function, we can ask whether the permutation sends a value k to itself. If so, we say the permutation *fixes* k. More generally, we can ask about the image of any element.

- The permutation p fixes 5 because it sends 5 to itself.

```
In[45]:= FixQ[p = {2, 3, 4, 1, 5}, 5]
```

```
Out[45]= True
```

■ The `PermutationImage` function confirms this.

In[46]:= **PermutationImage[p, 5]**

Out[46]= 5

FixQ[*p*, *k*]	give True if the permutation *p* (of {1, 2, ..., *n* }) fixes the element *k*, and False otherwise
FixQ[*S*, *p*, *el*]	give True if the permutation *p* of the elements in the set *S* fixes the element *el* (from *S*), and False otherwise
PermutationImage[*p*, *k*]	give the image of *k* under the permutation *p*
PermutationImage[*S*, *p*, *el*]	give the image of *el* under the permutation *p* of the elements in the set *S*

FixQ and PermutationImage variations.

■ Here we see that the element "c" is not fixed, since it is sent to "e".

In[47]:= **Sset = {"a", "b", "c", "d", "e"};**
       p = {"c", "b", "e", "d", "a"};
       {FixQ[Sset, p, "c"], PermutationImage[Sset, p, "c"]}

Out[47]= {False, e}

Given a group *G* of permutations on a set *S*, and given an element *x* ∈ *S*, we are often interested in the *stabilizer* and *orbit* of *x*.

Stabilizer[*G*, *S*, *x*]	give the stabilizer of the element *x* from the set *S*, where *G* is a group of permutations on *S*
Stabilizer[*S*, *x*]	give the stabilizer of the element *x* from the set *S*, where the understood group *G* is the full group of permutations on *S*
Orbit[*G*, *S*, *x*]	give the orbit of the element *x* from the set *S*, where *G* is a group of permutations on *S*
Orbit[*S*, *x*]	give the orbit of the element *x* from the set *S*, where the understood group *G* is the full group of permutations on *S*

How to work with orbits and stabilizers.

■ Let *G* be a group formed from cycles as follows. Note that $|G| = 6$ and *G* is a subgroup of S_8.

In[48]:= **G = FormGroupoidFromCycles[**
       {Cycle[1], Cycle[1, 3, 2] @ Cycle[4, 6, 5] @ Cycle[7, 8],
       Cycle[1, 3, 2] @ Cycle[4, 6, 5] ,

```
        Cycle[1, 2, 3] @ Cycle[4, 5, 6],
        Cycle[1, 2, 3] @ Cycle[4, 5, 6] @ Cycle[7, 8],
        Cycle[7, 8]}]
```

```
Out[48]= Groupoid[{{1, 2, 3, 4, 5, 6, 7, 8},
            {3, 1, 2, 6, 4, 5, 8, 7}, {3, 1, 2, 6, 4, 5, 7, 8},
            {2, 3, 1, 5, 6, 4, 7, 8}, {2, 3, 1, 5, 6, 4, 8, 7},
            {1, 2, 3, 4, 5, 6, 8, 7}}, -Operation-]
```

- The orbit of 4 under the group G is the set $\{4, 6, 5\}$.

```
In[49]:= Orbit[G, Range[8], 4]
```

```
Out[49]= {4, 6, 5}
```

- The orbit can also be obtained by directly applying the definition of the orbit, $\{\phi(s) \mid \phi \in G\}$.

```
In[50]:= Map[PermutationImage[Range[8], #, 4] &, Elements[G]]
```

```
Out[50]= {4, 6, 6, 5, 5, 4}
```

- The stabilizer of the same element with the same group is the set of the following two permutations.

```
In[51]:= Stabilizer[G, Range[8], 4]
```

```
Out[51]= {{1, 2, 3, 4, 5, 6, 7, 8}, {1, 2, 3, 4, 5, 6, 8, 7}}
```

- This set can also be obtained by directly applying the definition of the stabilizer, $\{\phi \in G \mid \phi(x) = x\}$.

```
In[52]:= Select[Elements[G], PermutationImage[Range[8], #, 4] == 4 &]
```

```
Out[52]= {{1, 2, 3, 4, 5, 6, 7, 8}, {1, 2, 3, 4, 5, 6, 8, 7}}
```

- Using the `FixQ` function, the following may be more natural.

```
In[53]:= Select[Elements[G], FixQ[#, 4] &]
```

```
Out[53]= {{1, 2, 3, 4, 5, 6, 7, 8}, {1, 2, 3, 4, 5, 6, 8, 7}}
```

▣ 5.3 Working in $\mathbb{Z}\left[\sqrt{d}\right]$

■ 5.3.1 Basic functions

It may be worthwhile to first preview the `Adjoin` function discussed in section 5.4.3, since it is related. Here we would like to focus on working with elements in $\mathbb{Z}\left[\sqrt{d}\right]$, which is the set obtained by adjoining the square root of some integer d to the integers.

(We want d to be square-free.) This set is of interest when considering a unique factorization domain (UFD). First we look at some basic functions, then we look at some functions for divisibility.

`ZdQ[x]`	give `True` if x can be viewed as an element in $\mathbb{Z}[\sqrt{d}]$ for some integer d, and `False` otherwise
`ZdConjugate[a + b √d]`	return $a - b\sqrt{d}$
`ZdUnitQ[d, x]`	give `True` if x is a unit in $\mathbb{Z}[\sqrt{d}]$, and `False` otherwise
`ZdDivide[x, y]`	return the quotient x/y in the form $r + s\sqrt{d}$ when x and y are both in $\mathbb{Z}[\sqrt{d}]$
`ZdAssociatesQ[d, x, y]`	give `True` if x and y are associates in $\mathbb{Z}[\sqrt{d}]$, and `False` otherwise
`ZdIrreducibleQ[d, x]`	for negative d, give `True` if x is irreducible in $\mathbb{Z}[\sqrt{d}]$, and `False` otherwise
`RandomElement[d, max]`	return $a + b\sqrt{d}$ where $a, b \in [-max, max]$ and are integers
`RandomElements[d, max, n]`	return n random elements in $\mathbb{Z}[\sqrt{d}]$

Basic functions.

■ Some numbers do not belong to any set $\mathbb{Z}[\sqrt{d}]$.

```
In[54]:= Map[ZdQ,
            {3 - √5, ∛5, 18, π, 2 + 17 √-3 , RandomElement[-3, 25]}]
```

Out[54]= {True, False, True, False, True, True}

■ The conjugate of $a + b\sqrt{d}$ is $a - b\sqrt{d}$.

```
In[55]:= Map[ZdConjugate, {3 - √5, 2 + √-8 , 18}]
```

Out[55]= {3 + √5 , 2 - 2 i √2 , 18}

■ We can ask whether an element is a unit in $\mathbb{Z}[\sqrt{d}]$.

```
In[56]:= Map[ZdUnitQ[2, #] &, {1 + √2 , -1, 2 + √2 , 1 - √2}]
```

Out[56]= {True, True, False, True}

■ Here we calculate a quotient between two elements in $\mathbb{Z}[\sqrt{2}]$.

In[57]:= **q = ZdDivide$\left[2 - 3\sqrt{2}, 1 + \sqrt{2}\right]$**

Out[57]= $-8 + 5\sqrt{2}$

- Since we just saw that $1 + \sqrt{2}$ is a unit in $\mathbb{Z}\left[\sqrt{2}\right]$, $2 - 3\sqrt{2}$ and the previous quotient should be associates.

In[58]:= **ZdAssociatesQ$\left[2, 2 - 3\sqrt{2}, q\right]$**

Out[58]= True

- Irreducibility can be tested as follows.

In[59]:= **Map$\left[\text{ZdIrreducibleQ}[-3, \#] \&, \{2 - 5\sqrt{-3}, 5, 7\}\right]$**

Out[59]= {True, True, False}

■ 5.3.2 Divisibility

DividesQ[*r*, *s*]	give True if the integer *s* divided by the integer *r* yields an integer, and False otherwise
IntegerDivisors[*n*]	equivalent to Divisors[*n*]
IntegerDivisors[*n, opts*]	equivalent to Divisors[*n*] with additional options specified below with ZdDivisors

Functions for working with the ordinary integers.

- It is the case that 4 | 8, but it is not true that 8 | 4.

In[60]:= **{DividesQ[4, 8], DividesQ[8, 4]}**

Out[60]= {True, False}

- Here we have the divisors of 28 paired up in a natural way.

In[61]:= **IntegerDivisors[28, Combine → Products]**

Out[61]= {{1, 28}, {2, 14}, {4, 7}}

- Here we contrast the divisors of 28 with the nontrivial divisors of 28.

In[62]:= **{Divisors[28], IntegerDivisors[28, NonTrivialOnly → True]}**

Out[62]= {{1, 2, 4, 7, 14, 28}, {2, 4, 7, 14}}

DividesQ$\left[\,a+b\,\sqrt{d}\,,\right.$ $\left. c+e\,\sqrt{d}\,,\ \text{Radical}\ \to d\,\right]$	give True if $a+b\,\sqrt{d}\mid c+e\,\sqrt{d}$, and False otherwise
ZdDividesQ$[d,\ r,\ s]$	equivalent to DividesQ$[r,\ s,\ \text{Radical}\to d]$
ZdDivisors$[d,\ x,\ opts]$	return the divisors of x in $\mathbb{Z}\big[\sqrt{d}\,\big]$ for negative d, giving only one divisor per class of associates, using the options specified in *opts*
ZdDivisors$[$ $d,\ x,\ max,\ opts]$	return the divisors of x in $\mathbb{Z}\big[\sqrt{d}\,\big]$ for positive d, whose norm is less than or equal to the norm of *max*, using the options specified in *opts*

Divisor-related functions for elements in $\mathbb{Z}\big[\sqrt{d}\,\big]$.

option name	value	
DivisorsComplete	False	include all the divisors if set to True, but only include one from each class of associates (typically one in the first quadrant or right half-plane), if set to False (default)
NonTrivialOnly	False	if set to True, exclude all associates of 1 and the number whose divisors we are seeking, while include them if set to False (default)
Combine	False	do not combine the divisors in any manner (default value)
Combine	Products	combine divisors, if possible, in pairs of the form $\{a,b\}$ where $a*b$ is the number whose divisors we are seeking
Combine	Negations	combine divisors, if possible, in pairs of the form $\{a,-a\}$
Combine	Associates	combine divisors, if possible, in lists where each element in a list is an associate of the others; for $d<-1$, this is the same as Negations

Options for IntegerDivisors and ZdDivisors.

- We use the following to determine if $1-3\sqrt{-5}$ divides 46 over $\mathbb{Z}\big[\sqrt{-5}\,\big]$.

In[63]:= **DividesQ$\left[\,1-3\,\sqrt{-5}\,,\ 46,\ \text{Radical}\ \to -5\,\right]$**

Out[63]= True

- This is an alternate way of accomplishing the same thing.

In[64]:= **ZdDividesQ$\left[-5,\ 1\ -\ 3\ \sqrt{-5}\ ,\ 46\right]$**

Out[64]= True

- Since we know that $1 - 3\sqrt{-5} \mid 46$, we may be interested in knowing the other divisor (yielding a product of 46).

In[65]:= **ZdDivide$\left[46,\ 1\ -\ 3\ \sqrt{-5}\ \right]$**

Out[65]= $1 + 3\ \mathbb{i}\ \sqrt{5}$

- In this case, we are asking for all the divisors of 24 to be paired so that the product is 24.

In[66]:= **IntegerDivisors[24,**
   DivisorsComplete → True, Combine → Products]

Out[66]= {{-24, -1}, {-12, -2}, {-8, -3},
   {-6, -4}, {1, 24}, {2, 12}, {3, 8}, {4, 6}}

- Now we are looking for the divisors of 28 in $\mathbb{Z}\left[\sqrt{-5}\right]$ in order to pair them by associates. Since we are not looking for the complete list of divisors, all the negations (of the ones listed) are excluded. Since the negations are the only associates, each class has a single element.

In[67]:= **ZdDivisors[-5, 28, Combine → Associates]**

Out[67]= {{1}, {2}, {4}, {7}, {14}, {28},
   {$3 - \mathbb{i}\ \sqrt{5}$ }, {$3 + \mathbb{i}\ \sqrt{5}$ }, {$6 - 2\ \mathbb{i}\ \sqrt{5}$ }, {$6 + 2\ \mathbb{i}\ \sqrt{5}$ }}

- When we set DivisorsComplete to True, the natural pairing of associates (negations in this case) is given. Note also that NonTrivialOnly set to True prevents ± 1 and ± 28 from being listed.

In[68]:= **ZdDivisors[-5, 28, Combine → Associates,**
   DivisorsComplete → True, NonTrivialOnly → True]

Out[68]= {{-14, 14}, {-7, 7}, {-4, 4}, {-2, 2},
   {$-3 - \mathbb{i}\ \sqrt{5}$, $3 + \mathbb{i}\ \sqrt{5}$ }, {$3 - \mathbb{i}\ \sqrt{5}$, $-3 + \mathbb{i}\ \sqrt{5}$ },
   {$-6 - 2\ \mathbb{i}\ \sqrt{5}$, $6 + 2\ \mathbb{i}\ \sqrt{5}$ }, {$6 - 2\ \mathbb{i}\ \sqrt{5}$, $-6 + 2\ \mathbb{i}\ \sqrt{5}$ }}

- Since the ring $\mathbb{Z}[i]$ has four units, $\{\pm 1, \pm i\}$, combining by associates yields classes of length 4.

In[69]:= **ZdDivisors[-1, 8, Combine → Associates,**
   DivisorsComplete → True, NonTrivialOnly → True]

Out[69]= {{-4, -4 $\mathbb{i}$, 4 $\mathbb{i}$, 4}, {-4 - 4 $\mathbb{i}$, -4 + 4 $\mathbb{i}$, 4 - 4 $\mathbb{i}$, 4 + 4 $\mathbb{i}$},
   {-2, -2 $\mathbb{i}$, 2 $\mathbb{i}$, 2}, {-2 - 2 $\mathbb{i}$, -2 + 2 $\mathbb{i}$, 2 - 2 $\mathbb{i}$, 2 + 2 $\mathbb{i}$},
   {-1 - $\mathbb{i}$, -1 + $\mathbb{i}$, 1 - $\mathbb{i}$, 1 + $\mathbb{i}$}}

■ Here are the same divisors combined by negations.

In[70]:= **ZdDivisors[-1, 8, Combine → Negations,**
DivisorsComplete → True, NonTrivialOnly → True]

Out[70]= {{-4, 4}, {-4 - 4 i, 4 + 4 i}, {-4 + 4 i, 4 - 4 i},
{-2, 2}, {-2 - 2 i, 2 + 2 i}, {-2 + 2 i, 2 - 2 i},
{-1 - i, 1 + i}, {-1 + i, 1 - i}, {-2 i, 2 i}, {-4 i, 4 i}}

ZdCombineAssociates [*d*, *lst*]	return the list *lst* by combining all the associates in $\mathbb{Z}\big[\sqrt{d}\,\big]$ together in sublists

Means of combining associates if given a list of elements.

In[71]:= **ZdCombineAssociates[-1,**
{3, 4, 3 I, -4, 5, -5 I, 6, 7 I, -5}]

Out[71]= {{3 i, 3}, {-5, -5 i, 5}, {-4, 4}, {7 i}, {6}}

■ 5.3.3 Norm-related functions

ZdNorm $\big[a + b\sqrt{d}\,\big]$	return the norm of an element in $\mathbb{Z}\big[\sqrt{d}\,\big]$, defined to be $\mid a^2 - d\,b^2 \mid$
ZdPossibleNormQ [*d*, *nrm*]	for negative *d*, give True if the value *nrm* can occur in $\mathbb{Z}\big[\sqrt{d}\,\big]$, and False otherwise
ZdPossibleNorms [*d*, *max*]	for negative *d*, give a list of all norms less than or equal to *max* that are possible in $\mathbb{Z}\big[\sqrt{d}\,\big]$
ValuesHavingGivenNorm [*d*, *nrm*]	for negative *d*, return all values in $\mathbb{Z}\big[\sqrt{d}\,\big]$ having norm *nrm*
ValuesHavingGivenNorm [*d*, *nrm*, *max*]	return a partial list of values in $\mathbb{Z}\big[\sqrt{d}\,\big]$ having norm *nrm*, for positive *d*, using up to *max* iterations of an algorithm in this pursuit (defaulting to 50 if omitted)

Working with norms.

■ After choosing five random elements from $\mathbb{Z}\big[\sqrt{-3}\,\big]$, we determine the norm of each.

In[72]:= **TableForm[**
Map[{#, ZdNorm[#]} &, RandomElements[-3, 25, 5]],
TableHeadings → {None, {"x", "N(x)\n"}}]

Out[72]//TableForm=

x	N(x)
$-13 + 8\,\mathrm{i}\,\sqrt{3}$	361
$11 + 21\,\mathrm{i}\,\sqrt{3}$	1444
$4 - 2\,\mathrm{i}\,\sqrt{3}$	28
$-16 + 10\,\mathrm{i}\,\sqrt{3}$	556
$-3 - 19\,\mathrm{i}\,\sqrt{3}$	1092

- If we have a norm value in mind, we can check to see if this is a viable value in the specified ring.

In[73]:= **Map[{#, ZdPossibleNormQ[-5, #]} &, Range[0, 10]]**

Out[73]= {{0, True}, {1, True}, {2, False},
          {3, False}, {4, True}, {5, True}, {6, True},
          {7, False}, {8, False}, {9, True}, {10, False}}

- The previous result can also be achieved in the following manner.

In[74]:= **ZdPossibleNorms[-5, 10]**

Out[74]= {0, 1, 4, 5, 6, 9}

- For the value of a norm n to be possible in $\mathbb{Z}\!\left[\sqrt{d}\,\right]$, there must be integer coordinates $(a,\ b)$ on the ellipse $a^2 - d\,b^2 = n$, when d is negative, and on one of two hyperbolas for positive d. In the following figure, points of the same color have the same norm and are on a common ellipse concentric with the ellipse shown (corresponding to norm 56).

In[75]:= **ZdPossibleNorms[-5, 56, Mode → Visual]**

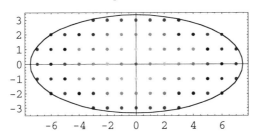

Out[75]= {0, 1, 4, 5, 6, 9, 14, 16, 20, 21,
          24, 25, 29, 30, 36, 41, 45, 46, 49, 54, 56}

- If you look carefully, you should be able to see four points on the ellipse shown in the figure. They have coordinates ($\pm\,6$, $\pm\,2$), which correspond to the following values in $\mathbb{Z}\!\left[\sqrt{-5}\,\right]$.

In[76]:= **ValuesHavingGivenNorm[-5, 56]**

Out[76]= $\{-6 - 2\,\mathfrak{i}\,\sqrt{5}\,,\ -6 + 2\,\mathfrak{i}\,\sqrt{5}\,,\ 6 - 2\,\mathfrak{i}\,\sqrt{5}\,,\ 6 + 2\,\mathfrak{i}\,\sqrt{5}\,\}$

- Here are some of the units (having norm 1) in $\mathbb{Z}\left[\sqrt{2}\right]$.

In[77]:= **ValuesHavingGivenNorm[2, 1, 35]**

Out[77]= $\{-41 - 29\,\sqrt{2}\,,\ -41 + 29\,\sqrt{2}\,,\ -17 - 12\,\sqrt{2}\,,\ -17 + 12\,\sqrt{2}\,,$
$-7 - 5\,\sqrt{2}\,,\ -7 + 5\,\sqrt{2}\,,\ -3 - 2\,\sqrt{2}\,,\ -3 + 2\,\sqrt{2}\,,\ -1 - \sqrt{2}\,,\ -1,$
$-1 + \sqrt{2}\,,\ 1 - \sqrt{2}\,,\ 1,\ 1 + \sqrt{2}\,,\ 3 - 2\,\sqrt{2}\,,\ 3 + 2\,\sqrt{2}\,,\ 7 - 5\,\sqrt{2}\,,$
$7 + 5\,\sqrt{2}\,,\ 17 - 12\,\sqrt{2}\,,\ 17 + 12\,\sqrt{2}\,,\ 41 - 29\,\sqrt{2}\,,\ 41 + 29\,\sqrt{2}\,\}$

5.4 Miscellaneous functions

5.4.1 Working with lists

Although *Mathematica* has many functions for working with lists, it is still wanting in a few areas. Sometimes we wish to take the union or complement of two sets but *not* have the elements returned already sorted. The Cartesian product of two sets is a natural construct in algebra, as is the notion of applying some operation to a Cartesian product. Finally, we frequently wish to consider relations between two sets, viewing two lists as mathematical sets.

- UnionNoSort does the work of Union by combining all the elements under one roof and removing duplicates, but it does not apply Sort to the final list.

In[78]:= **{UnionNoSort[{2,4,3,5,3}, {1,6,4,3,5}],**
    UnionNoSort[{2,4,3,5,3,1,6,4,3,5}],
    UnionNoSort[anyHead[2,4,3,5,3], anyHead[1,6,4,3,5]]}

Out[78]= {{2, 4, 3, 5, 1, 6},
    {2, 4, 3, 5, 1, 6}, anyHead[2, 4, 3, 5, 1, 6]}

- ComplementNoSort works in a similar fashion.

In[79]:= **{ComplementNoSort[{2,4,3,5,3,7}, {1,6,4,3,5}],**
    ComplementNoSort[{2,4,3,5,3,7}, {1,6,4,3,5}, {3,7}],
    ComplementNoSort[anyHead[2,4,3,5,3], anyHead[1,6,4,3,5]]}

Out[79]= {{2, 7}, {2}, anyHead[2]}

- The CartesianProduct function returns a list of *n*-tuples obtained by taking the Cartesian product of the lists given as arguments.

In[80]:= **CartesianProduct[{"a", "b", "c", "d"}, {1, 2, 3}]**

Out[80]= {{a, 1}, {a, 2}, {a, 3}, {b, 1}, {b, 2}, {b, 3},
    {c, 1}, {c, 2}, {c, 3}, {d, 1}, {d, 2}, {d, 3}}

UnionNoSort [*list₁*, *list₂*, ...]	equivalent to Union[*list₁*, *list₂*, ...] except the elements are not sorted
ComplementNoSort [*list₁*, *list₂*, ...]	equivalent to Complement[*list₁*, *list₂*, ...] except the elements are not sorted
CartesianProduct [*list₁*, *list₂*, ...]	give the Cartesian product of the lists *list₁*, *list₂*, ...
CartesianProduct [*list₁*, *list₂*, ..., Partition → True]	give the Cartesian product of the lists, partitioned according to the length of the last list
CloseSets [*list₁*, *list₂*, *op*]	give the list of all distinct elements obtained by applying the operation *op* on CartesianProduct[*list₁*, *list₂*]
Randomize [*list*]	randomly reorder the elements in *list*
SubsetQ [*child*, *parent*]	give True if *child* is a subset of *parent*, and False otherwise
ProperSubsetQ [*child*, *parent*]	give True if *child* is a proper subset of *parent*, and False otherwise
SameSetQ [*A*, *B*]	give True if *A* is mathematically the same set as *B*, and False otherwise

Functions to work with lists.

■ To have a Cartesian product partitioned according to the length of the last list, add the corresponding option.

```
In[81]:= TableForm[CartesianProduct[{"a", "b", "c", "d"},
           {1, 2, 3}, Partition → True], TableDepth → 1]
```

```
Out[81]//TableForm=
           {{a, 1}, {a, 2}, {a, 3}}
           {{b, 1}, {b, 2}, {b, 3}}
           {{c, 1}, {c, 2}, {c, 3}}
           {{d, 1}, {d, 2}, {d, 3}}
```

■ We can use more than two lists for arguments in the CartesianProduct function.

```
In[82]:= CartesianProduct[{0, 1}, {1, 2, 3}, {0, 1, 2}]
```

```
Out[82]= {{0, 1, 0}, {0, 1, 1}, {0, 1, 2}, {0, 2, 0},
          {0, 2, 1}, {0, 2, 2}, {0, 3, 0}, {0, 3, 1},
          {0, 3, 2}, {1, 1, 0}, {1, 1, 1}, {1, 1, 2}, {1, 2, 0},
          {1, 2, 1}, {1, 2, 2}, {1, 3, 0}, {1, 3, 1}, {1, 3, 2}}
```

■ If we want to apply an operation to a Cartesian product, the CloseSets function may be useful.

In[83]:= **{CartesianProduct[{2, 3, 5}, {0, 1}],**
      CloseSets[{2, 3, 5}, {0, 1}, Plus]} // ColumnForm

Out[83]= {{2, 0}, {2, 1}, {3, 0}, {3, 1}, {5, 0}, {5, 1}}
      {2, 3, 4, 5, 6}

- This function can also be used to help form a Boggle dictionary.

In[84]:= **words = CloseSets[**
      CloseSets[{"a", "c"}, {"a", "c", "p"}, StringJoin],
      {"t", "e"}, StringJoin]

Out[84]= {aat, aae, act, ace, apt, ape, cat, cae, cct, cce, cpt, cpe}

- Randomize does exactly what its name implies.

In[85]:= **Randomize[words]**

Out[85]= {aat, ape, ace, cpe, apt, cpt, cct, cat, act, cce, aae, cae}

- Note the differences between SubsetQ, ProperSubsetQ, and SameSetQ.

In[86]:= **{SubsetQ[{3, 4, 5}, {4, 3, 5}],**
      ProperSubsetQ[{3, 4, 5}, {4, 3, 5}],
      SameSetQ[{3, 4, 5}, {4, 3, 5}]}

Out[86]= {True, False, True}

▪ 5.4.2 Working with graphics

- DrawNgon[5] would have almost the same result as when using ShowFigure. as below.

In[87]:= **ShowFigure[5, {1, 2, 3, 4, 5}, "D"];**

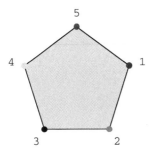

To illustrate some elementary group concepts, sometimes visual aids are useful. The following have been constructed with this in mind.

ShowCircle[*n*]	give a graph of the unit circle with *n* points uniformly placed and labeled 0 through *n* − 1
ShowCircle[*n, labels*]	give a graph of the unit circle with *n* points uniformly placed and labeled according to *labels*
DrawNgon[*n*]	give a graph of a regular *n*-gon with the vertices labeled 1 through *n* (*n* > 2)
ShowFigure[*n, perm, sym*]	give a graph of an *n*-gon possessing symmetry of type *sym*, with permutation *perm* applied to the vertices
ShowFigure[]	redraw the most recently drawn figure
ShowPermutation[*n, perm, sym*]	give the graphic of ShowFigure[*n, perm, sym*] with also the same *n*-gon in standard position so that the images before and after the permutation are shown
ShowPermutation[*perm*]	once a figure has been drawn via ShowFigure or ShowPermutation, show the before and after effects of the permutation *perm* on the *n*-gon

Functions for drawing some graphics.

■ After ShowFigure has been used, ShowPermutation can be used by simply giving a permutation to be applied to the last figure.

In[88]:= **ShowPermutation[{4, 3, 2, 1, 5}]**

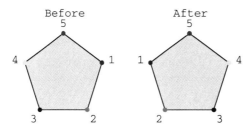

```
The figure on the left represents the original figure and
    the one on the right exhibits the effect of the
      transformation determined by the given permutation.
```

Out[88]= {4, 3, 2, 1, 5}

■ 5.4.3 Adjoin

The set of Gaussian integers is the set of all elements of the form {*a* + *b* *i* | *a, b* ∈ ℤ}. Although the polynomial $x^2 - 2$ does not have zeros in the rationals ℚ, it does have zeros in the set {*a* + *b* √2 | *a, b* ∈ ℚ}. Note the similarity between the two sets. In general, given √*d* for some integer *d* and some set *S* (usually a group or ring), we can construct the set $S[\sqrt{d}]$ defined as $\{a + b\sqrt{d} \mid a, b \in S\}$. The Gaussian integers are therefore denoted $\mathbb{Z}[i]$ (since $i = \sqrt{-1}$), while the second set is denoted $\mathbb{Q}[\sqrt{2}]$. In each case, we say we are *adjoining* an element to the underlying set. The same process can be applied to finite sets. Additionally, this process can be generalized to adjoining elements other than square roots, as well as generating lists of polynomials by adjoining an indeterminate.

Adjoin[*list, num*]	give the elements in *list* with *num* adjoined, where *num* needs to be of the form $n^{p/q}$, where *n*, *p*, and *q* are integers
Adjoin[*S, num*]	for *S* either a groupoid or ringoid, return Adjoin[Elements[*S*], *num*]
Adjoin[*list, ind, deg*]	give the set of polynomials in the indeterminate *ind* of degree less than or equal to *deg* with coefficients from *list*
Adjoin[*S, ind, deg*]	for *S* either a groupoid or ringoid, return Adjoin[Elements[*S*], *ind*, *deg*]

Uses of the Adjoin function.

- Here we adjoin $\sqrt{2}$ to the set $\{5, 6, 8\}$.

In[89]:= **Adjoin$\left[\{5, 6, 8\}, \sqrt{2}\,\right]$**

Out[89]= $\{5 + 5\sqrt{2}, 5 + 6\sqrt{2}, 5 + 8\sqrt{2}, 6 + 5\sqrt{2},$
   $6 + 6\sqrt{2}, 6 + 8\sqrt{2}, 8 + 5\sqrt{2}, 8 + 6\sqrt{2}, 8 + 8\sqrt{2}\}$

- We can adjoin a cube root to a list of symbols.

In[91]:= **Clear[a, t]**
      Adjoin[$\{a, t\}, 2^{1/3}\,$]

Out[92]= $\{a + 2^{1/3}\, a + 2^{2/3}\, a, a + 2^{1/3}\, a + 2^{2/3}\, t,$
   $a + 2^{2/3}\, a + 2^{1/3}\, t, a + 2^{1/3}\, t + 2^{2/3}\, t, 2^{1/3}\, a + 2^{2/3}\, a + t,$
   $2^{1/3}\, a + t + 2^{2/3}\, t, 2^{2/3}\, a + t + 2^{1/3}\, t, t + 2^{1/3}\, t + 2^{2/3}\, t\}$

- Adjoining can be nested. Note that here we use the group (or ring) $\mathbb{Z}_2$ instead of a list of elements.

In[93]:= **Adjoin$\left[$Adjoin$\left[Z[2], \sqrt{3}\,\right], \sqrt{2}\,\right]$**

Out[93]= $\{0, \sqrt{6}, \sqrt{2}, \sqrt{2} + \sqrt{6}, \sqrt{3}, \sqrt{3} + \sqrt{6}, \sqrt{2} + \sqrt{3},$
   $\sqrt{2} + \sqrt{3} + \sqrt{6}, 1, 1 + \sqrt{6}, 1 + \sqrt{2}, 1 + \sqrt{2} + \sqrt{6},$
   $1 + \sqrt{3}, 1 + \sqrt{3} + \sqrt{6}, 1 + \sqrt{2} + \sqrt{3}, 1 + \sqrt{2} + \sqrt{3} + \sqrt{6}\}$

- Here we adjoin the cube of a fifth root of 2 to $\{0, 1\}$.

In[94]:= **Adjoin[Z[2], $2^{3/5}$] // Short**

Out[94]//Short=
$$\{0, \ 4 \ 2^{2/5}, \ 2 \ 2^{4/5}, \ \ll 26 \gg, \ 1 + 2 \ 2^{1/5} + 4 \ 2^{2/5} + 2^{3/5},$$
$$1 + 2 \ 2^{1/5} + 2^{3/5} + 2 \ 2^{4/5}, \ 1 + 2 \ 2^{1/5} + 4 \ 2^{2/5} + 2^{3/5} + 2 \ 2^{4/5}\}$$

- We can also form a list of polynomials by providing the set of coefficients, the indeterminate, and the bound on the degree.

In[95]:= **Adjoin[Z[2], α, 2]**

Out[95]= $\{0, \ \alpha^2, \ \alpha, \ \alpha + \alpha^2, \ 1, \ 1 + \alpha^2, \ 1 + \alpha, \ 1 + \alpha + \alpha^2\}$

- Since we are interested in the elements for coefficients, it doesn't matter if we are working with a group or a ring. Here we have all third degree (and lower) polynomials over $\mathbb{Z}_3$ in the indeterminate x.

In[96]:= **Adjoin[Z[3, Structure → Ring], x, 3] // Short[#, 3] &**

Out[96]//Short=
$$\{0, \ x^3, \ 2 \ x^3, \ x^2, \ x^2 + x^3, \ x^2 + 2 \ x^3, \ 2 \ x^2, \ 2 \ x^2 + x^3,$$
$$2 \ x^2 + 2 \ x^3, \ x, \ x + x^3, \ \ll 59 \gg, \ 2 + x + 2 \ x^2 + x^3,$$
$$2 + x + 2 \ x^2 + 2 \ x^3, \ 2 + 2 \ x, \ 2 + 2 \ x + x^3, \ 2 + 2 \ x + 2 \ x^3,$$
$$2 + 2 \ x + x^2, \ 2 + 2 \ x + x^2 + x^3, \ 2 + 2 \ x + x^2 + 2 \ x^3,$$
$$2 + 2 \ x + 2 \ x^2, \ 2 + 2 \ x + 2 \ x^2 + x^3, \ 2 + 2 \ x + 2 \ x^2 + 2 \ x^3\}$$

■ 5.4.4 Disguising groups and rings

There are five groups of order 8. If they are presented as Cayley tables using only generic letters (*a* through *h*), is it easy to determine which table belongs to which group? The DisguiseGroupoid function can facilitate the task of presenting generic-looking groups.

DisguiseGroupoid[*G*]	return the groupoid *G* with the elements of *G* replaced with strings *a*, *b*, etc., in order to provide a context-free environment to explore group properties
DisguiseGroupoid[*G*, *rules*]	return the groupoid *G* disguised according to the *rules* given
DisguiseGroupoid[*G*, Randomize → True]	return the groupoid *G* disguised by first randomizing the elements before using the rule assignments; Randomize defaults to False
DisguiseRingoid[*R*]	return a disguised ringoid in the same fashion as DisguiseGroupoid does for groupoids

Disguising a groupoid or ringoid.

In[97]:= **CayleyTable[**
    {DisguiseGroupoid[G = DirectProduct[Z[2], Z[2]]],

```
DisguiseGroupoid[G, {{0, 0} → "cat",
    {0, 1} → "dog", {1, 1} → "hat", {1, 0} → "log"},
    Randomize → True]}, Mode → Visual];
```

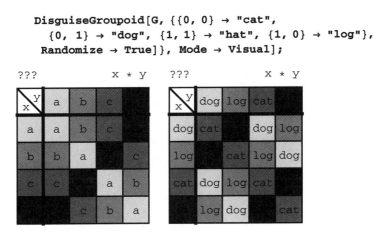

■ 5.4.5 A look at some functions in LabCode

■ This function organizes the data regarding the orders of the elements in $\mathbb{Z}_{15}$.

In[98]:= **CollectOrders[Orders[Z[15]]]**

Out[98]= {{1, {0}}, {3, {5, 10}}, {5, {3, 6, 9, 12}},
{15, {1, 2, 4, 7, 8, 11, 13, 14}}}

■ Here are two graphical ways of considering the same question.

In[99]:= **ShowGroupOrders[Z[15]];**

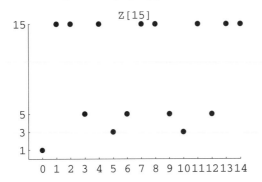

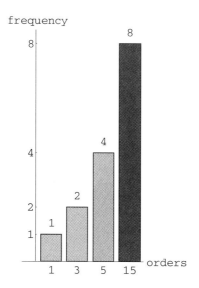

CollectOrders[Orders[*G*]]	given the output of the Orders function (equivalently, OrderOfAllElements), this function organizes the data in the form {*p*, *A*} where *p* is one of the possible orders and *A* is the set of elements in *G* with order *p*
ShowGroupOrders[*G*]	return a ListPlot showing pairs {*g*, |*g*|} and a bar chart indicating how many elements there are for each order
ShowColoredPermutation[*p*]	give a graphical illustration of the permutation *p*

Miscellaneous functions from LabCode that may be of interest.

■ 5.4.6 Potpourri

■ A complex number is a Gaussian integer if its real and imaginary parts are integers.

```
In[100]:= Map[GaussianIntegerQ, nums = {3, 3 - 2 I, 3.5 + 4.3 I, -8 I}]

Out[100]= {True, True, False, True}
```

■ Here are the real and imaginary parts of each of these numbers.

In[101]:= **Map[ComplexToPoint, nums]**

Out[101]= {{3, 0}, {3, -2}, {3.5, 4.3}, {0, -8}}

GaussianIntegerQ[z]	give True if the complex number z is a Gaussian integer (real and imaginary parts are integers), and False otherwise
ComplexToPoint[z]	given a complex number z, return {Re[z], Im[z]}
IntegerLatticeGrid[{a, b}, {c, d}, opts]	return a ListPlot of an integer lattice with domain $a \leq x \leq b$ and $c \leq y \leq d$ with options *opts* to be used by ListPlot
ESG[code]	return the group corresponding to *code* as found in the *Exploring Small Groups* software package

Miscellaneous functions.

■ Here are the two groups of order 4 given by the codes 0401 and 0402 used in the program *Exploring Small Groups*, written by Ladnor Geissinger.

In[102]:= **{ESG[0401], ESG[0402]}**

Out[102]= {Groupoid[{0, 1, 2, 3}, Mod[#1 + #2, 4] &],
        Groupoid[{1, Rot, Ref, Rot ** Ref}, -Operation-]}

Appendices

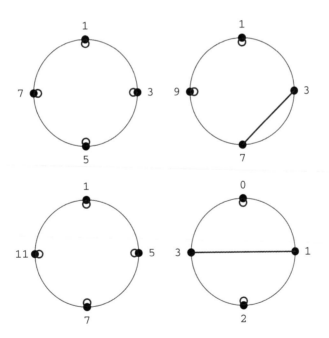

Appendix A

▓ Installation instructions

■ Version 2.x under Windows

From from the web site (`http://www.central.edu/eaam.html`) or from the CD, find the files appropriate for version 2.x under Windows. Place the entire `AbstractAlgebra` directory (called `Abstract` to limit the name to 8 characters) into the **Packages** directory (in the *Mathematica* directory). The GrpLabs and RngLabs directories can be placed wherever convenient.

As users of version 2.x under Windows are well aware, each cell contains only one possible style. Therefore, some of the emphases (italics, bold or otherwise) may be lost that are provided in other versions. Additionally, to discriminate input cells from text cells, all input cells have been colored yellow. (This can be changed by the user, if so inclined.) As you may be aware, there are a large number of advantages in upgrading to version 3.x, including a number of palettes available that complement *Exploring Abstract Algebra with Mathematica*.

■ Version 2.x under other platforms

From from the web site (`http://www.central.edu/eaam.html`) or from the CD, find the files appropriate for version 2.x under Macintosh or Unix. Place the entire `AbstractAlgebra` directory into the **Packages** directory (in the *Mathematica* directory). The GrpLabs and RngLabs directories can be placed wherever convenient.

As you may be aware, there are a large number of advantages in upgrading to version 3.x, including a number of palettes available that complement *Exploring Abstract Algebra with Mathematica*.

▪ Version 3.0 or higher

From from the web site (http://www.central.edu/eaam.html—you may wish to check here for updates to the files) or from the CD, find the files appropriate for version 3.0 or higher. (Note that these files work equally well on Macintosh, Unix or Windows platforms.) The files are located on the CD in such a way to imitate the layout of the main *Mathematica* directory. Here is a schematic of what is on the CD and on the web page (except the CD uses initial lower-case directory names that correspond to *Mathematica* directories to avoid an accidental overwrite of files). Note that the AddOns and Configuration directories are inside the main *Mathematica* 3.0 Files directory.

```
AddOns
  Applications
    AbstractAlgebra
      package files (.m files)
      Documentation
        English
          documentation files
      GroupLabs
        group lab notebooks from EAAM
      Kernel
        init.m (Master.m copied here and renamed to init.m)
      Palettes
        palette files from EAAM
        palette files for general distribution
      RingLabs
        ring lab notebooks from EAAM
Configuration
  FrontEnd
    StyleSheets
      EAAM.nb (needed if labs are moved from AbstractAlgebra)
    Palettes
      AbstractAlgebraPalette.nb (can also be found via the Help Browser)
```

Place the entire AbstractAlgebra directory into the Applications directory (in the AddOns directory in the *Mathematica* directory). From the above schematic, observe that this places the package files, labs, and Browser documentation in an accessible location for *Mathematica*. In particular, the group and ring labs are in the obvious directories. If it is more convenient, the lab directories can be moved to another location. (The only loss in doing so is that links related to these labs in the Help Browser will become broken. If the labs are not moved, they are easily accessible from the EAAM Info section of the accompanying help files.) After placing these files in the Applications directory, open *Mathematica* and choose Rebuild Help Index from the Help menu; this will enable the Help Browser to be aware of the documentation files accompanying AbstractAlgebra. (Choose the Add-ons button in the Help Browser and AbstractAlgebra will show up in the left-most column.)

In addition to copying the `AbstractAlgebra` directory into the Applications directory, copy the EAAM.nb style sheet into the StyleSheets directory (in the FrontEnd directory in the Configuration directory in the *Mathematica* directory). Similarly, the AbstractAlgebraPalette.nb palette should be copied into the Palettes directory (in the FrontEnd directory in the Configuration directory). From this palette, all the other palettes in the Palettes directory in the `AbstractAlgebra` directory will be accessible.

References

Anick, David, "A Model of Adams-Hilton Type for Fiber Squares," *Illinois J. Math*, 29 (3), 1985, pp. 463-502.

Dornhoff, L. and F. Hohn, *Applied Modern Algebra*, Macmillan, New York, 1978.

Fraleigh, J., *A First Course in Abstract Algebra*, Fifth Edition, Addison-Wesley, Reading, MA, 1994.

Gray, T. and J. Glynn, *Exploring Mathematics with Mathematica,* Addison-Wesley, Reading, MA, 1991.

Herstein, I., *Topics in Algebra*, Second Edition, Wiley, New York, 1975.

Herstein, I., *Abstract Algebra*, Third Edition, Prentice-Hall, Upper Saddle River, NJ, 1990.

Hungerford, T., *Abstract Algebra: An Introduction,* Second Edition, Saunders, New York, 1997.

Levasseur, K., "A Microworld for Elementary Group Theory," *Mathematica in Education*, 3 (Fall, 1994), pp. 5-10.

Lindl, R. and G. Pilz, *Applied Abstract Algebra,* Springer-Verlag, New York, 1984.

Lipson, J., *Elements of Algebra & Algebraic Computing,* Addison-Wesley, Reading, MA, 1981.

Pinter, C., *A Book of Abstract Algebra,* McGraw Hill, New York, 1990.

Rotman, J., *The Theory of Groups,* Allyn & Bacon, Boston, 1965.

Rotman, J., *A First Course in Abstract Algebra,* Prentice-Hall, Upper Saddle River, NJ, 1996.

Sims, C., *Abstract Algebra: a Computational Approach,* Wiley, New York, 1984.

Wolfram, S., *Mathematica, A System for Doing Mathematics by Computer*, Second Edition, Addison-Wesley, Redwood City, CA, 1991.

Wolfram, S., *The Mathematica Book*, Third Edition, Wolfram Media, Champaign, IL, 1997.

Wolfram Research, *Mathematica 3.0 Standard Add-on Packages*, Wolfram Media, Champaign, IL, 1996.

Appendix B—Lab 0

Getting Started with

Mathematica

This lab is intended to be an introduction to using *Mathematica*. There are several key skills and concepts that need to be mastered to ensure success with the labs in *Exploring Abstract Algebra with Mathematica*. They are:

1. how to recognize nested cells
2. how to open and close nested cells
3. how to evaluate an expression in an input cell
4. how to create an input cell in order to create one's own *Mathematica* input
5. learn some of the general principles of *Mathematica* syntax
6. learn where to get more help

Let's begin with the first two goals. For users of *Mathematica* 2.2, every cell has one or more brackets surrounding it on the right side of the window. For users of version 3.0 or higher, the brackets may be replaced by, or supplemented with, arrows (triangles) on the left that can be toggled. For instance, look at the next cell, headed by the title "0.0 Note regarding Exploring Abstract Algebra with *Mathematica*." (You may need to scroll down to see it. Note: This section occurs only in the notebook version, not in the printed version.) Note that this cell has a standard square bracket as well as another bracket to its right that looks somewhat like a harpoon. This indicates that there are cells nested inside. When you move the cursor over the bracket with the harpoon, the cursor takes on the shape of an arrow pointing left toward a vertical bar. When the cursor has this shape, you can double-click on the harpoon. This will open up to reveal the inner, nested cells.

Try this now with any of these first three sections (0.0, 0.1, or 0.2); after reading any or all of these, proceed to section 0.3 and open it up to continue.

▣ 0.1 Prerequisites

There are no prerequisites for this lab.

▣ 0.2 Goals for this lab

This lab is intended to introduce a few of the basics of *Mathematica*, as well as introduce a few rudimentary algebraic ideas. In particular, the user should master the six specific goals outlined at tthe outset of this lab.

▣ 0.3 The In's and Out's of evaluating *Mathematica* expressions

These nested cells are not too difficult to open up, once you get the hang of it, right?

> **Q1**. Go to section 0.4 (below) and open up the nested cells. What do you find inside at the inner-most nested level? Note: The printed version, contains only the section title, while the electronic version has the nested cells to be opened.

You may have noticed that some cells appear differently than others. For instance, this text is in a `Text` cell, while the cell heading this section (0.3) is called a `Section` cell. If you go up and click anywhere in the characters of the section 0.3 cell, you will see the word `Section` appear in the pop-up menu in the ruler at the top of the window. This menu indicates the current cell type; it can also be used to change one type of cell to another. To tell *Mathematica* to compute an evaluation, we need to use a special type of cell, called an `Input` cell. Consider the following cell.

$$\frac{26}{3}$$

Click the cursor anywhere in the body of the above cell containing $\frac{26}{3}$, or select its cell bracket. (Note that the name of the cell type becomes `Input` in the pop-up menu in the ruler.) To *evaluate* this cell (assuming you have clicked in the cell or selected its bracket), press the SHIFT/RETURN key combination (or just the ENTER key on a Macintosh). *Mathematica* will then evaluate this request and return its value. Do so.

Do not be too disappointed with the result; *Mathematica* always tries to return an exact result whenever possible. In this case, $\frac{26}{3}$ is more exact than 8.7 or 8.6667 or 8.666666667. Each of these approximations can be obtained as follows (evaluate this next cell).

$$\left\{\text{N}\left[\frac{26}{3}, 2\right], \text{N}\left[\frac{26}{3}, 5\right], \text{N}\left[\frac{26}{3}, 10\right]\right\}$$

Q2. Evaluate the cell below. (Remember to use the SHIFT/RETURN combination.) What do these results have in common with the above approximations of 26/3?

$$\text{N}\left[\left\{\frac{29}{3}, \frac{23}{3}, \frac{20}{3}\right\}\right]$$

The next thing to learn is how to make your own `Input` cell so that you can type in your own request. Move the cursor (slowly) right below the text in this paragraph. As you move it, you should see the cursor take on one or more of the following shapes: a horizontal I-beam, a down-arrow, a circle with a plus, a vertical I-beam or possibly other shapes. When it is just below this text, but yet above the next cell, you should see the cursor become a horizontal I-beam. On clicking the mouse when this cursor is present, a horizontal line appears in the window. You can now start typing and you are automatically in the `Input` style.

Q3. Above this cell, but below the preceding paragraph of text, create an input cell, type in 17/3, and evaluate the cell. Create another cell and determine the decimal approximations of 14/3 and 11/3. What do these results have in common with previous results?

Q4. Evaluate the following `Input` cell. Next, create a new input cell and either retype the line just evaluated or copy it and paste it in the new cell. (You are not expected to understand what these instructions in the `Input` cell mean.) Before evaluating, change the 2 to a 1. What do the results in the output have in common? How do they compare to the previous output?

$$\text{Map}\left[\left(\text{N}\left[\frac{\#1}{3}\right]\,\&\right), \text{Range}[2, 30, 3]\right]$$

◼ 0.4 What is inside?

(See question 1.)

🔳 0.5 Some syntax basics

Evaluate the following cell. (Recall that also saw this expression when working question two.)

$$N\left[\left\{\frac{29}{3}, \frac{23}{3}, \frac{20}{3}\right\}\right]$$

Mathematica is very particular in expecting the exact syntax for communicating requests. For example, the following are incorrect methods (for various reasons) of inputting the previous cell. Evaluate each of the following.

$$(* \text{ example 1 } *) \; n\left[\left\{\frac{29}{3}, \frac{23}{3}, \frac{20}{3}\right\}\right]$$

$$(* \text{ example 2 } *) \; N[(29/3, 23/3, 20/3)]$$

$$(* \text{ example 3 } *) \; N\left\{\frac{29}{3}, \frac{23}{3}, \frac{20}{3}\right\}$$

$$(* \text{ example 4 } *) \; N[\{29/3 \; 23/3 \; 20/3\}]$$
$$N[\{29/3; \; 23/3; \; 20/3\}]$$

$$(* \text{ example 5 } *) \; N[\{29/3, \; 23/3, \; 20/3]\}$$

In addition to the error messages that were given, here are additional reasons why these are incorrect.

1. All *Mathematica* commands or functions start with capital letters.

2. Lists are always enclosed with curly brackets {}; parentheses are used *only* for grouping expressions to alter the standard order of operations (such as $(a + b)^2$ in contrast to $a + b^2$).

3. Arguments for functions (such as the N function in this case) must always be enclosed in square brackets [], not parentheses (which are used only as grouping symbols).

4. Lists must always have their elements separated by commas. (Here the terms were multiplied, in the first instance, since a space implies multiplication.)

5. Of the three types of brackets, (), {}, and [], each type can only be used for the specific use for which it is intended. Furthermore every left bracket must have a matching right bracket of the same type.

> **Q5**. Evaluate the following Input cell. What error(s) does it have? Correct the error(s) to arrive at the intended input.

```
Expand[{x + y} ^ 2)
```

📖 0.6 Help

Mathematica has a sophisticated help facility called the Help Browser (or Function Browser for 2.2 users). The last item on *Mathematica*'s menubar is the Help menu. Choose this and select the first item (Help... in 3.0 or higher or Open Function Browser... in 2.2). In the browser for version 2.2, you have the choice of reviewing the use of any names in the Built-in Functions, Packages, or Loaded Packages by pressing the appropriate button. (For Loaded Packages, first press the Update button.) In version 3.0 or higher, the browser is much more sophisticated. In addition to the previous three categories, one can find the whole *Mathematica* book on-line, as well as an index to it and all the packages, in addition to other information. This browser is a *Mathematica* notebook and not just a static display device as in version 2.2. Try it out and find out for yourself. To learn more about the functions in *Exploring Abstract Algebra with Mathematica* in version 3.0 or higher, click on the Add-ons button and AbstractAlgebra should show up in the first column if a correct installation has been made (including a call to Rebuild Help Index, the last item on the Help menu).

> **Q6**. Using the help facility, learn more about the Expand function. What does this function do? (Hint: Expand is a built-in function used for algebraic computations in basic algebra.)

Another feature that occurs beginning in version 3.0 is the use of hyperlinks. By clicking on any underlined, blue text, one can be linked to another part of the notebook (or a different notebook altogether), the main book in the Help Browser, package documentation in the Help Browser, or even to a web page.

📖 0.7 Using *Mathematica* to learn a mod idea

Above we saw that 26/3, 29/3, 23/3 and several other fractions all had one thing in common: the quotient had a decimal part returning .666667. In other words, on division by three, the remainder was always two. Observe:

```
26 == 8 * 3 + 2
```

$29 == 9 * 3 + 2$

$23 == 7 * 3 + 2$

In each case, we see that the numbers 26, 29 and 23 have a remainder of two, on division by 3. We can see this visually as well. First, select all the cells in the following subsection. (You do not need to open up the subsection, simply *select* the outer harpoon bracket surrounding it.) Next, evaluate all these cells (type SHIFT/RETURN). This will define some functions we will be using. Now go to the next section.

■ Evaluate all cells in this subsection, if not done on opening

■ Continue on in this subsection

By evaluating the cell below, you can see why 2 is the remainder of 26 on division by 3.

```
IllustrateModReduction[26, 3];
```

Note that the word "mod" is used in the function name. We say 26 mod 3 is 2 because 2 is the remainder when 26 is divided by 3. *Mathematica* can compute this directly.

```
Mod[26, 3]
```

Note that each of 29, 26 and 23, when reduced modulo 3, results in the value 2.

```
{Mod[29, 3], Mod[26, 3], Mod[23, 3]}
```

In other words, 2 is the remainder when 29 is divided by 3 and when 23 is divided by 3. These can be seen visually as well; evaluate one or both of the following.

```
IllustrateModReduction[29, 3];
```

```
IllustrateModReduction[23, 3];
```

Q7. Determine the values of 22 mod 3, 25 mod 3 and 28 mod 3 using either the `Mod` function or the `IllustrateModReduction` function. What do these three values have in common? How are 22, 25 and 28 related to each other?

> **Q8**. Determine the values of 21 mod 3, 24 mod 3 and 27 mod 3 using either the Mod function or the `IllustrateModReduction` function. What do these three values have in common? How are 21, 24 and 27 related to each other?

▣ 0.8 Divide and conquer

Select the input cell below. After evaluating it, come back here and finish reading this text. There are some mathematical observations to make, as well some *Mathematica* observations. First, note that this is showing k mod 3 for k running from 18 to 28. As you look at the value of this (the value of the remainder of k divided by 3, which is also the height of the yellow rectangle), note how it cycles through 0 to 1 to 2 to 0 to 1 to 2 and so on. Do you know why?

Second, note that all the graphics are each in their own cell but there is also one cell that surrounds the whole collection of graphics. Put the cursor over this surrounding bracket and double-click. This will close up and hide the nested cells with only the first one showing. If you now double-click on the top graphic, an animation should take place. Since the motion may be going too fast, you may want to know how to adjust it. Typing any number from 1 to 9 adjusts the speed (1 slow and 9 fast), while pressing an up-arrow or down-arrow key causes one step to be taken in the specified direction. Try it.

```
Do[IllustrateModReduction[k, 3, showText → False],
  {k, 18, 28}]
```

For each of the cases where the remainder is zero (ie., k mod 3 $= 0$), observe that this happens because 3 is a divisor of k. Mathematically, we say "3 divides k" and denote this by 3 | k. Consider $k = 24$ and $j = 30$. In both cases we have 3 as a divisor. There are, however, other divisors of each. In previous course-work, you may have encountered the notion of the *greatest common divisor*, or gcd. In this case, the gcd of 24 and 30, frequently denoted gcd(24, 30) or simply (24, 30), is actually 6. Here is how we get *Mathematica* to confirm this. (Note the capital letters.)

```
GCD[24, 30]
```

> **Q9**. What is the gcd of 242652 and 3054876?

Suppose we consider the problem of determining the gcd of two fairly large numbers, say $a = 1234567891011121314$ and $b = 1413121110987654321$.

```
a = 1234567891011121314
b = 1413121110987654321
```

We will pursue the gcd in two ways. First, we will find *all* the divisors of *a* (and record how much time this took).

> `divisorsOfa = Timing[Divisors[a]]`

And then do the same for *b*.

> `divisorsOfb = Timing[Divisors[b]]`

Next we take the intersection of the lists of divisors and ask for the maximum value. (Note that `divisorsOfa` and `divisorsOfb` consist of a list containing both the time and the list of divisors. Since we want the second part, the list of divisors, we specify this by using the notation `divisorsOfa[[2]]`.)

> `max =`
> `    Timing[Max[Intersection[divisorsOfa[[2]], divisorsOfb[[2]]]]]`

We see that the gcd is 3. How much time did this take? We can add up the amounts found in the first part of each result.

> `totalTime = divisorsOfa[[1]] + divisorsOfb[[1]] + max[[1]]`

Now let's see how long it takes for the GCD function?

> `Timing[GCD[a, b]]`

Is this surprising? The GCD function did not use any previous results to enable it to be this fast, but rather the method of calculating the gcd happens to be a very fast algorithm called the Euclidean Algorithm. Hopefully, you will be learning more about this algorithm in this course; we will see it in this suite of labs when we study polynomial rings in Ring Lab 6. If you increase the number of digits in *a* and *b* by only a few, the time difference will be even more dramatic.

▥ 0.9 It all adds up

Earlier we saw how the values of *k* mod 3 (`Mod[k, 3]`) are 0, 1 or 2.

> `Table[Mod[k, 3], {k, 18, 28}]`

A question that algebraists ask in such a situation is "Does the set {0, 1, 2} have any other structure?" In other words, is there a natural operation that may exist between, for example, 14 mod 3 and 16 mod 3? In particular, does this answer have anything to do with 30 mod 3?

> `Mod[14, 3]`
> `Mod[16, 3]`
> `Mod[30, 3]`

This can be illustrated visually below.

```
IllustrateModAddition[14, 16, 3]
```

Note that $(14 \bmod 3)$ plus $(14 \bmod 3)$ (ie., 2 "plus" 2) yields 1, which is also 28 mod 3.

```
Mod[14, 3]
Mod[14, 3]
Mod[28, 3]
```

```
IllustrateModAddition[14, 14, 3]
```

The collection of values where the integers are reduced by some modulus (3 in this example) forms a very interesting structure for algebraists. Much more will be said about this in the future.

Index

EXPLORING ABSTRACT ALGEBRA WITH *MATHEMATICA*®

REGISTRATION CARD

Since this field is fast-moving, we expect updates and changes to occur that might necessitate sending you the most current pertinent information by paper, electronic media, or both, regarding *Exploring Abstract Algebra with Mathematica*®. Therefore, in order to not miss out on receiving your important update information, please fill out this card and return it to us promptly. Thank you.

Name: _____

Title: _____

Company: _____

Address: _____

City: _____ State: _____ Zip: _____

Country: _____ Phone: _____

E-mail: _____

Areas of Interest/Technical Expertise: _____

Comments on this Publication: _____

_____

_____

❑ Please check this box to indicate that we may use your comments in our promotion and advertising for this publication.

Purchased from: _____
Date of Purchase: _____

❑ Please add me to your mailing list to receive updated information on *Exploring Abstract Algebra with Mathematica*® and other TELOS publications.

❑ I have a ☐ IBM compatible ☐ Macintosh ☐ UNIX ☐ other

Designate specific model _____

TELOS ®
THE ELECTRONIC LIBRARY OF SCIENCE

NO POSTAGE
NECESSARY
IF MAILED
IN THE
UNITED STATES

BUSINESS REPLY MAIL
FIRST-CLASS MAIL PERMIT NO. 5863 NEW YORK, NY

POSTAGE WILL BE PAID BY ADDRESSEE

THE
ELECTRONIC
LIBRARY
OF
SCIENCE

TELOS PROMOTION
SPRINGER-VERLAG NEW YORK, INC.
ATTN: J. Roth
175 FIFTH AVENUE
NEW YORK NY 10160-0266